ADULT STEM CELL NICHES

Edited by **Sabine Wislet-Gendebien**

Adult Stem Cell Niches
http://dx.doi.org/10.5772/57290
Edited by Sabine Wislet-Gendebien

Contributors

Peter Siig Vestentoft, Helen ONeill, Thomas Meyer, Mirjana Kessler, Rike Zietlow, Adriana Bonomo, Fani Papagiannouli, Ali Dalloul, Christelle Freitas, Karl Balabanian, Alexandre Bignon, Agustín G Zapata, Andrew Zannettino, Jacqueline Noll, Kate Vandyke, Joshua Bernstock

Published by InTech

Janeza Trdine 9, 51000 Rijeka, Croatia

Publishing Process Manager Danijela Duric

Technical Editor InTech DTP team

Cover InTech Design team

First published August, 2014

Printed in Croatia

Additional hard copies can be obtained from orders@intechopen.com

Adult Stem Cell Niches, Edited by Sabine Wislet-Gendebien

p. cm.

ISBN 978-953-51-1718-6

For EU product safety concerns:
IN TECH d.o.o., Prolaz Marije Krucifikse Kozulić 3, 51000 Rijeka, Croatia,
info@intechopen.com or visit our website at intechopen.com.

Meet the editor

Dr. Sabine Wislet obtained her PhD from the Center of Cellular and Molecular Neurobiology at the University of Liège under the supervision of Dr. Bernard Rogister and Pr Gustave Moonen. During her PhD, she identified neurogenic stem cells in adult bone marrow that could potentially be used in cellular therapy for neurological disorders. She then completed her postdoctoral training at the Center for Research in Neurodegenerative Disorders directed by Professor St George-Hyslop, in Dr. Anurag Tandon's laboratory, at the university of Toronto, where she started to work on alpha-synuclein. She is currently a Senior Scientist at the GIGA Neurosciences and an Assistant Professor in the Department of Biomedical and Preclinical Sciences, at the University of Liège. Dr Wislet research focuses on cell replacement therapy in neurological disorders using adult bone marrow stromal cells as well as the molecular origins of Parkinson's disease.

Contents

Preface IX

Section 1 Hematopoietic Stem Cell Niches in Normal and Pathological Conditions 1

Chapter 1 **Delineation of Niches which Support Hematopoiesis 3**
Hong Kiat Lim and Helen C. O'Neill

Chapter 2 **The Adult Hematopoietic Niches — Cellular Composition, Histological Organization and Physiological Regulation 21**
Agustín G Zapata

Chapter 3 **CXCR4 in Central and Peripheral Lymphoid Niches - Physiology, Pathology and Therapeutic Perspectives in Immune Deficiencies and Malignancies 49**
Christelle Freitas, Alexandre Bignon, Karl Balabanian and Ali Dalloul

Chapter 4 **Hematopoietic Stem Cells, Tumor Cells and Lymphocytes — Party in the Bone Marrow 87**
Adriana Bonomo, Ana Carolina Monteiro and Alex Balduíno

Section 2 Stem Cell Niches and Reproductive Systems 111

Chapter 5 **Male stem Cell Niche and Spermatogenesis in the Drosophila testis — A Tale of Germline-Soma Communication 113**
Fani Papagiannouli

Chapter 6 **Adult Stem Cell Niches — Stem Cells in the Female Reproductive System 139**
Mirjana Kessler, Rike Zietlow and Thomas F. Meyer

Section 3 Hepatic Stem Cell Niches 171

Chapter 7 **Adult Hepatic Progenitor Cells 173**
Peter Siig Vestentoft

Section 4 Stem Cell Niches in Adult Nentral Nervous System 209

Chapter 8 **Typical and Atypical Stem Cell Niches of the Adult Nervous System in Health and Inflammatory Brain and Spinal Cord Diseases 211**
Joshua Bernstock, Jeroen Verheyen, Bing Huang, John Hallenbeck and Stefano Pluchino

Section 5 Stem Cell Niches and Cancer 289

Chapter 9 **The Role of the "Cancer Stem Cell Niche" in Cancer Initiation and Progression 291**
Jacqueline E. Noll, Kate Vandyke and Andrew C.W. Zannettino

Preface

Stem cells are characterized as cells endowed with continuous self-renewal ability and pluri- or multipotentiality, and could consequently give rise to a wide panel of cell types. Non-germinal stem cells are classified into different categories: Embryonic stem cells (ES) are found in the inner cell mass of blastocyst and are pluripotent stem cells that can generate any mature cell of each of the three germ layers; Somatic stem cells are tissue-specific and more restricted than ES cells in terms of differentiation capabilities. They can be isolated from various fetal and adult tissues, which make them an attractive supply of material for cell therapy.

To better understand and control stem cell proliferation and differentiation, scientists analyzed deeply the stem cells niches. The concept of stem cell niches was first described by Schofield and collaborators in 1978. Stem cell niche refers to a microenvironment where stem cells are found and regulate cell fate. During embryonic development, various niche factors act on embryonic stem cells to alter gene expression, and induce their proliferation or differentiation for the development of the fetus. Within the human body, stem cell niches maintain adult stem cells in a quiescent state, but after tissue injury, the surrounding micro-environment actively signals to stem cells to promote either self renewal or differentiation to form new tissues. Several important factors regulate stem cell characteristics within the niche. Among others, we can mention cell-cell interactions between stem cells or between stem cells and neighboring differentiated cells, interactions between stem cells and adhesion molecules, extracellular matrix components, etc.

Mammalian adult stem cell niches have been described in many tissues including the germinal (testis), hematopoietic, epidermal, intestinal system or central nervous system. The main objective of this book was therefore to highlight the molecular mechanisms that mediate the balanced response of stem cells to the needs of the organisms. Likewise, niches have also been linked to pathologies, by imposing aberrant function on stem cells or other targets. Therefore, the second objective of this book is to give a current view of the molecular dysregulation of niche biology leading to the disease. The third objective is to review the therapeutical targets described within stem cell niches.

Finally, I would like to conclude this preface by expressing my deepest gratitude to all authors who contributed to the elaboration of this book.

Sabine Wislet-Gendebien, PhD
Assistant Professor
GIGA-Neurosciences
University of Liège
Belgium

Section 1

Hematopoietic Stem Cell Niches in Normal and Pathological Conditions

Chapter 1

Delineation of Niches which Support Hematopoiesis

Hong Kiat Lim and Helen C. O'Neill

Additional information is available at the end of the chapter

http://dx.doi.org/10.5772/58715

1. Introduction

Hematopoiesis is the process of blood cell formation from self-renewing, multipotent hematopoietic stem cells (HSC) crucial for blood homeostasis of all living organisms [1]. This developmental process has two distinct phases, primitive and definitive hematopoiesis. In mice, the primitive phase of hematopoiesis occurs in the yolk sac at embryonic (E) day 7.5 [2-5]. It results in the production of primitive hematopoietic cells including mainly large, nucleated erythroblasts, some megakaryocytes, and primitive macrophages needed for embryonic growth [2-5]. The definitive phase of hematopoiesis in mice begins in the extra-embryonic yolk sac at E8.25 [2, 6]. Development in this phase is a hierarchical process, with two main lymphoid and myeloid lineages forming from which all blood cells develop. The lymphoid lineage comprises T and B lymphocytes and natural killer cells that arise from a common lymphoid progenitor (CLP). In contrast, the myeloid lineage comprises granulocytes (neutrophils, eosinophils, mast cells, and basophils), monocytes, erythrocytes and dendritic cells (DC), all developing from a common myeloid progenitor (CMP). The multipotent HSC resides at the apex of this hematopoietic hierarchy [1, 7].

Most of our present knowledge on hematopoiesis stems from decades of research on bone marrow, in part due to the clinical relevance of bone marrow as the main site for hematopoiesis in adults. However, there is cumulating evidence to suggest that the spleen can adopt extramedullary hematopoiesis in both the steady-state, and during disease and inflammation. While very little is understood about extramedullary hematopoiesis, there portends to be huge potential for regenerative medicine, if the hematopoietic capacity of spleen can be harnessed. For example, the hematopoietic output of spleen could be experimentally enhanced on bone marrow transplantation using either *in vivo* or *in vitro* means. In this report, information on hematopoiesis in bone marrow is reviewed and the current understanding of extramedullary hematopoiesis in spleen is considered.

2. Hematopoietic Stem Cells

HSC are multipotent adult stem cells, capable of differentiating to give all mature blood cell types, while still maintaining a pool of stem cells due to their unique ability to self-renew [8]. Definitive hematopoiesis occurs at E8.5 and is detected experimentally at E10.5 in the aorta-gonad-mesonephros region of the embryo [6, 9-11]. At E11.5, HSC appear in fetal liver where they exhibit high proliferative and differentiative potential with cell number expansion [12]. HSC appear in fetal spleen at E13-14 and undergo proliferation and differentiation to form mature blood cells [13]. At E17.5, HSC then proceed to bone marrow, the major hematopoietic site in adults where they are maintained during postnatal life [14].

HSC in murine bone marrow represent a heterogeneous population characterised phenotypically by the absence of lineage (Lin)-specific markers and high expression of Sca-1, also known as lymphocyte antigen 6A (Ly6A) and the c-Kit tyrosine kinase receptor (c-Kit). Within the $Lin^-Sca\text{-}1^+c\text{-}Kit^+$cell subset, HSC can be further delineated as $CD150^+CD34^-Flt3^-CD244^-CD48^-$ long-term reconstituting HSC (LT-HSC) and $Flt3^-CD34^+$short-term reconstituting HSC (ST-HSC) [15-17]. Functional studies involving adoptive transfer of HSC to reconstitute the hematopoietic system of lethally irradiated mice, remains the gold standard assay to distinguish ST-HSC from LT-HSC [18]. While LT-HSC are able to sustain reconstitution of the hematopoietic system over more than 25 weeks, and even a lifetime [19], ST-HSC provide short-term reconstitution for only ~6 weeks [17].

3. The fate of HSC

HSC can adopt several fates throughout the lifetime of an organism, including quiescence, dormancy, self-renewal and differentiation. Quiescence refers to a state of inactivity, whereby most HSC are in the G_0/G_1 phase of cell cycle and are not dividing [20]. It is believed that keeping HSC in the quiescent state is necessary for long-term maintenance of the HSC compartment, in part due to reduced stress associated with cellular respiration and DNA replication [20]. Signalling involving the TGF-β/Smad pathway has been implicated in the maintenance of quiescence in HSC [21], and the addition of TGF-β to *in vitro* cultures of LT-HSC can inhibit cell proliferation. The state which involves HSC in long-term quiescence is called dormancy [22]. Self-renewal is the process by which HSC undergo symmetrical or asymmetrical cell division to produce one or more daughter stem cells [23]. This process leads to expansion of stem cell numbers during development, and restores the stem cell pool after injury [23]. Self-renewal, along with quiescence, prevents the depletion of stem cells. Differentiation then involves specialisation of multipotent HSC to give mature blood cells, and is needed to replenish the hematopoietic system since most mature hematopoietic cell types are short-lived.

While the hematopoietic system has a high daily rate of cell turnover, most HSC are quiescent and divide very rarely. This raises the perplexing question of how the hematopoietic system can produce such large numbers of mature blood cells from slowly dividing HSC. The answer lies in the balance between the various HSC fates of quiescence, dormancy, self-renewal and differentiation, such that tight regulation occurs between these fates to ensure that HSC are both maintained for the life of the animal, and sufficient mature blood cells are produced to meet the demands of their development. The importance of this balance is highlighted in disease states such as leukaemia, which result from abnormal HSC development.

4. Niches for HSC

In the 1970s, Schofield introduced the concept of the HSC 'niche' after observing that once HSC were removed from the bone marrow microenvironment, they quickly lost the capacity to self-renew and to reconstitute the hematopoietic system. The HSC 'niche' involves a microenvironment comprising non-hematopoietic stromal cells, extracellular matrix and soluble regulatory factors that regulate the different fates of HSC. To date, three stromal cell types have been found to contribute to the HSC niche, namely endosteal, vascular, and perivascular cells [24-26]. However, it is still not known to what extent these distinct niches are truly independent of each other, and the extent of hematopoietic support contributed by each cell type is not clear.

The endosteal niche comprises a heterogeneous group of osteoblastic cells residing close to the endosteal lining of trabecular bone where they interact with HSC. The role of osteoblastic cells in HSC maintenance has been demonstrated in studies which vary the number of osteoblastic cells experimentally [27]. In experimental mouse models, it was shown that expression of a constitutively active form of parathyroid hormone (PTH) or the PTH-related protein receptor (PPR) important for calcium regulation, under control of the type 1 collagen $\alpha 1$ promoter, gave a marked increase not only in number of osteoblastic cells but also in HSC [27]. Osteoblastic cells maintain and regulate HSC through secretion of cytokines like angiopoietin-1 (ANGPT1), thrombopoietin (THPO) and osteopontin (SPP1), which bind to cell surface receptors on HSC, namely endothelial-specific receptor tyrosine kinase (TEK), myeloproliferative leukemia virus oncogene (MPL), and either CD44 or integrins including $\alpha_4\beta_1$, $\alpha_5\beta_1$,$\alpha_9\beta_1$, and $\alpha_4\beta_7$ [28-30]. A role for THPO and ANGPT1 in HSC maintenance was demonstrated in studies which showed a reduction in the number of HSC in the bone marrow of *Thpo*$^{-/-}$and *Angpt*$^{-/-}$mutant mice [28-30]. Similarly, *Spp1*$^{-/-}$mice showed a marked increase in the number of HSC cycling, consistent with osteopontin (SPP1) being a negative regulator of HSC proliferation [30]. In addition, osteoblastic cells express Jagged 1, which leads to the inhibition of HSC differentiation, and the enhancement of HSC self-renewal when engaged with Notch receptors on HSC [27].

There is increasing evidence now that HSC niches are located in the vicinity of blood vessels, and in close proximity with sinusoids in bone marrow, so allowing rapid mobilisation of HSC into the bloodstream after administration of granulocyte colony-stimulating factor (G-CSF) [16]. Vascular niches also play an important role during embryogenesis since HSC self-renew and differentiate at a stage of foetal development when bone marrow cavities are not yet formed [31]. Endothelial cells in the vascular niche express cell surface molecules that allow HSC and immune cells to move between the bone marrow and the periphery [32]. Vascular niches also rely on the organisation of sinusoids or capillaries similar to fenestrated endothelium. When hematopoietic cells enter tissues, the slow blood flow in the larger sinusoids allows blood-borne cells to interact with sinusoidal endothelial cells via adhesion molecules, resulting in movement of cells through the sinus wall and into the extravascular space within bone marrow [33-35]. Sinusoidal endothelial cells express adhesion molecules like E-selectin and vascular cell adhesion molecule 1 (VCAM1) [32]. The importance of vascular niches for hematopoiesis was demonstrated by conditionally deleting the signalling protein vascular endothelial growth factor receptor 2 (VEGFR2) in adult mice [36]. This resulted in an inability of sinusoidal endothelial cells to develop after irradiation of mice, so preventing reconstitution of the hematopoietic system [36]. In addition, sinusoidal endothelial cells express gp130, a cytokine receptor essential for HSC self-renewal [37]. The importance of endothelial cells expressing gp130 in hematopoiesis was also demonstrated in reconstitution studies in mice where bone marrow cells from *gp130*$^{-/-}$mice could reconstitute the hematopoietic system of irradiated wild-type mice, while wild-type bone marrow cells were unable to reconstitute the hematopoietic system of *gp130*$^{-/-}$mice [38].

Recently, three populations of perivascular reticular cells expressing high levels of CXCL12, were identified as important niche elements for HSC in bone marrow. They have been described as CXCL12-abundant reticular (CAR) cells [39], nestin-GFP^{+}mesenchymal stem cells [40] and leptin receptor^{+}stromal cells [41]. CAR cells were characterised as bipotent adipo-osteogenic progenitors, which develop around sinusoids [39, 42]. By conditionally ablating CAR cells using transgenic mice with the diphtheria toxin receptor gene inserted into the *cxcl12* locus, CAR cells were shown to promote HSC proliferation, while maintaining them in an undifferentiated state [42]. The absence of CAR cells resulted in a reduction in HSC and progenitors, and of early myeloid differentiation of HSC.

Perivascular niches are also described by nestin^{+}mesenchymal stem cells situated near the larger blood vessels in BM [40]. These cells were identified in close association with HSC by their nestin expression. When nestin^{+}mesenchymal stem cells were conditionally ablated from mice, the frequency of HSC decreased, indicating the importance of these cells as a perivascular niche element in bone marrow [40]. Leptin receptor^{+}stromal cells have also been identified as perivascular cells surrounding sinusoids. This subset expresses high levels of CXCL12 and stem cell factor (SCF), and may overlap with other previously described CAR cell subsets. All three described perivascular subsets are an important source of SCF [41], a cytokine that binds to the c-Kit tyrosine kinase receptor expressed on hematopoietic stem/progenitor cells [43]. A recent study demonstrated depletion of the HSC pool in *Scf*$^{-/-}$mice, highlighting the importance of SCF as well as perivascular cells in HSC maintenance [41].

5. Spleen as a hematopoietic organ

The spleen is the largest blood-filtering organ. It also contains structures which support lymphoid cell development and function, and in this respect resembles a lymph node. The spleen is central to both hematopoiesis and immunity. Anatomically, it is composed of red pulp which is responsible for blood cell removal, and white pulp, which houses T and B lymphocytes in discrete regions [44]. Red and white pulp of murine spleen are separated by a marginal zone comprising marginal zone macrophages, B cells, and DC, and a marginal sinus where the smallest arterial branches terminate [44]. The white pulp is further divided into T cell zones and B cell follicles, with germinal centres containing B cells [44]. While murine and human spleens are largely similar in terms of their architecture, there are some notable differences [44]. Unlike murine spleen, the marginal sinus is not present in the human spleen, although human spleen does contain a perifollicular region, external to the marginal zone in the red pulp, which is absent in the murine spleen.

5.1. Development of murine spleen

Spleen organogenesis involves formation of the splanchnic mesodermal plate from lateral plate mesoderm at E9.5 [45, 46]. The splanchnic mesodermal plate comprises epithelial-like cells which are arranged as an organized plate of cells at the right and left side of the foregut [45]. The initial condensation of mesenchyme within the dorsal pancreatic mesenchyme occurs adjacent to the splanchnic mesodermal plate at E10.5 [46]. By E10.5, there is progressive loss of thickness on the right side of splanchnic mesodermal plate, causing it to be thicker at the left side [46]. In addition, the cells constituting the splenic condensing mesenchyme also acquire the splenic cell fate [45]. At the same time, leftward growth of pancreatic mesenchyme and spleen anlagen occur [45]. By E11.5, the spleen anlage proper is formed laterally to the developing stomach [45, 47].

While the exact cellular requirements governing development of spleen remain poorly understood, it is widely known that the interaction between the non-hematopoietic $CD45^-VCAM1^+ICAM1^+LT\beta R^+$lymphoid tissue organiser (LTo) cells and hematopoietic $CD45^+CD3^-CD4^+cKit^+LT^+$lymphoid tissue inducer (LTi) cells is crucial for the development of secondary lymphoid organs such as lymph nodes and Peyer's patches [48]. This is due to lymphotoxin LTαβ signaling, which occurs as a result of binding between lymphotoxin-α1β2 ($Lt\alpha_1\beta_2$) expressed on LTi cells and the lymphotoxin-β receptor (LTβR) on LTo cells [49]. Engagement of $Lt\alpha_1\beta_2$ and LTβR induces the activation of two NFκB pathways which promote mesenchymal and endothelial cell differentiation [50]. In addition, increased expression of adhesion molecules like VCAM1, intracellular adhesion molecule 1 (ICAM1) and mucosal vascular addressin cell adhesion molecule 1 (MAdCAM1), and homeostatic chemokines such as CXCL13, CCL19 and CCL21, ensues as part of the differentiation program of mesenchymal cells [51]. These adhesion molecules and chemokines attract and retain additional hematopoietic cells at the formation site.

The importance of $Lt\alpha_1\beta_2$ signaling in secondary lymphoid organ development was first documented after the initial observation that lymph nodes and Peyer's patches failed to develop in $LT\alpha^{-/-}$mice [52]. Identification of what are currently known as LTi cells began with the search for the LTαβ-expressing cell subset which was first found in developing mesenteric lymph nodes [53]. These LTαβ-expressing cells were later characterised as $CD45^+CD3^-CD4^+$cells. Insight into the role of LTαβ-expressing $CD45^+CD3^-CD4^+$cells in secondary lymphoid organ development was gained with the generation of two mutant mice that lacked

$CD45^{+}CD3^{-}CD4^{+}$cells [54, 55]. A detailed analysis of these two mutant mice revealed an absence of lymph node and Peyer's patch development. However, the architecture of the T and B cell areas in spleen was unaffected [55, 56], suggesting the existence of multiple cellular subsets, which provide instructions to the splenic stromal cells.

5.2. The red pulp regulates erythrocyte turnover

The red pulp region has a very specialised function as a blood filter. This region contains many splenic cords formed by red pulp fibroblasts and reticular fibres where blood first enters via the afferent splenic artery [44]. Blood then enters venous sinuses lined by endothelium with a discontinuous structure [44, 57]. Slits formed in the endothelium of the sinuses allow the passage of most blood cells but not aged or damaged platelets and erythrocytes. These are phagocytosed by red pulp macrophages in the cords [44, 58].

In addition to the filtration of blood, the red pulp is involved in recycling iron. Phagocytosis of aged erythrocytes by splenic macrophages results in release of hem from hemoglobin, which is converted into biliverdin and ferrous iron [44, 59]. The resultant iron is either stored or released as ferritins from cells. Plasmablasts and plasma cells are also present in the red pulp for antibody production [44].

5.3. The white pulp facilitates T and B lymphoid responses

The white pulp of spleen contains the T cell zones and B cell follicles. In T cell zones, fibroblastic reticular cells encircle the central arterioles and form a network that connects the T cell zones to the marginal zone [60, 61]. This network serves to guide T cells in their migration from marginal zone to T cell zones [60, 61]. It is in the T cell zone where T cells interact with DC and passing B cells. In contrast, B cell follicles comprise activated B cells undergoing clonal expansion, isotype switching and somatic hypermutation. They also house specialised antigen handling cells called follicular dendritic cells, which form the architectural framework of the follicles, and are involved in retaining and presenting antigens to activated B cells in the germinal centre.

5.4. Multiple subsets of DC exist in spleen

Dendritic cells are specialised antigen presenting cells capable of inducing an adaptive immune response [62]. They do so by acquiring, processing and presenting antigens on major histocompatibility complex (MHC) molecules to naïve T cells. While all DC have capacity for antigen acquisition, processing and presentation, they are heterogeneous and their subtypes differ in location, migratory pathway, cell surface marker expression and immunological function [62]. The definition of a DC is therefore not straightforward, although most DC are phenotypically marked by high expression of CD11c and MHC-II [63].

The commonly described DC subsets in murine spleen include conventional DC (cDC), plasmacytoid DC (pDC), regulatory DC (DCreg) and monocyte-derived DC (mo-DC) [64, 65]. Conventional DC represent steady-state, mature DC with $CD11c^{hi}MHC\text{-}II^{hi}$ marker expression that can be further classified into $CD8\alpha^{+}$and $CD8\alpha^{-}$subsets [65]. $CD8\alpha^{+}$cDC are phenotypically distinguishable as CD11c^{+}CD11b^{-}CD8α^{+}MHC-II^{+}B220^{-}cells, while CD8α^{-}cDC are CD11c $^{+}$CD11b^{+}CD8α^{-}MHC-II^{+}B220^{-}cells [65, 66]. Plasmacytoid DC exist as less mature DC precursors in the steady-state, producing natural type I interferon (IFN-α) upon viral exposure. They can be distinguished from other DC subsets by expression of Ly6C, B220, and low expression level of the CD11c marker [65, 67]. Under certain inflammatory conditions, monocytes which are phagocytic myeloid cells, are induced to differentiate into mo-DC for antigen presentation [68]. These cells can be distinguished from other DC by CD64 (FcR) expression.

5.5. Lineage origin of splenic DC

With high heterogeneity within the DC lineage, an important question is whether each subset is developmentally distinct. Early studies provided direct evidence for a myeloid origin of DC following reconstitution of cDC and pDC in the spleen and thymus when mouse bone marrow-derived CMP were transplanted into irradiated recipients [69]. It is now well established that all leukocytes originate from bone marrow-derived HSC. In the early stages of hematopoiesis, successive commitment steps result in the divergence of lymphoid and myeloid lineages, generating CLP and CMP [70]. The CLP give rise to B, T and natural killer cells, whilst CMP develop into macrophage/DC progenitors (MDP), identified as Lin^{-}CX3CR1^{+}CD11b^{-} CD115^{+}cKit^{+}CD135^{+}cells [70]. These subsequently give rise to either common DC progenitors (CDP) or two monocyte subsets distinguishable as Ly-6C^{+}and Ly-6C^{-}cells [64, 71]. In contrast to the developmental flexibility of MDP in terms of ability to produce cDC, pDC and monocytes, CDP are restricted to producing cDC and pDC [71]. It has also been shown that adoptive transfer of either CMP or MDP can give rise to CDP and monocytes, indicating that CDP are downstream of CMP or MDP [70].

While monocytes and pDC have been found to exit bone marrow as mature cells, cDC leave as immature precursors that further differentiate and mature within lymphoid organs [71]. In mouse blood, two populations of DC precursors have been described as CD11c^{int}CD11b $^{+}$CD45RA^{-}and CD11c^{lo}CD11b^{-}CD45RA^{hi} cells that differentiate to give mature CD8α^{-}cDC in the presence of TNF-α and GM-CSF, and IFN-α producing pDC in the presence of GM-CSF and CpG [72]. CD8α^{-}cDC are able to stimulate T cells to produce IL-2 in response to microbial stimuli, while pDC weakly stimulate T cells by producing large quantities of natural type I interferon [69]. A recent study reported the identification of the most immediate cDC precursors (pre-DC) in spleen with a CD11c^{int}CD45RA^{lo}CD43^{int}SIRP-α^{int}CD4^{-}CD8α^{-}phenotype [73]. Progenitors in bone marrow give rise to pre-DC which enter the spleen where they differentiate further to give cDC and p-pre-DC.

6. Extramedullary hematopoiesis in spleen

The spleen can support hematopoiesis, and this finding is based on early evidence documenting the recovery of mice lethally irradiated while the spleen was shielded [74]. This process is termed extramedullary hematopoiesis, since it occurs in sites other than the bone medullary cavity. Extramedullary hematopoiesis can be viewed as an active or a passive process. While extramedullary hematopoiesis occurs as an active natural process during fetal development and also during infection, passive extramedullary hematopoiesis occurs due to impairment of hematopoiesis in the bone marrow [75]. Occurrence of an active process is evidenced by studies, showing that the low number of hematopoietic stem/progenitor cells present in murine spleen in the steady-state increases quickly following inflammation [76, 77]. The presence of hematopoietic stem/progenitor cells in the steady-state is however not restricted to murine spleen as spleens of pigs, baboons and humans was also found to retain a low number of hematopoietic stem/progenitor cells under steady-state conditions [78]. Moreover, in cell tracing experiments, spleen cells derived from both neonatal and adult mice were able to provide hematopoietic reconstitution of lethally irradiated host mice following adoptive transfer [79, 80]. These findings confirm that spleen can adopt a role in extramedullary hematopoiesis, at least during times of stress or inflammation. Evidence, which suggests a role for spleen in hematopoiesis, also raises the possibility of a splenic HSC niche that supports the maintenance of HSC in the resting state.

6.1. Splenic stromal cell network

Thus far, HSC niches in spleen have not been well investigated, although at least six distinct stromal cell types have been mapped to different regions of the spleen. These include $gp38^+$fibroblastic reticular cells in the T cell zones, $CD35^+$follicular dendritic cells in the B cell follicles, $MAdCAM1^+$marginal reticular cells in the marginal zone, and red pulp fibroblasts, lymphatic endothelial cells and vascular endothelial cells in the red pulp [81].

Fibroblastic reticular cells are mesenchymal cells, which along with reticular fibres and fibrous extracellular matrix bundles are required for the formation of a reticular network that serves as a scaffolding for the three-dimensional structure of secondary lymphoid organs [82]. In addition to creating a network in the T cell areas of spleen, which serves to guide T lymphocytes in their migration from marginal zone to the T cell zones [60, 61], fibroblastic reticular cells are also involved in creating a conduit system that selectively allows molecules of low molecular mass such as chemokines and antigens to enter the T cell zone [82-84]. Large molecules are trapped in the cortical sinuses by subcapsular sinus macrophages [82-84]. Such a conduit system offers an opportunity for different types of signals to be delivered in secondary lymphoid organs for optimisation of immunity to different pathogens [84].

In addition to their structural role, fibroblastic reticular cells are also involved in regulation of the immune response. They secrete homeostatic chemokines like CCL21, CCL19 and CXCL12 which attract naïve T cells expressing the chemokine receptors CCR7 and CXCR4 [85]. Fibroblastic reticular cells also enhance the survival of naïve T cells in the steady-state by producing interleukin-7. This is important in increasing the probability that naïve T cells meet cognate antigen presented on antigen presenting cells. A recent study revealed that fibroblastic reticular cells stimulate the activation of cytotoxic T cells via alarmin IL-33 secretion during viral infection [86].

Marginal reticular cells represent a unique subset of stromal cells located in the marginal zone of the spleen. Despite expressing many common stromal cell surface markers like ER-TR7, desmin, laminin, VCAM1, and MAdCAM1, marginal reticular cells appear to also specifically express RANKL, the receptor activator of NF-κB ligand, and secrete CXCL13 [87]. While RANKL has been found to be essential for lymph node development [88], its function in spleen remains elusive. These stromal cells form a conduit system which can capture and deliver antigens to the B cell follicles [87].

In the red pulp region, red pulp fibroblasts form the splenic cords necessary for filtration of blood [81]. They also control splenic blood flow and assist in removal of dead or dying red blood cells [81]. Additionally, red pulp fibroblasts are involved in the localisation of different cells in the red pulp of spleen through expression of the cell adhesion molecule ICAM1 that binds to lymphocyte function-associated antigen 1 (LFA1), a heterodimeric receptor protein found on lymphocytes [89, 90]. For instance, ICAM1-LFA1 interaction between red pulp fibroblasts and plasma cells results in movement of plasma cells into the red pulp region for secretion of antibodies into the circulation [91].

7. Evidence that splenic stroma supports hematopoiesis

Evidence suggesting that the spleen contains stromal cells, which support *in vitro* hematopoiesis has been previously reported by this lab [92-95]. In particular, it was demonstrated that the stromal monolayer forming in long-term cultures (LTC) of 8-day old murine spleen supports the maintenance of small progenitors and the production of a distinct class of large, immature DC, coined 'LTC-DC'. These can be produced continuously in LTC for years in the absence of added growth factors and cytokines [92-95]. When the small progenitors maintained in LTC were specifically sorted and co-cultured over STX3 stroma, large immature dendritic-like cells were produced [96]. STX3 stroma had been isolated as a stromal cell line from a splenic LTC which had ceased production of dendritic-like cells after multiple passages, apparently due to the loss of progenitors [97]. Interestingly, similar dendritic-like cells, termed 'L-DC' were also produced when STX3 was overlaid with lineage-negative (Lin^-) cells derived from bone marrow [98]. L-DC, like LTC-DC, are large cells which express CD11b and CD11c but not MHC-II and CD8α[99, 100]. These cells are highly efficient in endocytosis and cross presentation of antigen for $CD8^+$T cell activation, particularly after exposure to lipopolysaccharide [99, 100].

In order to better understand the cellular composition of the heterogeneous STX3 stroma, the line was cloned to form 102 splenic stromal lines [98]. These include the 5G3 clone, which supports *in vitro* hematopoiesis, and the 3B5 clone, which is a non-supporter. The 5G3 clone supports production of 'L-DC' from hematopoietic progenitors in a highly reproducible and contact-dependent manner, similar to the parent line, STX3 [94, 99]. Transcriptome analysis of 5G3 stroma has revealed high expression of genes including *Sca-1*, *Vcam1*, *Pdgfrα/β* and *CXCL12*, which are associated with perivascular reticular cells described in the bone marrow (data in preparation). In addition, 5G3 was shown to have osteogenic but not adipogenic differentiative capacity (data in preparation). Such evidence raises the hypothesis that spleen contains a unique perivascular niche comprising mesenchymal stromal cells resembling osteoprogenitors that supports extramedullary hematopoiesis.

The highly reproducible nature of DC production in splenic LTC and in stromal co-cultures, suggests that a potential equivalent process may exist *in vivo*. Already there is some evidence for an *in vivo* equivalent L-DC subset in spleen, which was originally termed 'IVL-DC' [101]. However, the *in vivo* equivalent stromal supporter cell line has not yet been identified. Moreover, since DC production in splenic stromal co-cultures can be maintained for a long time, it seems likely that the progenitors, which are maintained in cultures, are self-renewing. One hypothesis is that the particular splenic stromal cells maintain self-renewing progenitors through close contact. To test this hypothesis, various progenitor subsets from bone marrow and spleen were sorted and tested for their capacity to seed 5G3 stroma with production of L-DC. 5G3 stromal co-cultures were established with MDP, CDP, ST-HSC or LT-HSC. Since no L-DC production was observed in co-cultures overlaid with MDP and CDP, it was evident that L-DC did not derive from the same progenitors which gave rise to cDC, pDC and monocytes [102]. When the Flt3$^-$c-Kit$^+$Lin$^-$Sca-1$^+$subset of LT-HSC from bone marrow, and the Flt3$^+$c-Kit$^+$Lin$^-$Sca-1$^+$subset of ST-HSC were overlaid on 5G3 stroma, L-DC production was clearly supported. In some co-cultures, contact between non-adherent cells and stroma was prevented using a Transwell membrane [92]. L-DC production occurred for up to 35 days from overlaid HSC, but only in co-cultures where overlaid cells maintained contact with 5G3 stroma [92]. The development of L-DC is therefore dependent on the interaction between primitive HSC and the competent 5G3 stroma. The progenitor of L-DC therefore appears to be a subset of HSC, some cells of which express Flt3 and resemble ST-HSC and the less primitive multi-potent progenitors (MPP) [92, 102].

Indeed, it has become clear that L-DC progenitors *in vivo* reflect HSC endogenous to spleen, and that the process of L-DC development reflects extramedullary hematopoiesis. L-DC progenitors first appear in murine spleen at E18.5, while progenitors of cDC appear after 4 days [103]. This raises the possibility that hematopoietic progenitors in spleen are laid down during ontogeny, and that hematopoiesis in spleen and the differentiation of dendritic-like 'L-DC' occurs as an active process, independent of inflammatory signaling. One model is that the splenic stromal microenvironment supports the restricted differentiation of endogenous progenitors to give antigen presenting cells unique to the spleen microenvironment. Indeed, studies to date on the *in vivo* tissue distribution of cells equivalent to L-DC, confirm that L-DC reflect a novel subset which is limited in its tissue distribution to spleen [100]. Such a model does not discount the possibility that spleen can also act as a site for extramedullary hemato-poiesis under inflammatory conditions.

While it is known that hematopoiesis can be driven by inflammation *in vivo* [104], it is important to consider whether *in vitro* hematopoiesis in 5G3 cocultures involves the same mechanism. Whether the same stromal microenvironment contributes to both steady-state and inflamma-tory processes remains to be determined. This laboratory has studied the role of toll-like receptor signaling in the production of L-DC in *in vitro* co-cultures involving 5G3 stroma. Toll-like receptor binding of pathogen components triggers an inflammatory response. When co-cultures were established with bone marrow progenitors derived from mutant mice lacking the adapter proteins MyD88 and TRIF crucial for toll-like receptor signaling, theywere found to be producers of L-DC [92]. This suggests that L-DC production must occur independently of toll-like receptor signaling and inflammation. L-DC production therefore appears to reflect an active process of extramedullary hematopoiesis, dependent on primitive HSC endogenous to spleen.

8. Conclusion

While spleen has been traditionally viewed as an organ of immunity, recent evidence has shed light on its involvement in hematopoiesis. In this report, we have advanced the existence of niches for hematopoiesis in spleen, comprising a perivascular reticular cell type that supports extramedullary hematopoiesis from primitive HSC. Previous evidence in mouse and humans suggests that perivascular reticular cells reflect mesenchymal stem/progenitor cells located in close proximity with endothelial cells associated with the vasculature. Since the perivascular cells in bone marrow which constitute the HSC niche surround the sinusoids, it is quite possible that a similar type of cell exists in spleen, particularly in the red pulp region, where copious numbers of sinusoids are found. Moreover, transcriptome analysis of spleen-derived 5G3 stroma has revealed a close genetic resemblance between 5G3 stroma and perivascular cells in the bone marrow (data in preparation). Also consistent with this is the osteogenic capacity of 5G3 cells, suggesting the presence of mesenchymal osteoprogenitors in spleen with ability to support *in vitro* hematopoiesis.

The hypothesis that spleen contains perivascular cells as niches for hematopoiesis is not without functional precedence. Perivascular reticular cells present in the red pulp of spleen could provide a niche for maintenance of HSC in the steady-state. Furthermore, restricted hematopoiesis leading to development of only L-DC raises the possibility that the spleen microenvironment may support the production of tissue-specific antigen presenting cells. Their location in red pulp could reflect a specific role in monitoring blood-borne antigens and interacting with migrating lymphoid cells.

The lack of understanding of spleen as a hematopoietic organ has limited clinical application to date. A history of work from this lab has however considered spleen as a site for extramedullary hematopoiesis. Since spleen is a secondary site for hematopoiesis, it offers potential for reengineering niches to increase hematopoietic cell production. For example, if these unique stromal cells could be isolated and used to expand HSC *in vitro*, or provided as an ectopic niche *in vivo* for the same purpose, then the potential exists to enhance hematopoiesis during HSC transplantation. Regeneration or expansion of these niches could represent future therapy for patients undergoing myeloablative treatment, involution of lymphoid tissue with ageing, or HSC transplantation.

Acknowledgements

This work was supported by grants from the National Health and Medical Research Council of Australia, and the Australian Research Council.

Author details

Hong Kiat Lim and Helen C. O'Neill*

*Address all correspondence to: Helen.ONeill@anu.edu.au

Research School of Biology, Australian National University, Canberra, Australia

References

[1] Seita, J. and I.L. Weissman, *Hematopoietic stem cell: self-renewal versus differentiation.* Wiley Interdiscip Rev Syst Biol Med, 2010. 2(6): p. 640-53.

[2] Hirschi, K.K., *Hemogenic endothelium during development and beyond.* Blood, 2012. 119(21): p. 4823-7.

[3] Ueno, H. and I.L. Weissman, *Clonal Analysis of Mouse Development Reveals a Polyclonal Origin for Yolk Sac Blood Islands.* Developmental Cell, 2006. 11(4): p. 519-533.

[4] Baron, M.H., *Embryonic origins of mammalian hematopoiesis.* Exp Hematol, 2003. 31(12): p. 1160-9.

[5] Haar, J.L. and G.A. Ackerman, *A phase and electron microscopic study of vasculogenesis and erythropoiesis in the yolk sac of the mouse.* Anat Rec, 1971. 170(2): p. 199-223.

[6] Medvinsky, A. and E. Dzierzak, *Definitive Hematopoiesis Is Autonomously Initiated by the AGM Region.* Cell, 1996. 86(6): p. 897-906.

[7] Doulatov, S., et al., *Hematopoiesis: A Human Perspective.* Cell Stem Cell, 2012. 10(2): p. 120-136.

[8] Trumpp, A., M. Essers, and A. Wilson, *Awakening dormant hematopoietic stem cells.* Nature Reviews Immunology, 2010. 10(3): p. 201-209.

[9] Li, Z., et al., *Mouse Embryonic Head as a Site for Hematopoietic Stem Cell Development.* Cell Stem Cell, 2012. 11(5): p. 663-675.

[10] Palis, J., et al., *Development of erythroid and myeloid progenitors in the yolk sac and embryo proper of the mouse.* Development, 1999. 126(22): p. 5073-84.

[11] Müller, A.M., et al., *Development of hematopoietic stem cell activity in the mouse embryo.* Immunity, 1994. 1(4): p. 291-301.

[12] Ema, H. and H. Nakauchi, *Expansion of hematopoietic stem cells in the developing liver of a mouse embryo.* Blood, 2000. 95(7): p. 2284-8.

[13] Godin, I., et al., *Stem Cell Emergence and Hemopoietic Activity Are Incompatible in Mouse Intraembryonic Sites.* The Journal of Experimental Medicine, 1999. 190(1): p. 43-52.

[14] Dzierzak, E. and N.A. Speck, *Of lineage and legacy: the development of mammalian hematopoietic stem cells.* Nat Immunol, 2008. 9(2): p. 129-136.

[15] Christensen, J.L. and I.L. Weissman, *Flk-2 is a marker in hematopoietic stem cell differentiation: A simple method to isolate long-term stem cells.* Proceedings of the National Academy of Sciences, 2001. 98(25): p. 14541-14546.

[16] Kiel, M.J., et al., *SLAM Family Receptors Distinguish Hematopoietic Stem and Progenitor Cells and Reveal Endothelial Niches for Stem Cells.* Cell, 2005. 121(7): p. 1109-1121.

[17] Yang, L., et al., *Identification of Lin–Sca1+kit+CD34+Flt3– short-term hematopoietic stem cells capable of rapidly reconstituting and rescuing myeloablated transplant recipients.* Blood, 2005. 105(7): p. 2717-2723.

[18] Muller-Sieburg, C., et al., *Stem cell heterogeneity: implications for aging and regenerative medicine.* Blood, 2012.

[19] Spangrude, G.J., S. Heimfeld, and I.L. Weissman, *Purification and characterization of mouse hematopoietic stem cells.* Science, 1988. 241(4861): p. 58-62.

[20] Eliasson, P. and J.I. Jonsson, *The hematopoietic stem cell niche: low in oxygen but a nice place to be.* J Cell Physiol, 2010. 222(1): p. 17-22.

[21] Sitnicka, E., et al., *Transforming growth factor beta 1 directly and reversibly inhibits the initial cell divisions of long-term repopulating hematopoietic stem cells.* Blood, 1996. 88(1): p. 82-8.

[22] Wilson, A., et al., *Hematopoietic Stem Cells Reversibly Switch from Dormancy to Self-Renewal during Homeostasis and Repair.* Cell, 2008. 135(6): p. 1118-1129.

[23] He, S., D. Nakada, and S.J. Morrison, *Mechanisms of stem cell self-renewal.* Annu Rev Cell Dev Biol, 2009. 25: p. 377-406.

[24] Bianco, P., *Bone and the hematopoietic niche: a tale of two stem cells.* Blood, 2011.

[25] Kiel, M.J. and S.J. Morrison, *Uncertainty in the niches that maintain hematopoietic stem cells.* Nature Reviews Immunology, 2008. 8(4): p. 290-301.

[26] Nagasawa, T., Y. Omatsu, and T. Sugiyama, *Control of hematopoietic stem cells by the bone marrow stromal niche: the role of reticular cells.* Trends in Immunology, 2011. 32(7): p. 315-320.

[27] Calvi, L.M., et al., *Osteoblastic cells regulate the hematopoietic stem cell niche.* Nature, 2003. 425(6960): p. 841-846.

[28] Qian, H., et al., *Critical role of thrombopoietin in maintaining adult quiescent hematopoietic stem cells.* Cell Stem Cell, 2007. 1(6): p. 671-84.

[29] Lilly, A.J., W.E. Johnson, and C.M. Bunce, *The Hematopoietic Stem Cell Niche: New Insights into the Mechanisms Regulating Hematopoietic Stem Cell Behaviour.* Stem Cells International, 2011. 2011.

[30] Nilsson, S.K., et al., *Osteopontin, a key component of the hematopoietic stem cell niche and regulator of primitive hematopoietic progenitor cells.* Blood, 2005. 106(4): p. 1232-9.

[31] Huber, T.L., et al., *Hemangioblast commitment is initiated in the primitive streak of the mouse embryo.* Nature, 2004. 432(7017): p. 625-630.

[32] Sipkins, D.A., et al., *In vivo imaging of specialized bone marrow endothelial microdomains for tumour engraftment.* Nature, 2005. 435(7044): p. 969-973.

[33] Quesenberry, P.J. and P.S. Becker, *Stem cell homing: Rolling, crawling, and nesting.* Proceedings of the National Academy of Sciences, USA, 1998. 95(26): p. 15155-15157.

[34] Mazo, I.B., et al., *Hematopoietic Progenitor Cell Rolling in Bone Marrow Microvessels: Parallel Contributions by Endothelial Selectins and Vascular Cell Adhesion Molecule 1.* The Journal of Experimental Medicine, 1998. 188(3): p. 465-474.

[35] Frenette, P.S., et al., *Endothelial selectins and vascular cell adhesion molecule-1 promote hematopoietic progenitor homing to bone marrow.* Proceedings of the National Academy of Sciences, USA, 1998. 95(24): p. 14423-14428.

[36] Hooper, A.T., et al., *Engraftment and reconstitution of hematopoiesis is dependent on VEGFR2-mediated regeneration of sinusoidal endothelial cells.* Cell Stem Cell, 2009. 4(3): p. 263-74.

[37] Audet, J., et al., *Distinct role of gp130 activation in promoting self-renewal divisions by mitogenically stimulated murine hematopoietic stem cells.* Proceedings of the National Academy of Sciences, USA, 2001. 98(4): p. 1757-1762.

[38] Yao, L., et al., *Bone marrow dysfunction in mice lacking the cytokine receptor gp130 in endothelial cells.* Blood, 2005. 106(13): p. 4093-4101.

[39] Sugiyama, T., et al., *Maintenance of the hematopoietic stem cell pool by CXCL12-CXCR4 chemokine signaling in bone marrow stromal cell niches.* Immunity, 2006. 25(6): p. 977-88.

[40] Mendez-Ferrer, S., et al., *Mesenchymal and hematopoietic stem cells form a unique bone marrow niche.* Nature, 2010. 466(7308): p. 829-834.

[41] Ding, L., et al., *Endothelial and perivascular cells maintain hematopoietic stem cells.* Nature, 2012. 481(7382): p. 457-462.

[42] Omatsu, Y., et al., *The Essential Functions of Adipo-osteogenic Progenitors as the Hematopoietic Stem and Progenitor Cell Niche.* Immunity, 2010. 33(3): p. 387-399.

[43] McNiece, I.K. and R.A. Briddell, *Stem cell factor.* Journal of Leukocyte Biology, 1995. 58(1): p. 14-22.

[44] Mebius, R.E. and G. Kraal, *Structure and function of the spleen.* Nat Rev Immunol, 2005. 5(8): p. 606-616.

[45] Brendolan, A., et al., *Development and function of the mammalian spleen.* Bioessays, 2007. 29(2): p. 166-77.

[46] Hecksher-Sorensen, J., et al., *The splanchnic mesodermal plate directs spleen and pancreatic laterality, and is regulated by Bapx1/Nkx3.2.* Development, 2004. 131(19): p. 4665-75.

[47] Burn, S.F., et al., *The dynamics of spleen morphogenesis.* Dev Biol, 2008. 318(2): p. 303-11.

[48] Randall, T.D., D.M. Carragher, and J. Rangel-Moreno, *Development of secondary lymphoid organs.* Annu Rev Immunol, 2008. 26: p. 627-50.

[49] Tan, J.K.H. and T. Watanabe, *Chapter 5-Artificial Engineering of Secondary Lymphoid Organs*, in *Advances in Immunology*, W.A. Frederick, Editor. 2010, Academic Press. p. 131-157.

[50] Dejardin, E., et al., *The lymphotoxin-beta receptor induces different patterns of gene expression via two NF-kappaB pathways.* Immunity, 2002. 17(4): p. 525-35.

[51] Honda, K., et al., *Molecular basis for hematopoietic/mesenchymal interaction during initiation of Peyer's patch organogenesis.* J Exp Med, 2001. 193(5): p. 621-30.

[52] De Togni, P., et al., *Abnormal development of peripheral lymphoid organs in mice deficient in lymphotoxin.* Science, 1994. 264(5159): p. 703-7.

[53] Mebius, R.E., P. Rennert, and I.L. Weissman, *Developing lymph nodes collect CD4+CD3-LTbeta+cells that can differentiate to APC, NK cells, and follicular cells but not T or B cells.* Immunity, 1997. 7(4): p. 493-504.

[54] Sun, Z., et al., *Requirement for RORgamma in thymocyte survival and lymphoid organ development.* Science, 2000. 288(5475): p. 2369-73.

[55] Yokota, Y., et al., *Development of peripheral lymphoid organs and natural killer cells depends on the helix-loop-helix inhibitor Id2.* Nature, 1999. 397(6721): p. 702-706.

[56] Zhang, N., J. Guo, and Y.W. He, *Lymphocyte accumulation in the spleen of retinoic acid receptor-related orphan receptor gamma-deficient mice.* J Immunol, 2003. 171(4): p. 1667-75.

[57] Drenckhahn, D. and J. Wagner, *Stress fibers in the splenic sinus endothelium in situ: molecular structure, relationship to the extracellular matrix, and contractility.* The Journal of Cell Biology, 1986. 102(5): p. 1738-1747.

[58] Bratosin, D., et al., *Cellular and molecular mechanisms of senescent erythrocyte phagocytosis by macrophages. A review.* Biochimie, 1998. 80(2): p. 173-195.

[59] Knutson, M. and M. Wessling-Resnick, *Iron metabolism in the reticuloendothelial system.* Crit Rev Biochem Mol Biol, 2003. 38(1): p. 61-88.

[60] Bajénoff, M., N. Glaichenhaus, and R.N. Germain, *Fibroblastic Reticular Cells Guide T Lymphocyte Entry into and Migration within the Splenic T Cell Zone.* The Journal of Immunology, 2008. 181(6): p. 3947-3954.

[61] Anderson, A.O. and S. Shaw, *T cell adhesion to endothelium: the FRC conduit system and other anatomic and molecular features which facilitate the adhesion cascade in lymph node.* Semin Immunol, 1993. 5(4): p. 271-82.

[62] Shortman, K. and S.H. Naik, *Steady-state and inflammatory dendritic-cell development.* Nat Rev Immunol, 2007. 7(1): p. 19-30.

[63] Miloud, T., G.J. Hämmerling, and N. Garbi, *Review of Murine Dendritic Cells: Types, Location, and Development*. 2009. p. 21-42.

[64] Geissmann, F., et al., *Development of Monocytes, Macrophages, and Dendritic Cells.* Science, 2010. 327(5966): p. 656-661.

[65] Merad, M., et al., *The dendritic cell lineage: ontogeny and function of dendritic cells and their subsets in the steady state and the inflamed setting.* Annu Rev Immunol, 2013. 31: p. 563-604.

[66] Vremec, D., et al., *CD4 and CD8 Expression by Dendritic Cell Subtypes in Mouse Thymus and Spleen.* The Journal of Immunology, 2000. 164(6): p. 2978-2986.

[67] Asselin-Paturel, C., et al., *Mouse type I IFN-producing cells are immature APCs with plasmacytoid morphology.* Nat Immunol, 2001. 2(12): p. 1144-1150.

[68] Cheong, C., et al., *Microbial Stimulation Fully Differentiates Monocytes to DC-SIGN/CD209+Dendritic Cells for Immune T Cell Areas.* Cell, 2010. 143(3): p. 416-429.

[69] Wu, L. and Y.-J. Liu, *Development of Dendritic-Cell Lineages.* Immunity, 2007. 26(6): p. 741-750.

[70] Liu, K. and M.C. Nussenzweig, *Origin and development of dendritic cells.* Immunological Reviews, 2010. 234(1): p. 45-54.

[71] Liu, K., et al., *In Vivo Analysis of Dendritic Cell Development and Homeostasis.* Science, 2009. 324(5925): p. 392-397.

[72] Naik, S.H., et al., *Development of plasmacytoid and conventional dendritic cell subtypes from single precursor cells derived in vitro and in vivo.* Nat Immunol, 2007. 8(11): p. 1217-1226.

[73] Wu, L. and Y.J. Liu, *Development of dendritic-cell lineages.* Immunity, 2007. 26(6): p. 741-50.

[74] Rugh, R. and E. Grupp, *Splenic Radioprotective Agent: Particulate or Diffusible?* Radiation Research, 1960. 13(5): p. 657-660.

[75] Kim, C.H., *Homeostatic and pathogenic extramedullary hematopoiesis.* Journal of Blood Medicine, 2010. 1: p. 13-19.

[76] Massberg, S., et al., *Immunosurveillance by Hematopoietic Progenitor Cells Trafficking through Blood, Lymph, and Peripheral Tissues.* Cell, 2007. 131(5): p. 994-1008.

[77] Wolber, F.M., et al., *Roles of spleen and liver in development of the murine hematopoietic system.* Experimental Hematology, 2002. 30(9): p. 1010-1019.

[78] Dor, F.J., et al., *Primitive hematopoietic cell populations reside in the spleen: Studies in the pig, baboon, and human.* Exp Hematol, 2006. 34(11): p. 1573-82.

[79] Tan, J.K. and H.C. O'Neill, *Myelopoiesis in spleen-producing distinct dendritic-like cells.* J Cell Mol Med, 2012. 16(8): p. 1924-33.

[80] Tan, J.K., P. Periasamy, and H.C. O'Neill, *Delineation of precursors in murine spleen that develop in contact with splenic endothelium to give novel dendritic-like cells.* Blood, 2010. 115(18): p. 3678-85.

[81] Mueller, S.N. and R.N. Germain, *Stromal cell contributions to the homeostasis and functionality of the immune system.* Nature Reviews Immunology, 2009. 9(9): p. 618-629.

[82] Gretz, J.E., A.O. Anderson, and S. Shaw, *Cords, channels, corridors and conduits: critical architectural elements facilitating cell interactions in the lymph node cortex.* Immunological Reviews, 1997. 156(1): p. 11-24.

[83] Lokmic, Z., et al., *The extracellular matrix of the spleen as a potential organizer of immune cell compartments.* Seminars in Immunology, 2008. 20(1): p. 4-13.

[84] Nolte, M.A., et al., *A conduit system distributes chemokines and small blood-borne molecules through the splenic white pulp.* J Exp Med, 2003. 198(3): p. 505-12.

[85] Forster, R., et al., *CCR7 coordinates the primary immune response by establishing functional microenvironments in secondary lymphoid organs.* Cell, 1999. 99(1): p. 23-33.

[86] Bonilla, W.V., et al., *The alarmin interleukin-33 drives protective antiviral CD8(+) T cell responses.* Science, 2012. 335(6071): p. 984-9.

[87] Katakai, T., et al., *Organizer-Like Reticular Stromal Cell Layer Common to Adult Secondary Lymphoid Organs.* The Journal of Immunology, 2008. 181(9): p. 6189-6200.

[88] Dougall, W.C., et al., *RANK is essential for osteoclast and lymph node development.* Genes Dev, 1999. 13(18): p. 2412-24.

[89] Williams, A.F. and A.N. Barclay, *The Immunoglobulin Superfamily—Domains for Cell Surface Recognition.* Annual Review of Immunology, 1988. 6(1): p. 381-405.

[90] Hynes, R.O., *Integrins: A family of cell surface receptors.* Cell, 1987. 48(4): p. 549-554.

[91] Ellyard, J.I., et al., *Contribution of stromal cells to the migration, function and retention of plasma cells in human spleen: potential roles of CXCL12, IL-6 and CD54.* European Journal of Immunology, 2005. 35(3): p. 699-708.

[92] Periasamy, P., S. Petvises, and H.C. O'Neill, *Development of two distinct dendritic-like APCs in the context of splenic stroma.* Front Immunol, 2013. 4: p. 73.

[93] O'Neill, H.C., et al., *Spleen as a Site for Hematopoiesis of a Distinct Antigen Presenting Cell Type.* Stem Cells International, 2011. 2011.

[94] Periasamy, P., et al., *Splenic stromal niches support hematopoiesis of dendritic-like cells from precursors in bone marrow and spleen.* Experimental Hematology, 2009. 37(9): p. 1060-1071.

[95] Ni, K. and H. O'Neill, *Spleen stromal cells support hemopoiesis and in vitro growth of dendritic cells from bone marrow.* Br J Hematol, 1999. 105(1): p. 58-67.

[96] Wilson, H.L., K. Ni, and H.C. O'Neill, *Identification of progenitor cells in long-term spleen stromal cultures that produce immature dendritic cells.* Proc Natl Acad Sci U S A, 2000. 97(9): p. 4784-9.

[97] Ni, K. and H.C. O'Neill, *Hemopoiesis in long-term stroma-dependent cultures from lymphoid tissue: production of cells with myeloid/dendritic characteristics.* In Vitro Cell Dev Biol Anim, 1998. 34(4): p. 298-307.

[98] Despars, G. and H.C. O'Neill, *Splenic endothelial cell lines support development of dendritic cells from bone marrow.* Stem Cells, 2006. 24(6): p. 1496-504.

[99] Periasamy, P. and H.C. O'Neill, *Stroma-dependent development of two dendritic-like cell types with distinct antigen presenting capability.* Experimental Hematology, 2013. 41(3): p. 281-292.

[100] Tan, J.K., et al., *Identification of a novel antigen cross-presenting cell type in spleen.* J Cell Mol Med, 2011. 15(5): p. 1189-99.

[101] Tan, J.K.H. and H.C. O'Neill, *Hematopoietic stem cells in spleen have distinct differentiative potential for antigen presenting cells.* Journal of Cellular and Molecular Medicine, 2010. 14(8): p. 2144-2150.

[102] Petvises, S. and H.C. O'Neill, *Distinct progenitor origin distinguishes a lineage of dendritic-like cells in spleen.* Front Immunol, 2014. 4: p. 501.

[103] Petvises, S. and H.C. O'Neill, *Characterisation of dendritic cells arising from progenitors endogenous to mu*rine spleen. PLoS One, 2014. 9(2): p. e88311.

[104] Nagai, Y., et al., Toll-like receptors on hematopoietic progenitor cells stimulate innate immune system replenishment. Immunity, 2006. 24(6): p. 801-12.

Chapter 2

The Adult Hematopoietic Niches — Cellular Composition, Histological Organization and Physiological Regulation

Agustín G Zapata

Additional information is available at the end of the chapter

http://dx.doi.org/10.5772/58634

1. Introduction

For many years the adult bone marrow (BM) was described as a highly vascularised organ consisting histologically of a network of reticular cells, some of them intimately associated with blood vessel walls, so-called adventitial reticular cells, in whose holes mature and developing hematopoietic cells can be found. In long bones, central longitudinal arteries give rise to radial arteries that branch into arterioles near the endosteum where they connect to venous sinusoids that extend back toward the central region where they unite to form a large central sinus. The sinusoidal network occupies 30 ± 5 % of BM volume whereas arterioles comprise a smaller volume, 1.2 ± 0.1 % of total BM [1]. Macrophages, adipocytes, endosteal bone-lining cells and nerve endings also form part of the haematopoietic microenvironment of BM.

Twenty years ago, it was proposed that hematopoietic stem cells (HSCs) and their progeny occupied specific places (niches) in the BM network, a topological arrangement necessary to create a gradient of maturation from the bone endosteum, in which primitive B progenitor cells were identified, to central blood vessels through which mature B lymphocytes migrated into the blood circulation [2]. Thus, the relevance of bone lining cell (endosteal cells)-HSC interface was emphasized for the control of haematopoiesis. However, more recently based on the phenotypic definition of HSCs as CD150+CD48-CD41-Lin-cells, most early hematopoietic progenitors appear to really be in contact with sinusoidal endothelia, defining the so-called vascular niche [3]. Despite this evidence, the cells that organize these different BM microenvironments, the physiological relationships between them and molecules governing their functioning remain elusive. In the present chapter, we will be trying to clarify these issues providing current evidence on the functional niches operating in the adult bone marrow, their main cell components and the mechanisms governing their homeostasis. Niches involved in the homing and maintenance of distinct lymphoid cell subsets will be also addressed as well as the available information on the organization of hematopoietic niches occurring in primitive vertebrates that lack a functional bone marrow.

2. The existence of an endosteal niche in the adult BM is controversial

Bone development and the appearance of hematopoietic niches occur concomitantly during development. In foetal mice, the vascular invasion of specific chondrogenic sites favours bone formation, organizes the BM and seeds it with HSCs. Prior to the appearance of a bone microenvironment, HSCs occurring in foetal liver cannot home into the BM. In addition, injected HSCs disappear from the blood circulation to home rapidly in the endosteal niche [4]. On the other hand, E14.5 bone tissue grafted under the kidney capsule of syngeneic mice produces bone containing bone marrow that houses HSCs, and grafted CD105+Thy-1-cells isolated from E14.5 bone recruit blood vessels, produce ectopic bone through endochondral ossification and generate a BM capable of recruiting long term (LT) HSCs. In any experimental condition, the presence of bone was mandatory for BM organization [5]. Mixed multicellular spheroids containing CD45-CD105+CD31-Ter119-Sca-1+CD51+osteoprogenitors form a 3D hematopoietic niche that produces CXCL12 and osteopontin, retaining HSCs [6]. Nevertheless, bone marrowless vertebrates contain numerous hematopoietic loci in which HSCs develop in the absence of bone. Remarkably, the stroma of these organs morphologically resembles that of mammalian BM [7].

The existence of an endosteal niche gained considerable support several years ago when it was shown that endosteal cells produced several molecules involved in HSC maintenance, including G-CSF [8], thrombopoietin, angiopoietin [9, 10], CXCL12 [11], Jagged 1, Wnt ligands, among others [12, 13]. They, also, expressed adhesion molecules (VCAM, ICAM-1; N-cadherin, CD44, CD164) that could mediate endosteum-HSC interactions [4]. HSCs express Ca^{++}-sensing receptors that recognize Ca^{++}concentration in endosteum. In mice defective in these receptors HSC migrate into the BM in response to CXCL12 but do not anchor to the endosteum [4]. On the other hand, an increase in osteoblasts or osteolinage cells through enforced expression of parathyroid hormone receptor 1 [14] or by conditional inactivation of BMP receptor I [15] resulted in increased numbers of HSCs. Conversely, conditional depletion of osteolinage cells is concomitant with a loss of HSCs in the BM and extramedullary haematopoiesis [16].

However, recently some of these results have been questioned. Mice deficient in biglycan, a leucine-rich repeat proteoglycan of the connective tissue, have reduced osteoblasts but contain normal numbers of HSCs [17]. Osteoblast depletion in transgenic mice expressing thymidine kinase under the collagen $\alpha 1$ promotor courses with important alterations in haematopoiesis but only slight changes in the Lin-Sca-1+c-Kit+fraction of hematopoietic progenitors [16]. On the other hand, increased numbers of short term HSCs after parathyroid treatment could be mediated through Wnt ligand production by T cells [18] rather than through osteoblast expansion. Moreover, an increase in osteoblasts is not sufficient to increase HSCs. Thus, the treatment of mice with strontium, a bone anabolic agent, induces expansion of mature osteoblasts but does not affect the number or function of HSCs [19]. Likewise, in mice suffering chronic arthritis that produces osteoblast suppression, HSCs exhibit a normal behaviour [20] and conditional deletion of CXCL12 [21, 22] or SCF [23] from mature osteoblasts has no effect on HSCs, while osteoblasts would promote proliferation and differentiation of lymphoid progenitors [21].

A good example of the controversy about the significance of endosteal niche for haematopoiesis is the role played by the N-cadherin positive cells. As mentioned above, N-cadherin was pointed out to be an important adhesion molecule for endosteal cell-HSC interactions. These interactions were considered to be occurring between osteoblasts and HSCs, but other studies indicated that N-cadherin was largely expressed on immature osteolinage cells called spindle-shaped N-cadherin+osteoblasts [15]. These cells also express high levels of SCF and CXCL12 and support long term repopulating activity of HSCs [24]. In this respect, N-cadherin knock-down or expression of a dominant negative molecule in HSCs inhibited their repopulating capacity [25, 26] but conditional deletion of N-cadherin encoding gene in HSC had no effects on HSC number, mobilization, proliferation or repopulating capacity [25, 27].

In fact, a key issue to understanding these controversial results is the different cell composition attributed to the endosteum in various studies: whereas the earliest studies assigned any effect of endosteal niches to osteoblasts, more recent studies consider that, apart from osteoblasts, osteoclasts, osteoprogenitor cells, blood vessels and reticular cells provide a more accurate idea of the endosteal cell components and their function in the BM microenvironment. It has been proposed that osteoprogenitors rather than osteoblasts provide a functional niche for HSCs [28], whereas osteoclasts contribute to endosteal niches governing osteoblast maturation [29]. In this respect, macrophages interspersed between endosteal osteoblasts, called osteo-macrophages, support osteoblast function because their depletion in vivo leads to osteoblast disappearance and HSP mobilization [30], and RANK (receptor activator of nuclear factor κ B) ligand, essential for osteoclast maturation, induces mobilization of hematopoietic progenitors to the circulation [31]. In fact, osteoprogenitors, osteoblasts and osteoclasts are in a dynamic balance on the endosteal surface. Furthermore, recent evidence also supports a role for osteocytes in the function of the endosteal niche [32]. An analysis of the kinetics of gene expression in mice treated subcutaneously with injections of G-CSF each 12 hours showed that osteocytes responded more rapidly than osteoblasts to G-CSF. Thus, from the 4th dose onwards a rapid suppression of specific osteocyte genes was observed from the 1st or 2nd dose before HSC mobilization. After 8 doses of G-CSF, osteocytes showed fewer cell processes that connected them with endosteal osteoblasts. Remarkably, the specific elimination of osteocytes that disrupted the bone cell network, resulted in severe impairment of G-CSF-mediated HSC mobilization, as previously observed in other experimental models [33-35]. The authors suggest that these defects in HSC mobilization are due to impairment of the endosteal niche via altered communication between osteocytes and osteoblasts. Thus, changes in osteoblasts would be secondary to osteocyte depletion.

Nevertheless, results that emphasize a close association between HSCs and vascular endothelia, and those emphasizing the relevance of endosteal niche are not discordant because both HSCs and early committed pluripotent progenitors are perivascular but preferentially occupy the highly vascular endosteum [1]. However, it is more difficult to demonstrate that endosteal and vascular niches represent functionally different environments housing quiescent HSC and cycling hematopoietic progenitors, respectively, as suggested by some authors [12, 36]. The endosteal niche would house dormant, primitive HSCs due to its presumptive hypoxic condition [37, 38].It has been reported that disruption of HIF 1α (hypoxia-inducible factor) coursed with loss of HSC quiescence and repopulating activity whereas stabilization of HIF-1α induced opposite effects [39, 40], but CFU-S associated with endosteum has been described as non-quiescent cells and, indeed, endosteum is a highly vascularised and, therefore, presumably well-oxygenated area of the BM. Thus, it has been proposed that hematopoietic progenitors exhibit an intrinsic hypoxic profile independently of their location in the BM [41, 42]. On the other hand, cycling hematopoietic progenitors present in the vascular niches would be ready to respond to the physiological demand of blood cells. However, a recent study which combines three-dimensional BM imaging with computational modelling to carefully analyze relationships between HSCs and BM stromal component, including blood vessels, shows that different vascular niches house quiescent or proliferating stem cells. Whereas arterioles that occur predominantly close to bone [1], contribute to a niche that maintains quiescent HSCs, sinusoids could represent a niche which houses proliferating HSCs [43]. In addition, E-selectin expressed in certain BM sinusoids [44] promotes HSC proliferation and its blockade protects HSCs following chemotherapy or γ-irradiation [45].

3. The existence of a vascular niche

As described above, a restricted analysis of the topological position of HSC identified by the expression of SLAM family molecules CD150, CD244, CD48 and CD41 [46], together with experiments that indicate the relevance of vascular meshwork for the biology of hematopoietic progenitors led to a hypothesis of the occurrence of a so-called vascular niche [36]. Thus, vascular endothelial cells promote haematological recovery in mice following total body irradiation [47] inducing HSCs to enter in cycle after stimulation with different injury factors [48] and conditional deletion of VEGFR2, which inhibits the regeneration of sinusoidal endothelia, impedes the hematopoietic recovery following sublethal irradiation [47]. Also, specific activation of BMPR1A in endothelial cells increases the number of HSCs and hematopoietic recovery after myeloablation [49]. HSC mobilization induced by different agents (cyclophosphamide, G-CSF) that results in extramedullar haematopoiesis is associated with splenic sinusoids [36] and thrombocytopenic mice only recover thrombopoiesis in the presence of BM endothelial cells [50].

4. The cellular components of distinct BM microenvironments

For many years, the characterization of BM stromal cells was merely morphologic identifying a supporting network consisting largely of reticular cells, adventitial reticular cells and blood vessels whose functions were unknown. An important part of our knowledge of the presumptive functions of these BM stromal cells comes from studies in which genes important for HSC maintenance (i.e., CXCL12, SCF, TGFβ) are selectively deleted from different niche cell types [42, 51]. Unfortunately, results arising from these studies are controversial partially due to the disparity of these experimental models from physiological conditions and different degrees of specific deletion of the genes studied.

CXCL12, a chemokine implicated in the retention of CXCR4 (the CXCL12 receptor) positive hematopoietic multipotent progenitors is produced by different cell types of BM stroma, including bone-lining cells, but particularly by reticular cells that were named CAR cells (CXCL12 abundant reticular cells) [36]. A strong CXCL12 expression was demonstrated in a small $VCAM^{+}$reticular cell population scattered throughout the BM. It constitutes a relatively homogeneous CD45-CD31-Sca-1-PDGFRβ+cell population that in human BM could correspond to CD146-expressing subendothelial cells [31]. Selective depletion of CAR cells results in a drastic reduction of hematopoietic progenitors as well as of both cycling B lymphocytes and erythroid progenitors. On the other hand, CAR cells express adipogenic and osteogenic genes and are capable of differentiating in vitro to adipocytes and bone lineage cells.

On the other hand, nestin, a neuroectoderm stem cell marker, has been used to identify another stromal cell type in murine BM [52]. Nestin+cells have also been reported in human foetal and adult BM [53]. Previously, other authors identified self-renewing osteoprogenitors from stromal cultures containing all the human BM CFU-F activity, although their phenotype ($CD45^{-}CD146^{-}$or $CD45^{-}CD271^{+}CD146^{-/lo}$) is a matter of discussion [54, 55]. PDGFRα $^{+}CD51^{+}$stromal cells isolated from human BM form self-renewing clonal mesenpheres and support maintenance and expansion of human hematopoietic progenitors in a dose-dependent manner [56]. Remarkably, these authors indicate that these $PDGFR\alpha^{+}CD51^{+}$stromal cells comprise a subset of $CD146^{+}$cells associated with hematopoietic and MSC activities in foetal human BM. Nestin-positive cells are morphologically similar to pericytes; in fact, they appear to be largely restricted to the perivascular areas in intimate association with HSCs and to heavily express genes known to be involved in HSC maintenance, such as CXCL12, SCF, angiopoietin 1, IL7, V-CAM and osteopontin. Nestin+cells are innervated by the sympathetic nervous system, express β3-adrenergic receptors, mediate G-CSF-induced HSC mobilization and regenerate in vivo a BM stroma capable of supporting haematopoiesis. Furthermore, β3-adrenergic stimulation reduces the expression of the above-mentioned genes involved in HSC biology [52]. Depletion of nestin+cells reduces by half the HSC numbers in BM increasing concomitantly in the spleen. Relationships between nestin+cells and CAR cells are unclear. They could represent two overlapping BM stromal cells, nestin+cells being more primitive because they are less abundant, contain all CFU-F activity of bone marrow, exhibit autorenewal capacity and are able to differentiate into osteoblasts, chondrocytes and adipocytes.

On the other hand, it has been proposed that the BM stromal network, largely formed by CAR cells and nestin-expressing cells, would constitute a functional syncytium through gap junctions capable of affecting the behaviour of hematopoietic progenitors modulating CXCL12 secretion [57]. Nestin+MSCs heavily express connexin (Cx) 43 and Cx45, and G-CSF mediated reduction of CXCL12 expression is intimately related to decreased expression of both Cx43 and Cx45. On the contrary, the pharmacological blockade of gap junctions reduces the CXCL12 expression preventing homing into BM of mononuclear cells. In addition, a tight correlation occurs in confluent cultures of human BM-MSCs between the functional modulation of gap junctions and CXCL12 production. Gap junctions could modulate CXCL12 mediating gene transcription or directly acting on CXCL12 secretion. Loss of function of the Cx45 gene courses with reduced CXCL12 transcription particularly if Cx 43 transcription is also disrupted. On the other hand, Ca^{++}release, due to osteoclast activity in bone, necessary for AMPc-dependent PKA activation involved in CXCL12 secretion, is transmitted by gap junctions [58]. However, other studies have proven that Cx43 deficiency increases CXCL12 secretion, although it compromises HSC homing and hematopoietic recovery following myeloablation [59].

Cre-recombinase expressed under the control of leptin receptor (LepR) regulatory elements in the mouse BM [23] identified another perivascular stromal cell subset that, like the CAR cells and nestin+cells, expresses high levels of CXCL12 and SCF (c-Kit ligand).

A fourth element, in this case related to BM innervation, completes, for the moment, the myriad of stromal cells identified in the mouse BM. GFAP+integrin β8+(a Schwann cell marker), non-myelinated Schwann cells have been identified surrounding sympathetic nerves in BM. They produce important factors for HSC maintenance and appear to be in close contact with high numbers of hematopoietic progenitors [60]. Presumably, BM GFAP+cells are the main cell type processing latent TGFβ into an active form of the molecule, key for controlling the HSC niche. In fact, surgical sympathectomy performed by ligation of the sympathetic trunk reduces the HSC frequency and provokes HSC differentiation [33].

On the other hand, at least a subset of the total nestin-expressing cells present in the BM stroma could correspond to adventitial reticular cells identified in the EM studies and in vivo equivalents of the so-called mesenchymal stem cells (MSCs), an in vitro characterized BM stromal cell subset that exhibits interesting immunomodulatory properties and is extensively used in cell therapy [61]. Currently, it is assumed that MSCs represent approximately 0.001% to 0.01% of the BM nucleated cells, 10 fold less than the percentage of HSCs [62]. In 2007, [54] and later other authors [63-65] pointed out that MSC could be identical to, or derived from, pericytes, as previously indicated for nestin+cells. In addition, the CD146+subendothelial cell population of BM stroma, after ex vivo culture, produces bone, adipocytes and a stroma capable of supporting haematopoiesis. Other authors have identified a non-hematopoietic perivascular cell population that expresses both PDGFRα and Sca-1 and, after intravenous injection, produces perivascular cells, but also osteoblasts and adipocytes [66]. In human BM, Stro-1+stromal cells support haematopoiesis, exhibit capacity to differentiate into multiple mesenchymal lineages [67, 68] and after long-term culture express αsM-actin, as vascular smooth muscle cells. In vivo, asM-actin+cells occur in both foetal and adult bone marrow and exhibit long cell processes in intimate contact with HSCs [69], similarly to the alkaline phosphatase adventitial reticular (AdR) cells described previously [70]. The AdR cells express CD271 and appear in the foetal BM before the presence of HSCs [71]. CD146+cells, another BM-MSC subset [55], are also morphologically and physiologically similar to AdR cells and, after ex vivo expansion, express alkaline phosphatase, α-SM actin and CXCL12 [54].

Although results need further confirmation, [43] two types of nestin+cells have recently been reported in the murine BM. Nestin^{hi} expressing cells appeared along arterioles in both cancellous and long bone BM, closely associated with tyroxine hydroxylase-positive sympathetic nerves and GFAP+Schwann cells. Nestin^{weak} cells were reticular in shape and appeared associated to blood sinusoids. Transcriptome analysis also revealed differences between the two nestin+cell types: periarteriolar nestin^{hi} cells expressed mainly genes related with HSC biology whereas reticular nestin^{weak} cells were particularly enriched in genes involved in DNA replication and cell cycle.

On the other hand, clonal analysis revealed that BM nestin+cells represent a mixed population composed of neural crest (NC) stem cells plus a few MSCs [72]. Classically nestin+MSC of adult BM were considered to be mesodermal in origin [73], but recent studies have conclusively demonstrated that some adult BM-MSCs derive from the NC [74-78]. The vertebrate NC cells are a multipotent population that gives rise to multiple types of derivatives [79]. In chick embryos, a common mesenchyme-neural progenitor derived from the neural crest has been described [80] and tracing studies using the Sox1 neuroepithelial marker and PO (protein O) NC marker indicate that the first wave of trunk MSC come from the neuroepithelium rather than from the mesoderm, although MSC derived from Sox1+PO+cells decrease gradually and appear to be substituted by MSCs from other unknown sources [74]. Murine NC stem cells migrate in the blood to the BM via the AGM region between E12.5 and E15.5 similar to HSCs and form spheres that express nestin as BM-MSCs [75]. It has been pointed out that TGF-β signalling confers mesenchymal potential to NC cells and suppresses their neurogenic potential [81] supporting the hypothesis that GFAP+Schwann cells intimately associated with nestin+cells in the murine BM produce TGF-β that could affect the BM-MSC properties. On the other hand, the neuroectodermal origin of some BM-MSC could confer them a neural potentiality useful for Cell Therapy. In this respect, cranial NC cell-derived MSCs from gingivae show a high capacity to differentiate into neural cells [82] and BM stem cells have been described to differentiate into both neurons and glial cells in vivo and in vitro [83, 84]. However, other studies demonstrate that NC derived BM-MSCs are not more competent than mere MSCs or whole bone MSCs at differentiating into neurons and *in vivo* recovery of the experimentally-damaged dopaminergic system [85].

As mentioned above, it is difficult to establish clear relationships between all these cell types of BM stroma. It has been proposed that they are overlapping cell subsets because LepR is among the top 1% most highly expressed genes in sorted nestin+cells [52] and also in CAR cells [22]. However, conditioned deletion of SCF gene in either osteoblasts, nestin+cells, endothelial cells or perivascular LepR+cells only negatively affects the HSC frequency in the two last experimental conditions suggesting that LepR+cells and nestin+cells are different components of BM stroma [23]. [51] have recently proposed that CAR+cells, which represent the highest BM stromal cell subset (0.26% of total BM nucleated cells), would encompass nestin

+cells (0.08%), GFAP+cells (0.004%) and LepR+cells (0.012%) whereas the nestin+cell subpopulation would partially include LepR+cells. On the contrary, [86] found only a partial overlapping of nestin+MSCs with the other three stromal cell types. In fact, all GFAP+cells express nestin but not PDGFRα, although nestin+cells and PDGFRα cells largely overlap. Thus, BM GFAP+cells seem to differ from PDGFRα MSCs, as further supported by the finding that they do not express smooth muscle actin [60]. Recently, [43] reported that LepR+cells and reticular nestin^{weak} cells largely overlap but they did not find overlapping between LepR+cells and periarteriolar nestin^{hi} cells that these authors consider to be related to pericytes that express Ng2 and α-smooth muscle actin.

Some evidence supports reciprocal interactions between HSCs and the niche. After acute bleeding, HSCs secrete BMP2 and BMP6 that differentiate MSCs to osteoblasts [87]. Also, megakaryocytes located close to the endosteum stimulate osteoblasts by secreting BMP2, 4 and 6 [50, 88].

5. Rhythms and haematopoiesis

HSCs, like other stem cells, are affected by systemic signals, including those associated with nutrition, seasonal and circadian rhythms, exercise, mating, pregnancy, etc. [89]. Recently, [90] have reported that HSCs suffer sexual dimorphism due to estrogens that directly affect HSC biology. More interestingly within the context of this chapter are results that suggest that certain physiological biorhythms known to occur in the BM could be mediated through its stromal components. The close association between nerve endings, GFAP+Schwann cells and some nestin+cells (ultrastructurally identified as adventitial reticular cells [91]), as well as the above mentioned possible neuroectodermic origin of part of this BM cell population suggested by some authors support this hypothesis.

A central clock in the central nervous system [92] orchestrates a uniform temporal programme by synchronizing multiple peripheral clocks that occur in practically all cell types [93]. The haematopoietic system exhibits circadian rhythms that oscillate according to the rest-activity phase that, by turn, depend on whether the species is diurnal or nocturnal. Thus, the numbers of HSCs, committed progenitor cells and most mature leukocytes, except CD8+T lymphocytes [94] peak in the circulation during the resting phase (night for humans and day for rodents) and decrease during the active period [95, 96].

Haematopoietic progenitors and mature immune cells leave the BM to enter the blood circulation at the beginning of the resting phase [97], migrating to tissues predominantly during the active phase [98]. The mobilization from BM is dependent on local sympathetic innervation which down-regulates the CXCL12 expression in the BM in close association with a reduced CXCR4 expression of haematopoietic progenitors [99] and both CD4+and CD8+T cell subsets [94]. On the contrary, the onset of the active phase is related to a peak of glucocorticoids, adrenaline, noradrenaline, TNF and IL1β [95, 96].

Numerous data support a role for the sympathetic nervous system (SNS) in regulating HSC trafficking. Noradrenalin administration stimulates proliferation and migration of human

HSCs [100], and $Cgt^{-/-}$mice, deficient in the UDP-galactose ceramide galactosyl transferase necessary for sulfatide synthesis, a key component of myelin, have compromised G-CSF-mediated HSC mobilization [33]. Nevertheless, the most remarkable information comes from studies using G-CSF, a cytokine that mobilizes HSCs from the bone marrow to peripheral blood by stimulating adrenergic receptors on target cells resulting in decreased secretion of CXCL12 [33]. The first studies identified osteoblasts as the target stromal cell for the effects observed on HSCs because β2 adrenergic receptor activation on osteoblasts increases the expression of the vitamin D receptor necessary for G-CSF induced HSC mobilization [34, 101]. Recently, it has been proposed that osteocytes, which indirectly affect osteoblasts on the endosteal surface, could also mediate G-CSF dependent HSC mobilization through a sympathetic tone [32]. This possibility was discarded because adrenergic receptors involved in these processes were largely β3 receptors not expressed on osteoblasts but on CAR cells and nestin-expressing perivascular cells [33, 52]. Presumably, both adrenergic receptors contribute to regulate haematopoiesis in BM [102].

The circadian rhythms could therefore be explained as a consequence of transmitted information from the central pacemaker to the BM through SNS that rhythmically secretes noradrenalin activating β receptors expressed on different stromal cells. Neural signals could be propagated through the BM stroma via gap junctions whose effects on HSC mobilization have already been commented. Chemical sympathectomy, by using 6-OH dopamine, abolishes the rhythms of circulating blood cell precursors and surgical denervation of mouse tibiae courses with the disappearance of rhythmic CXCL12 oscillations in the denervated limbs [86]. However, recent results suggest that the process could be more complex and the effects of innervation would vary rhythmically and affect different processes in the BM. Systemic activation of β adrenergic receptors does not increase circulating haematopoietic progenitors unless HSC homing is also blocked [97] suggesting that adrenergic stimulation might not only affect mobilization but also homing of haematopoietic progenitors in the BM and/or peripheral tissues [102]. Indeed, when mice defective in E-and P-selectins showing additional $\alpha4$ integrin blockade, which inhibited the homing to BM, received isoproterenol, an adrenergic antagonist, during the morning phase they showed considerable HSC mobilization. On the contrary, if isoproterenol administration occurred 1 hour earlier than the onset of light the HSC mobilization was prevented [86].

In any case, it is not easy to establish a direct relationship between sympathetic innervation and mobilizing agents (i.e., G-CSF) that induce haematopoietic progenitor egress from the BM, although some evidence for this is available. Administration of either G-CSF or β_3 adrenergic agonists has similar effects on the haematopoietic niche and HSC mobilization [52]. Tyrosine hydroxylase neurons from the murine superior cervical sympathetic ganglion express G-CSF receptors and this cytokine promotes neuron survival after brain damage [103]. On this basis, [104] have shown that G-CSF increases the sympathetic tone modifying the capacity of neurons to uptake noradrenalin and, consequently, increasing available catecholamine for BM target cells. Furthermore, in vivo administration of G-CSF plus noradrenalin re-uptake inhibitors, such as desipramine or reboxetine, induce significantly higher numbers of haematopoietic progenitors in the blood circulation, without changes in BM cell content, than those of animals receiving only cytokine. These studies support a direct effect of G-CSF on the sympathetic terminals through signalling transmitted by specific receptors on neurons and noradrenalin uptake inhibition.

6. The role of myeloid cells in the hematopoietic niche

We have extensively reported the organization of the main components of BM stroma and their relevance for regulating haematopoiesis. However, many other cells present in the BM that do not form part of the stromal reticular network are also important for this regulation. A clear example is the response of BM to G-CSF administration that is known to have provided such important information on the functioning of the haematopoietic niche. It is obvious that, apart from the BM stromal cells, monocytes/macrophages (Mo/MØ) and granulocytes are the main targets of its action. On the other hand, numerous inflammatory molecules produced mainly by leukocytes affect the behaviour of HSCs in both cell autonomous and non-cell autonomous fashion [105]. Could the response to C-CSF administration between stromal cells and leukocytes be coordinated? or, by contrast, does G-CSF administration induce specific, independent effects on each BM cell type that expresses receptors for this growth factor? Moreover, if this presumptive cooperation exists, how is it established?

Various results support the involvement of monocytes (Mo) and macrophages (MØ) in the functional regulation of the HSC niche [106]. In mice, the loss of MØ results in decreased self-renewal capacity and retention of HSC [107, 108], and it has been reported that G-CSF signalling in a Mo cell line is sufficient to induce HSC mobilization [109]. Is it possible to reconcile these results in which G-CSF mediated HSC mobilization depend on Mo/MØ with those previously described in which HSC egress seems to be mediated by changes in sympathetic tone that affects CXCL12 secretion by CAR cells and/or nestin+cells ? Some results try to do this although stromal cell types affected by Mo/MØ activity are controversial. MØ elimination by *in vivo* administration of clodronate-loaded liposomes or use of Fas-induced apoptosis transgenic mice courses with loss of endosteal osteoblasts and reduced HSC mobilization into the blood [108]. In these experiments, endosteal osteomacrophages were particularly affected. In this same study [107] and others, it has been shown that CD169+macrophages directly modulated the activity of nestin+cells by promoting the expression of HSC maintenance molecules, including angiopoietin, CXCL12, SCF and VCAM-1. Factors concerned with these effects are unknown, although some inflammatory molecules have been implicated [107].

Another key question on the regulation of haematopoiesis is how the BM "knows" the necessities of different blood cell subsets at any time and responds to the peripheral demand in each case. Every day there is an enormous production of blood cells and it is necessary to control precisely the number of each blood cell type in the periphery. Some recent results suggest an important role of Mo/MØ and neutrophils (Neu) in this control and show the high plasticity exhibited by the BM stromal components to address these processes. It was reported that BM responds to *Listeria monocytogenes* infection inducing migration of inflammatory Mo in a process that involves the chemokine receptor CCR2 [110]. CCR2+Mo also migrate from the BM stroma into peripheral blood through vascular sinusoids in response to low concentrations of LPS, a TLR ligand. In these experimental conditions, there is also a significant increase in CCL2 production by perivascular CXCL12+nestin+cells [111]. These same authors showed that conditional deletion of the CCL2 gene under the control of nestin promoter coursed with an important reduction of circulating inflammatory Ly6C^{hi} Mo. The reduction was less severe when CCL2 gene was deleted in endothelial cells. These results show that CCL12 gene up-regulation in nestin+cells, CAR cells and endothelial cells is necessary for inducing migration from the BM into the blood circulation of inflammatory Mo in response to low concentrations of LPS [111]. A similar behaviour was detected after *L. monocytogenes* infection. In mice infected which had deleted the CCL2 gene from nestin+cells, the bacterial clearance was significantly low [111]. It is unclear, however, how the reticular cells of BM stroma govern Mo migration and "perceive" the occurrence of TLR ligands (i.e., LPS) in the circulation, but it is evident that distinct chemokine-chemokine receptor pairs regulate trafficking of haematopoietic progenitors or their mature progeny between BM and the peripheral circulation. Nor is it clear how the migration of a specific cell type affects another blood cell lineage but stimulation of BM-MSC through TLR4 down-regulates the expression of Jagged, a Notch ligand, involved in HSC self-renewal [112].

These results support the existence of mutual influences between BM stromal cells and Mo/MØ but also open interesting concerns about the mechanisms of response of the BM to the peripheral demands of specific blood cell subsets and the possible existence of different stromal cells that define different niches in the BM microenvironment or, alternatively, a unique nestin +cell population capable of adapting its function (i.e., production of different chemokines) in time for the specific demands of the hematopoietic system (see later). The neutrophil is a blood cell type with a short life-span that reflects a fine-tuned turnover between elimination and production, particularly in inflammatory situations. Thus, mice with impaired Neu extravasation, a process necessary for their efficient clearance, results in an imbalance of G-CSF levels and enhanced granulopoiesis suggesting that Neu elimination is part of a homeostatic loop for controlling their level in blood [113]. In this respect, it has been proposed that physiological clearance of Neu in the BM triggers signals that modulate the hematopoietic niche and promote cycles of hematopoietic progenitor mobilization [114]. Neu during aging lose CD62L expression and increase that of CXCR4, that facilitates their migration into the BM [115]. Although the loss of CD62L^{lo} does not contribute to BM homing, within BM CD62L^{lo} Neu are preferentially engulfed by MØ as compared to CD62L^{hi} Neu. On the other hand, the a priori surprising release of haematopoietic progenitors from BM to peripheral blood during Neu clearance could be related to increased levels of G-CSF occurring in that condition [113] that favour HSC mobilization. It may be hypothesized, therefore, that circulating Neu could function as a sensor of the organism status.

7. The other niches of adult bone marrow

Until now we have almost exclusively paid attention to the regulation through the BM niche of early hematopoietic progenitors, their mechanisms of survival, proliferation and differentiation. However, BM stroma houses many developing cells belonging to different blood cell lineages that also need to receive signals from the BM microenvironment for maintenance and maturation. This raises the question of whether the BM constitutes a unique niche capable of numerous functions that vary in their timing and physiological demands or if, instead, the so-called BM niche should be considered as a myriad of different specific "subniches". In the following pages we will describe some of these "subniches" and their possible relationships to the previously described microenvironments that house and maintain early hematopoietic progenitors.

We have previously emphasized the mutual influences between Mo/MØ and BM stromal cells and their relevance in regulating hematopoietic activity. Erythroblastic islands that are "classical" BM niches, consisting of a central MØ surrounded by developing erythroid cells, and are considered essential for normal erythropoiesis [116] represent other distinct environments in the BM not associated with HSCs but with erythroblast maturation. Recent results suggest a dual role for central MØ contributing to both red blood cell production and clearance in the steady state but playing a supporting role for erythropoiesis during the erythroid recovery that follows haemolytic anaemia, acute blood loss, myeloablation and JAK2 induced polycythemia vera [117, 118]. Thus, during the myeloablative process the depletion of radioresistant CD169+MØ impaired the recovery of BM and splenic erythroblasts. A similar condition occurred when antibody blockade of V-CAM was performed suggesting that adhesion molecule-mediated interactions between erythroblasts and MØ are important in this process.

Lymphoid cells constitute another important component of the BM. B lymphocytes mature in the BM whereas other lymphoid cell subsets (i.e., plasmablasts, naïve and memory T-cells) migrate into the BM at any time of their life cycle [13]. The movement of B lymphocytes throughout the central and peripheral lymphoid organs is a good model to evaluate the presumptive occurrence of distinct subniches in the BM. As mentioned above, a pioneer study [2] proposed that primitive B progenitor cells that occurred close to the endosteum progressed during the maturation of distinct cell subsets along the BM reticular network toward the central blood vessels where mature B lymphocytes migrated into the blood circulation. Information on the niches that developing B cells occupy during this journey throughout BM stroma is scarce. The existence of $IL7^{hi}$-and $IL7^{lo}$-niches has been reported but their precise location in the BM stroma remains unresolved [119]. On the other hand, pre-BcR signalling necessary for the maturation of pre-pro B cells induces CXCR4 expression and its interaction with CXCL12-expressing stromal cells induces downstream activation of the focal adhesion kinase that could facilitate the movement of pre B cells into the $IL7^{lo}$ niches [120].

In the periphery (i.e., lymph nodes, spleen), naïve B lymphocytes activated by antigens, largely T-dependent ones, become plasmablasts, after passing through the named germinal centres. The conversion of activated B lymphocytes in plasmablasts courses with down-regulation of

http://dx.doi.org/10.5772/58634

both CXCR5 and CCR7 and up-regulation of the S1P1 (sphingosine-1-phosphate) receptor necessary for the exit of plasmablasts from the peripheral lymphoid organs [121]. Plasmablasts that leave the peripheral lymphoid organs migrate predominantly to the BM in a process mediated in part by CXCR4 because sinusoidal endothelial cells of BM express its ligand, CXCL12 [119]. However, other chemokines, including CXCL9, CXCL10 and CXCL11 are also involved in the homing to murine BM [122]. In humans, other chemokines, such as CXCL16 and CCL28 are implicated [123]. Once in the BM, plasmablasts need to home to appropriate niches that guarantee their retention, survival and differentiation to mature plasma cells. CXCL12, some adhesion molecules principally, VCAM1 constitutively expressed by BM stromal cells, and extracellular matrix molecules, such as fibronectin and hyaluronic acid [124] contribute to their retention, but remarkably many arriving plasmablasts fail to establish in the BM. Possible mere competence between arriving plasmablasts, maturing pre-or pro-B cells and/or pre-existing long-lived plasma cells by the space or, more specifically, by the CXCL12-expressing niches could explain these results [125]. Another remarkable finding is that about 50% of the recent immigrant plasmablasts interact with eosinophils (Eos), a smaller number with MØ and some with megakariocytes [126, 127]. In peripheral lymphoid organs most plasmablasts contact with MØ. The relevance of these cell types for plasmablast colonization of the BM was evidenced in Eos deficient ΔdbI GATA-1 mice in which plasmablasts migrate normally, but only a few are retained and mature to plasma cells. Importantly, the BM stroma in these mutants is normal indicating that VCAM-1+CXCL12+stromal cells are not sufficient to retain and presumptively to allow plasmablast differentiation [128, 129].

At this point it is important to remark that in vitro studies have shown IL5, IL6, TNFα, BAFF (B-cell activating factor), APRIL (a proliferation-inducing ligand) and CXCL12 as well as fibronectin, CD44 and CD28 as plasma cell survival factors [125]. However, in vivo studies have particularly pointed out the importance of APRIL. APRIL-deficient mice or mice deficient in BCMA, the high affinity receptor of APRIL, show a significantly lower number of plasma cells in the BM than WT mice [130, 131]. In addition, plasmablasts transferred into APRIL-deficient mice home normally in the BM but the maintenance of long-lived plasma cells is impaired [129] and blockade of BAFF and APRIL by fusion proteins significantly reduced the plasma cell numbers in the BM [130]. The relevance of APRIL for the survival of plasmablasts and plasma cells is also shown because SLE patients contain numerous plasma cells whose survival is controlled by autocrine expression of APRIL in the self plasma cells [132]. Plasmablasts retained in the BM move from the sinusoids into the stroma where clusters of Eos appear scattered at random throughout the BM parenchyma. These areas containing stromal cells and Eos become evident late after secondary immunization and could constitute the survival niche for mature plasma cells. Remarkably, the number of Eos in BM significantly increases during secondary immune responses when the production of plasma cells peaks. These are activated producing increased amount of cytokines, including APRIL and IL6. It, therefore, appears that a positive feedback governed by the plasma cells activate Eos to induce high levels of plasma cell survival factors.

These results report a new niche in the BM in which stromal cells need APRIL+Eos to home and maintain both plasmablasts and plasma cells. Nothing is known about the stromal cells associated with Eos, if they are similar to those that support the differentiation of early B lymphocyte progenitors or, by contrast, represent another phenotypic and functional subniche within the BM microenvironment.

8. Adipocytes as negative regulators of haematopoiesis

Adipocytes, other cell component of BM stroma, that were regarded as a passive element in the haematopoietic activity occurring in the BM, in fact appear to be active suppressors of haematopoiesis [133]. Thus, adipocyte-rich BM contains lower numbers of haematopoietic progenitors and fat cells impede their expansion preserving the HSC pool in a process partially mediated through increased expression of neurophilin 1, a semaphorine receptor, involved in angiogenesis and axon guidance [134]. In addition, in lipoathrophic A-ZIP/F1 mice, deficient in adipocytes, or after pharmacological inhibition of adipogenesis, BM exhibits massive osteogenesis and rapid recovery of haematopoiesis [135].

All these data suggest that the BM stromal cells, particularly CAR cells and nestin+cells, represent a highly plastic cell population capable under different physiological conditions, of differentiating either to bone cell lineage or adipocytes, or constitute a niche for homing and maintaining hematopoietic progenitors [136].

9. The haematopoietic niches in the lower vertebrates

When we analyze the characteristics, components and functional relevance of the haematopoietic niches we forget that many vertebrates do not contain a haematopoietic BM but exhibit highly efficient haematopoiesis raising the question of whether they use other mechanisms for controlling blood production, or if blood-forming organs in lower vertebrates have well organized haematopoietic niches equivalent to those described here in mammalian BM. A BM functionally implicated in haematopoiesis appears for the first time in the most evolved urodeles of the *Pletodontidae* family [137]. Earlier in phylogeny, particularly in fish, there are numerous haematopoietic organs which share a structural and functional resemblance to the BM of higher vertebrates [7]. Apparently, in these lower vertebrates any organ that contains a stroma consisting of a reticular network arranged between sinusoidal blood vessels could home and supports differentiation of blood cell progenitors. Obviously, alterations in the histological organization of these organs would result in the loss of their haematopoietic capacities.

The variety of organs involved in haematopoiesis in primitive vertebrates is impressive. In most elasmobranches, masses of haematopoietic tissue can be found in the oesophageal (Leydig organ) and gonad walls (Epigonal organs) [138]. The kidney is the main haematopoietic organ in many lower vertebrates, including cyclostomes, chondrostei, holostei, dipnoi, polypteriformi, teleostei, some urodelans and embryonic anurans. In this case, cell cords containing developing blood cells occur arranged among the renal tubules and enlarged blood sinusoids.

The brain and cranium are also suitable sites for haematopoiesis. It occurs in the meninx primitive of some elasmobranches [139] and in the orbit and subcranium of the holocephali, *Chimaera monstrosa* [140, 141]. An analogous tissue occurs in the meninges of ganoids [142] and in the brain of the urodeles *Ambystoma* [143] *and Megalobatrachus japonica* [144]. The importance of niches for determining the haematopoietic capacity of distinct organs is evidenced during the ontogeny of amphibians and throughout the complex life cycle of lampreys. In amphibians which do not have haematopoietic BM, the kidney retains the blood-forming function. In contrast, in the urodeles of the *Plethodontidae* family which already contain a haematopoietic BM, the kidney is not capable of producing blood cells [137]. In adult anurans with a functional BM, the kidney which had formed blood cells during embryonic and foetal life also loses its hematopoietic capacity [145], although some primitive families (i.e., *Pipidae*) maintain a granulopoietic perihepatic cell layer [146]. In larval lampreys, the typhlosole, a fold of the midgut, and the nephric fold consisting of the larval opistonephros and a little closely-associated adipose tissue are the main haematopoietic organs. All of these have a histological organization similar to that of mammalian BM. During metamorphosis, these organs, particularly the typhlosole, regress completely because the reticular network that constituted the haematopoietic niche is substituted by fibroblasts and masses of collagen fibres. After metamorphosis, when the adult opistonephros is organized, the haematopoietic tissue re-appears there, although the most remarkable adult haematopoietic organ in lampreys is the supraneural body. In larvae, this fat column along the central nervous system only contains adipocytes but just after metamorphosis it begins to home blood cell progenitors in cell cords arranged among big fat cells resembling the organization of BM of higher vertebrates [147].

10. Concluding remarks

It is evident that in a few years the data on the organization, cell and molecular components of adult BM have increased exponentially. However, this mass of information has generated new, important questions that reflect the enormous complexity of the haematopoietic niche and its physiology. The haematopoietic niche is more complex and heterogeneous than the current picture we have owing to the limited cell markers and experimental approaches available. To determine whether this heterogeneity is real, or rather reflects a high plasticity that permits a specific cell type to change its functionality throughout time depending on physiological demands is an important matter that needs an urgent answer. Before this global issue can be resolved we must first address some specific points:

1. An accurate, fine anatomical analysis of the BM is required which permits the precise location of the different components and the relationships between bone, blood vessels (arterioles and sinusoids) and reticular cells to be conclusively established.

2. The relationships between CAR cells, nestin+cells and LepR+cells are unclear. Although they presumably overlap considerably it will be important to define precisely which cell

type (and its possible subsets) is included in each cell population and what is its function in the haematopoietic niche.

3. The physiological significance of the homeostatic changes in the biology of haematopoietic niches also needs clarification and the appearance and maturation of distinct stromal components of BM during ontogeny, aging and disease must be determined.

4. It is also important to define the genes/signals involved in the maintenance and differentiation of various haematopoietic progenitor subsets as well as those that specifically differentiate BM stromal cells to either adipocytes or bone cells.

More importantly, we need to have an integrative model that allows to us to understand the complex interactions that govern haematopoiesis through the BM microenvironment in physiological and pathological conditions. Most self-renewing organs are supported by a myriad of stem cell niche units, distributed throughout the tissue. Moreover, growing evidence points to high heterogeneity in terms of molecular profiles, division patterns and population sizes of stem cells and niches within a given tissue [148]. Organs, therefore, face the considerable challenge of regulating not only each individual niche but also critically, the collective output of all the niches in a given organ. It is, therefore, important to define how multiple, heterogeneous and spatially dispersed units are coordinated in a tissue,

Acknowledgements

This work was supported by grants BFU2010-18250 from the Spanish Ministry of Education and Science, RD012/0019/0007 from the Spanish Ministry of Health (TerCell network) and S2010/BMD-2420 from the Regional Government of Madrid

Author details

Agustín G Zapata

Department of Cell Biology, Faculty of Biology, Complutense University, Madrid, Spain

References

[1] Nombela-Arrieta C, Pivarnik G, Winkel B, Canty KJ, Harley B, Mahoney JE, et al. Quantitative imaging of haematopoietic stem and progenitor cell localization and hypoxic status in the bone marrow microenvironment. Nat Cell Biol. 2013 May;15(5): 533-43.

[2] Jacobsen K, Osmond DG. Microenvironmental organization and stromal cell associations of B lymphocyte precursor cells in mouse bone marrow. Eur J Immunol. 1990 Nov;20(11):2395-404.

[3] Trumpp A, Essers M, Wilson A. Awakening dormant haematopoietic stem cells. Nat Rev Immunol. 2010 Mar;10(3):201-9.

[4] Zapata AG, Alfaro D, Garcia-Ceca J. Biology of stem cells: the role of microenvironments. Adv Exp Med Biol. 2012;741:135-51.

[5] Chan CK, Chen CC, Luppen CA, Kim JB, DeBoer AT, Wei K, et al. Endochondral ossification is required for haematopoietic stem-cell niche formation. Nature. 2009 Jan 22;457(7228):490-4.

[6] de Barros AP, Takiya CM, Garzoni LR, Leal-Ferreira ML, Dutra HS, Chiarini LB, et al. Osteoblasts and bone marrow mesenchymal stromal cells control hematopoietic stem cell migration and proliferation in 3D in vitro model. PLoS One. 2010;5(2):e9093.

[7] Zapata AG, Torroba M, Vicente A, Varas A, Sacedon R, Jimenez E. The relevance of cell microenvironments for the appearance of lympho-haemopoietic tissues in primitive vertebrates. Histol Histopathol. 1995 Jul;10(3):761-78.

[8] Taichman RS, Emerson SG. Human osteoblasts support hematopoiesis through the production of granulocyte colony-stimulating factor. J Exp Med. 1994 May 1;179(5): 1677-82.

[9] Qian H, Buza-Vidas N, Hyland CD, Jensen CT, Antonchuk J, Mansson R, et al. Critical role of thrombopoietin in maintaining adult quiescent hematopoietic stem cells. Cell Stem Cell. 2007 Dec 13;1(6):671-84.

[10] Yoshihara H, Arai F, Hosokawa K, Hagiwara T, Takubo K, Nakamura Y, et al. Thrombopoietin/MPL signaling regulates hematopoietic stem cell quiescence and interaction with the osteoblastic niche. Cell Stem Cell. 2007 Dec 13;1(6):685-97.

[11] Jung Y, Wang J, Schneider A, Sun YX, Koh-Paige AJ, Osman NI, et al. Regulation of SDF-1 (CXCL12) production by osteoblasts; a possible mechanism for stem cell homing. Bone. 2006 Apr;38(4):497-508.

[12] Wilson A, Trumpp A. Bone-marrow haematopoietic-stem-cell niches. Nat Rev Immunol. 2006 Feb;6(2):93-106.

[13] Mercier FE, Ragu C, Scadden DT. The bone marrow at the crossroads of blood and immunity. Nat Rev Immunol. 2011 Jan;12(1):49-60.

[14] Calvi LM, Adams GB, Weibrecht KW, Weber JM, Olson DP, Knight MC, et al. Osteoblastic cells regulate the haematopoietic stem cell niche. Nature. 2003 Oct 23;425(6960):841-6.

[15] Zhang J, Niu C, Ye L, Huang H, He X, Tong WG, et al. Identification of the haematopoietic stem cell niche and control of the niche size. Nature. 2003 Oct 23;425(6960): 836-41.

[16] Visnjic D, Kalajzic Z, Rowe DW, Katavic V, Lorenzo J, Aguila HL. Hematopoiesis is severely altered in mice with an induced osteoblast deficiency. Blood. 2004 May 1;103(9):3258-64.

[17] Kiel MJ, Radice GL, Morrison SJ. Lack of evidence that hematopoietic stem cells depend on N-cadherin-mediated adhesion to osteoblasts for their maintenance. Cell Stem Cell. 2007 Aug 16;1(2):204-17.

[18] Li JY, Adams J, Calvi LM, Lane TF, DiPaolo R, Weitzmann MN, et al. PTH expands short-term murine hemopoietic stem cells through T cells. Blood. 2012 Nov 22;120(22):4352-62.

[19] Lymperi S, Horwood N, Marley S, Gordon MY, Cope AP, Dazzi F. Strontium can increase some osteoblasts without increasing hematopoietic stem cells. Blood. 2008 Feb 1;111(3):1173-81.

[20] Ma YD, Park C, Zhao H, Oduro KA, Jr., Tu X, Long F, et al. Defects in osteoblast function but no changes in long-term repopulating potential of hematopoietic stem cells in a mouse chronic inflammatory arthritis model. Blood. 2009 Nov 12;114(20): 4402-10.

[21] Ding L, Morrison SJ. Haematopoietic stem cells and early lymphoid progenitors occupy distinct bone marrow niches. Nature. 2013 Mar 14;495(7440):231-5.

[22] Greenbaum A, Hsu YM, Day RB, Schuettpelz LG, Christopher MJ, Borgerding JN, et al. CXCL12 in early mesenchymal progenitors is required for haematopoietic stem-cell maintenance. Nature. 2013 Mar 14;495(7440):227-30.

[23] Ding L, Saunders TL, Enikolopov G, Morrison SJ. Endothelial and perivascular cells maintain haematopoietic stem cells. Nature. 2012 Jan 26;481(7382):457-62.

[24] Nakamura Y, Arai F, Iwasaki H, Hosokawa K, Kobayashi I, Gomei Y, et al. Isolation and characterization of endosteal niche cell populations that regulate hematopoietic stem cells. Blood. 2010 Sep 2;116(9):1422-32.

[25] Hosokawa K, Arai F, Yoshihara H, Iwasaki H, Hembree M, Yin T, et al. Cadherin-based adhesion is a potential target for niche manipulation to protect hematopoietic stem cells in adult bone marrow. Cell Stem Cell. 2010 Mar 5;6(3):194-8.

[26] Hosokawa K, Arai F, Yoshihara H, Iwasaki H, Nakamura Y, Gomei Y, et al. Knockdown of N-cadherin suppresses the long-term engraftment of hematopoietic stem cells. Blood. 2010 Jul 29;116(4):554-63.

[27] Greenbaum AM, Revollo LD, Woloszynek JR, Civitelli R, Link DC. N-cadherin in osteolineage cells is not required for maintenance of hematopoietic stem cells. Blood. 2012 Jul 12;120(2):295-302.

[28] Raaijmakers MH, Mukherjee S, Guo S, Zhang S, Kobayashi T, Schoonmaker JA, et al. Bone progenitor dysfunction induces myelodysplasia and secondary leukaemia. Nature. 2010 Apr 8;464(7290):852-7.

[29] Mansour A, Abou-Ezzi G, Sitnicka E, Jacobsen SE, Wakkach A, Blin-Wakkach C. Osteoclasts promote the formation of hematopoietic stem cell niches in the bone marrow. J Exp Med. 2012 Mar 12;209(3):537-49.

[30] Chang MK, Raggatt LJ, Alexander KA, Kuliwaba JS, Fazzalari NL, Schroder K, et al. Osteal tissue macrophages are intercalated throughout human and mouse bone lining tissues and regulate osteoblast function in vitro and in vivo. J Immunol. 2008 Jul 15;181(2):1232-44.

[31] Nagasawa T, Omatsu Y, Sugiyama T. Control of hematopoietic stem cells by the bone marrow stromal niche: the role of reticular cells. Trends Immunol. 2011 Jul;32(7): 315-20.

[32] Asada N, Katayama Y, Sato M, Minagawa K, Wakahashi K, Kawano H, et al. Matrix-embedded osteocytes regulate mobilization of hematopoietic stem/progenitor cells. Cell Stem Cell. 2013 Jun 6;12(6):737-47.

[33] Katayama Y, Battista M, Kao WM, Hidalgo A, Peired AJ, Thomas SA, et al. Signals from the sympathetic nervous system regulate hematopoietic stem cell egress from bone marrow. Cell. 2006 Jan 27;124(2):407-21.

[34] Kawamori Y, Katayama Y, Asada N, Minagawa K, Sato M, Okamura A, et al. Role for vitamin D receptor in the neuronal control of the hematopoietic stem cell niche. Blood. 2010 Dec 16;116(25):5528-35.

[35] Ferraro F, Lymperi S, Mendez-Ferrer S, Saez B, Spencer JA, Yeap BY, et al. Diabetes impairs hematopoietic stem cell mobilization by altering niche function. Sci Transl Med. 2011 Oct 12;3(104):104ra1.

[36] Kiel MJ, Morrison SJ. Uncertainty in the niches that maintain haematopoietic stem cells. Nat Rev Immunol. 2008 Apr;8(4):290-301.

[37] Levesque JP, Winkler IG, Hendy J, Williams B, Helwani F, Barbier V, et al. Hematopoietic progenitor cell mobilization results in hypoxia with increased hypoxia-inducible transcription factor-1 alpha and vascular endothelial growth factor A in bone marrow. Stem Cells. 2007 Aug;25(8):1954-65.

[38] Parmar K, Mauch P, Vergilio JA, Sackstein R, Down JD. Distribution of hematopoietic stem cells in the bone marrow according to regional hypoxia. Proc Natl Acad Sci U S A. 2007 Mar 27;104(13):5431-6.

[39] Takubo K, Goda N, Yamada W, Iriuchishima H, Ikeda E, Kubota Y, et al. Regulation of the HIF-1alpha level is essential for hematopoietic stem cells. Cell Stem Cell. 2010 Sep 3;7(3):391-402.

[40] Forristal CE, Winkler IG, Nowlan B, Barbier V, Walkinshaw G, Levesque JP. Pharmacologic stabilization of HIF-1alpha increases hematopoietic stem cell quiescence in vivo and accelerates blood recovery after severe irradiation. Blood. 2013 Jan 31;121(5):759-69.

[41] Wang LD, Wagers AJ. Dynamic niches in the origination and differentiation of haematopoietic stem cells. Nat Rev Mol Cell Biol. 2011 Oct;12(10):643-55.

[42] Anthony BA, Link DC. Regulation of hematopoietic stem cells by bone marrow stromal cells. Trends Immunol. 2014 Jan;35(1):32-7.

[43] Kunisaki Y, Bruns I, Scheiermann C, Ahmed J, Pinho S, Zhang D, et al. Arteriolar niches maintain haematopoietic stem cell quiescence. Nature. 2013 Oct 31;502(7473): 637-43.

[44] Sipkins DA, Wei X, Wu JW, Runnels JM, Cote D, Means TK, et al. In vivo imaging of specialized bone marrow endothelial microdomains for tumour engraftment. Nature. 2005 Jun 16;435(7044):969-73.

[45] Winkler IG, Barbier V, Nowlan B, Jacobsen RN, Forristal CE, Patton JT, et al. Vascular niche E-selectin regulates hematopoietic stem cell dormancy, self renewal and chemoresistance. Nat Med. 2012 Nov;18(11):1651-7.

[46] Kiel MJ, Yilmaz OH, Iwashita T, Terhorst C, Morrison SJ. SLAM family receptors distinguish hematopoietic stem and progenitor cells and reveal endothelial niches for stem cells. Cell. 2005 Jul 1;121(7):1109-21.

[47] Hooper AT, Butler JM, Nolan DJ, Kranz A, Iida K, Kobayashi M, et al. Engraftment and reconstitution of hematopoiesis is dependent on VEGFR2-mediated regeneration of sinusoidal endothelial cells. Cell Stem Cell. 2009 Mar 6;4(3):263-74.

[48] Essers MA, Offner S, Blanco-Bose WE, Waibler Z, Kalinke U, Duchosal MA, et al. IFNalpha activates dormant haematopoietic stem cells in vivo. Nature. 2009 Apr 16;458(7240):904-8.

[49] Kobayashi H, Butler JM, O'Donnell R, Kobayashi M, Ding BS, Bonner B, et al. Angiocrine factors from Akt-activated endothelial cells balance self-renewal and differentiation of haematopoietic stem cells. Nat Cell Biol. 2010 Nov;12(11):1046-56.

[50] Kacena MA, Gundberg CM, Horowitz MC. A reciprocal regulatory interaction between megakaryocytes, bone cells, and hematopoietic stem cells. Bone. 2006 Nov; 39(5):978-84.

[51] Frenette PS, Pinho S, Lucas D, Scheiermann C. Mesenchymal stem cell: keystone of the hematopoietic stem cell niche and a stepping-stone for regenerative medicine. Annu Rev Immunol. 2013;31:285-316.

[52] Mendez-Ferrer S, Michurina TV, Ferraro F, Mazloom AR, Macarthur BD, Lira SA, et al. Mesenchymal and haematopoietic stem cells form a unique bone marrow niche. Nature. 2010 Aug 12;466(7308):829-34.

[53] Isern J, Martin-Antonio B, Ghazanfari R, Martin AM, Lopez JA, del Toro R, et al. Self-renewing human bone marrow mesenspheres promote hematopoietic stem cell expansion. Cell Rep. 2013 May 30;3(5):1714-24.

[54] Sacchetti B, Funari A, Michienzi S, Di Cesare S, Piersanti S, Saggio I, et al. Self-renewing osteoprogenitors in bone marrow sinusoids can organize a hematopoietic microenvironment. Cell. 2007 Oct 19;131(2):324-36.

[55] Tormin A, Li O, Brune JC, Walsh S, Schutz B, Ehinger M, et al. CD146 expression on primary nonhematopoietic bone marrow stem cells is correlated with in situ localization. Blood. 2011 May 12;117(19):5067-77.

[56] Pinho S, Lacombe J, Hanoun M, Mizoguchi T, Bruns I, Kunisaki Y, et al. PDGFRalpha and CD51 mark human nestin+sphere-forming mesenchymal stem cells capable of hematopoietic progenitor cell expansion. J Exp Med. 2013 Jul 1;210(7):1351-67.

[57] Schajnovitz A, Itkin T, D'Uva G, Kalinkovich A, Golan K, Ludin A, et al. CXCL12 secretion by bone marrow stromal cells is dependent on cell contact and mediated by connexin-43 and connexin-45 gap junctions. Nat Immunol. 2011 May;12(5):391-8.

[58] Milsom MD, Trumpp A. Bridging the information gap. Nat Immunol. 2011 May; 12(5):377-9.

[59] Gonzalez-Nieto D, Li L, Kohler A, Ghiaur G, Ishikawa E, Sengupta A, et al. Connexin-43 in the osteogenic BM niche regulates its cellular composition and the bidirectional traffic of hematopoietic stem cells and progenitors. Blood. 2012 May 31;119(22): 5144-54.

[60] Yamazaki S, Ema H, Karlsson G, Yamaguchi T, Miyoshi H, Shioda S, et al. Nonmyelinating Schwann cells maintain hematopoietic stem cell hibernation in the bone marrow niche. Cell. 2011 Nov 23;147(5):1146-58.

[61] Zapata A. Bone marrow stromal cells: a source of MSC for Cell Therapy. In Simon C, Pellicer A (eds) Stem Cells in Human Reproduction Informa Healthcare. 2013:117-87.

[62] Pittenger MF, Martin BJ. Mesenchymal stem cells and their potential as cardiac therapeutics. Circ Res. 2004 Jul 9;95(1):9-20.

[63] da Silva Meirelles L, Caplan AI, Nardi NB. In search of the in vivo identity of mesenchymal stem cells. Stem Cells. 2008 Sep;26(9):2287-99.

[64] Pontikoglou C, Delorme B, Charbord P. Human bone marrow native mesenchymal stem cells. Regen Med. 2008 Sep;3(5):731-41.

[65] Crisan M, Yap S, Casteilla L, Chen CW, Corselli M, Park TS, et al. A perivascular origin for mesenchymal stem cells in multiple human organs. Cell Stem Cell. 2008 Sep 11;3(3):301-13.

[66] Morikawa S, Mabuchi Y, Kubota Y, Nagai Y, Niibe K, Hiratsu E, et al. Prospective identification, isolation, and systemic transplantation of multipotent mesenchymal stem cells in murine bone marrow. J Exp Med. 2009 Oct 26;206(11):2483-96.

[67] Dennis JE, Carbillet JP, Caplan AI, Charbord P. The STRO-1+marrow cell population is multipotential. Cells Tissues Organs. 2002;170(2-3):73-82.

[68] Gang EJ, Bosnakovski D, Figueiredo CA, Visser JW, Perlingeiro RC. SSEA-4 identifies mesenchymal stem cells from bone marrow. Blood. 2007 Feb 15;109(4):1743-51.

[69] Charbord P, Tavian M, Humeau L, Peault B. Early ontogeny of the human marrow from long bones: an immunohistochemical study of hematopoiesis and its microenvironment. Blood. 1996 May 15;87(10):4109-19.

[70] Westen H, Bainton DF. Association of alkaline-phosphatase-positive reticulum cells in bone marrow with granulocytic precursors. J Exp Med. 1979 Oct 1;150(4):919-37.

[71] Quirici N, Soligo D, Bossolasco P, Servida F, Lumini C, Deliliers GL. Isolation of bone marrow mesenchymal stem cells by anti-nerve growth factor receptor antibodies. Exp Hematol. 2002 Jul;30(7):783-91.

[72] Wislet-Gendebien S, Laudet E, Neirinckx V, Alix P, Leprince P, Glejzer A, et al. Mesenchymal stem cells and neural crest stem cells from adult bone marrow: characterization of their surprising similarities and differences. Cell Mol Life Sci. 2012 Aug; 69(15):2593-608.

[73] Dennis JE, Charbord P. Origin and differentiation of human and murine stroma. Stem Cells. 2002;20(3):205-14.

[74] Takashima Y, Era T, Nakao K, Kondo S, Kasuga M, Smith AG, et al. Neuroepithelial cells supply an initial transient wave of MSC differentiation. Cell. 2007 Jun 29;129(7): 1377-88.

[75] Nagoshi N, Shibata S, Kubota Y, Nakamura M, Nagai Y, Satoh E, et al. Ontogeny and multipotency of neural crest-derived stem cells in mouse bone marrow, dorsal root ganglia, and whisker pad. Cell Stem Cell. 2008 Apr 10;2(4):392-403.

[76] Morikawa S, Mabuchi Y, Niibe K, Suzuki S, Nagoshi N, Sunabori T, et al. Development of mesenchymal stem cells partially originate from the neural crest. Biochem Biophys Res Commun. 2009 Feb 20;379(4):1114-9.

[77] Glejzer A, Laudet E, Leprince P, Hennuy B, Poulet C, Shakhova O, et al. Wnt1 and BMP2: two factors recruiting multipotent neural crest progenitors isolated from adult bone marrow. Cell Mol Life Sci. 2011 Jun;68(12):2101-14.

[78] Wislet-Gendebien S, Laudet E, Neirinckx V, Rogister B. Adult bone marrow: which stem cells for cellular therapy protocols in neurodegenerative disorders? J Biomed Biotechnol. 2012;2012:601560.

[79] Bronner-Fraser M, Fraser SE. Cell lineage analysis reveals multipotency of some avian neural crest cells. Nature. 1988 Sep 8;335(6186):161-4.

[80] Le Douarin NM, Calloni GW, Dupin E. The stem cells of the neural crest. Cell Cycle. 2008 Apr 15;7(8):1013-9.

[81] John N, Cinelli P, Wegner M, Sommer L. Transforming growth factor beta-mediated Sox10 suppression controls mesenchymal progenitor generation in neural crest stem cells. Stem Cells. 2011 Apr;29(4):689-99.

[82] Xu X, Chen C, Akiyama K, Chai Y, Le AD, Wang Z, et al. Gingivae contain neural-crest-and mesoderm-derived mesenchymal stem cells. J Dent Res. 2013 Sep;92(9): 825-32.

[83] Jiang Y, Jahagirdar BN, Reinhardt RL, Schwartz RE, Keene CD, Ortiz-Gonzalez XR, et al. Pluripotency of mesenchymal stem cells derived from adult marrow. Nature. 2002 Jul 4;418(6893):41-9.

[84] Fernandez CI, Alberti E, Mendoza Y, Martinez L, Collazo J, Rosillo JC, et al. Motor and cognitive recovery induced by bone marrow stem cells grafted to striatum and hippocampus of impaired aged rats: functional and therapeutic considerations. Ann N Y Acad Sci. 2004 Jun;1019:48-52.

[85] Neirinckx V, Marquet A, Coste C, Rogister B, Wislet-Gendebien S. Adult bone marrow neural crest stem cells and mesenchymal stem cells are not able to replace lost neurons in acute MPTP-lesioned mice. PLoS One. 2013;8(5):e64723.

[86] Arranz L, Mendez-Ferrer S. Network anatomy and in vivo physiology of mesenchymal stem and stromal cells. Inflammation and Regeneration. 2013;33(1):38-47.

[87] Jung Y, Song J, Shiozawa Y, Wang J, Wang Z, Williams B, et al. Hematopoietic stem cells regulate mesenchymal stromal cell induction into osteoblasts thereby participating in the formation of the stem cell niche. Stem Cells. 2008 Aug;26(8):2042-51.

[88] Walkley CR, Olsen GH, Dworkin S, Fabb SA, Swann J, McArthur GA, et al. A microenvironment-induced myeloproliferative syndrome caused by retinoic acid receptor gamma deficiency. Cell. 2007 Jun 15;129(6):1097-110.

[89] Leeman DS, Brunet A. Stem cells: Sex specificity in the blood. Nature. 2014 Jan 23;505(7484):488-90.

[90] Nakada D, Oguro H, Levi BP, Ryan N, Kitano A, Saitoh Y, et al. Oestrogen increases haematopoietic stem-cell self-renewal in females and during pregnancy. Nature. 2014 Jan 23;505(7484):555-8.

[91] Yamazaki K, Allen TD. Ultrastructural morphometric study of efferent nerve terminals on murine bone marrow stromal cells, and the recognition of a novel anatomical unit: the "neuro-reticular complex". Am J Anat. 1990 Mar;187(3):261-76.

[92] Ralph MR, Foster RG, Davis FC, Menaker M. Transplanted suprachiasmatic nucleus determines circadian period. Science. 1990 Feb 23;247(4945):975-8.

[93] Scheiermann C, Kunisaki Y, Frenette PS. Circadian control of the immune system. Nat Rev Immunol. 2013 Mar;13(3):190-8.

[94] Dimitrov S, Benedict C, Heutling D, Westermann J, Born J, Lange T. Cortisol and epinephrine control opposing circadian rhythms in T cell subsets. Blood. 2009 May 21;113(21):5134-43.

[95] Haus E, Lakatua DJ, Swoyer J, Sackett-Lundeen L. Chronobiology in hematology and immunology. Am J Anat. 1983 Dec;168(4):467-517.

[96] Haus E, Smolensky MH. Biologic rhythms in the immune system. Chronobiol Int. 1999 Sep;16(5):581-622.

[97] Mendez-Ferrer S, Lucas D, Battista M, Frenette PS. Haematopoietic stem cell release is regulated by circadian oscillations. Nature. 2008 Mar 27;452(7186):442-7.

[98] Scheiermann C, Kunisaki Y, Lucas D, Chow A, Jang JE, Zhang D, et al. Adrenergic nerves govern circadian leukocyte recruitment to tissues. Immunity. 2012 Aug 24;37(2):290-301.

[99] Lucas D, Battista M, Shi PA, Isola L, Frenette PS. Mobilized hematopoietic stem cell yield depends on species-specific circadian timing. Cell Stem Cell. 2008 Oct 9;3(4): 364-6.

[100] Spiegel A, Shivtiel S, Kalinkovich A, Ludin A, Netzer N, Goichberg P, et al. Catecholaminergic neurotransmitters regulate migration and repopulation of immature human CD34+cells through Wnt signaling. Nat Immunol. 2007 Oct;8(10):1123-31.

[101] Jeanson NT, Scadden DT. Vitamin D receptor deletion leads to increased hematopoietic stem and progenitor cells residing in the spleen. Blood. 2010 Nov 18;116(20): 4126-9.

[102] del Toro R, Mendez-Ferrer S. Autonomic regulation of hematopoiesis and cancer. Haematologica. 2013 Nov;98(11):1663-6.

[103] Schneider A, Kruger C, Steigleder T, Weber D, Pitzer C, Laage R, et al. The hematopoietic factor G-CSF is a neuronal ligand that counteracts programmed cell death and drives neurogenesis. J Clin Invest. 2005 Aug;115(8):2083-98.

[104] Lucas D, Bruns I, Battista M, Mendez-Ferrer S, Magnon C, Kunisaki Y, et al. Norepinephrine reuptake inhibition promotes mobilization in mice: potential impact to rescue low stem cell yields. Blood. 2012 Apr 26;119(17):3962-5.

[105] Schuettpelz LG, Link DC. Regulation of hematopoietic stem cell activity by inflammation. Front Immunol. 2013;4:204.

[106] Ehninger A, Trumpp A. The bone marrow stem cell niche grows up: mesenchymal stem cells and macrophages move in. J Exp Med. 2011 Mar 14;208(3):421-8.

[107] Chow A, Lucas D, Hidalgo A, Mendez-Ferrer S, Hashimoto D, Scheiermann C, et al. Bone marrow CD169+macrophages promote the retention of hematopoietic stem and progenitor cells in the mesenchymal stem cell niche. J Exp Med. 2011 Feb 14;208(2): 261-71.

[108] Winkler IG, Sims NA, Pettit AR, Barbier V, Nowlan B, Helwani F, et al. Bone marrow macrophages maintain hematopoietic stem cell (HSC) niches and their depletion mobilizes HSCs. Blood. 2010 Dec 2;116(23):4815-28.

[109] Christopher MJ, Rao M, Liu F, Woloszynek JR, Link DC. Expression of the G-CSF receptor in monocytic cells is sufficient to mediate hematopoietic progenitor mobilization by G-CSF in mice. J Exp Med. 2011 Feb 14;208(2):251-60.

[110] Serbina NV, Pamer EG. Monocyte emigration from bone marrow during bacterial infection requires signals mediated by chemokine receptor CCR2. Nat Immunol. 2006 Mar;7(3):311-7.

[111] Shi C, Jia T, Mendez-Ferrer S, Hohl TM, Serbina NV, Lipuma L, et al. Bone marrow mesenchymal stem and progenitor cells induce monocyte emigration in response to circulating toll-like receptor ligands. Immunity. 2011 Apr 22;34(4):590-601.

[112] Liotta F, Angeli R, Cosmi L, Fili L, Manuelli C, Frosali F, et al. Toll-like receptors 3 and 4 are expressed by human bone marrow-derived mesenchymal stem cells and can inhibit their T-cell modulatory activity by impairing Notch signaling. Stem Cells. 2008 Jan;26(1):279-89.

[113] Stark MA, Huo Y, Burcin TL, Morris MA, Olson TS, Ley K. Phagocytosis of apoptotic neutrophils regulates granulopoiesis via IL-23 and IL-17. Immunity. 2005 Mar;22(3): 285-94.

[114] Casanova-Acebes M, Pitaval C, Weiss LA, Nombela-Arrieta C, Chevre R, N AG, et al. Rhythmic modulation of the hematopoietic niche through neutrophil clearance. Cell. 2013 May 23;153(5):1025-35.

[115] Martin C, Burdon PC, Bridger G, Gutierrez-Ramos JC, Williams TJ, Rankin SM. Chemokines acting via CXCR2 and CXCR4 control the release of neutrophils from the bone marrow and their return following senescence. Immunity. 2003 Oct;19(4): 583-93.

[116] Bessis M. [Erythroblastic island, functional unity of bone marrow]. Rev Hematol. 1958 Jan-Mar;13(1):8-11.

[117] Chow A, Huggins M, Ahmed J, Hashimoto D, Lucas D, Kunisaki Y, et al. CD169(+) macrophages provide a niche promoting erythropoiesis under homeostasis and stress. Nat Med. 2013 Apr;19(4):429-36.

[118] Ramos P, Casu C, Gardenghi S, Breda L, Crielaard BJ, Guy E, et al. Macrophages support pathological erythropoiesis in polycythemia vera and beta-thalassemia. Nat Med. 2013 Apr;19(4):437-45.

[119] Tokoyoda K, Egawa T, Sugiyama T, Choi BI, Nagasawa T. Cellular niches controlling B lymphocyte behavior within bone marrow during development. Immunity. 2004 Jun;20(6):707-18.

[120] Park SY, Wolfram P, Canty K, Harley B, Nombela-Arrieta C, Pivarnik G, et al. Focal adhesion kinase regulates the localization and retention of pro-B cells in bone marrow microenvironments. J Immunol. 2013 Feb 1;190(3):1094-102.

[121] Kabashima K, Haynes NM, Xu Y, Nutt SL, Allende ML, Proia RL, et al. Plasma cell S1P1 expression determines secondary lymphoid organ retention versus bone marrow tropism. J Exp Med. 2006 Nov 27;203(12):2683-90.

[122] Hauser AE, Debes GF, Arce S, Cassese G, Hamann A, Radbruch A, et al. Chemotactic responsiveness toward ligands for CXCR3 and CXCR4 is regulated on plasma blasts during the time course of a memory immune response. J Immunol. 2002 Aug 1;169(3):1277-82.

[123] Nakayama T, Hieshima K, Izawa D, Tatsumi Y, Kanamaru A, Yoshie O. Cutting edge: profile of chemokine receptor expression on human plasma cells accounts for their efficient recruitment to target tissues. J Immunol. 2003 Feb 1;170(3):1136-40.

[124] Cyster JG. Homing of antibody secreting cells. Immunol Rev. 2003 Aug;194:48-60.

[125] Chu VT, Berek C. The establishment of the plasma cell survival niche in the bone marrow. Immunol Rev. 2013 Jan;251(1):177-88.

[126] Chu VT, Frohlich A, Steinhauser G, Scheel T, Roch T, Fillatreau S, et al. Eosinophils are required for the maintenance of plasma cells in the bone marrow. Nat Immunol. 2011 Feb;12(2):151-9.

[127] Rozanski CH, Arens R, Carlson LM, Nair J, Boise LH, Chanan-Khan AA, et al. Sustained antibody responses depend on CD28 function in bone marrow-resident plasma cells. J Exp Med. 2011 Jul 4;208(7):1435-46.

[128] Yu C, Cantor AB, Yang H, Browne C, Wells RA, Fujiwara Y, et al. Targeted deletion of a high-affinity GATA-binding site in the GATA-1 promoter leads to selective loss of the eosinophil lineage in vivo. J Exp Med. 2002 Jun 3;195(11):1387-95.

[129] Belnoue E, Tougne C, Rochat AF, Lambert PH, Pinschewer DD, Siegrist CA. Homing and adhesion patterns determine the cellular composition of the bone marrow plasma cell niche. J Immunol. 2012 Feb 1;188(3):1283-91.

[130] O'Connor BP, Raman VS, Erickson LD, Cook WJ, Weaver LK, Ahonen C, et al. BCMA is essential for the survival of long-lived bone marrow plasma cells. J Exp Med. 2004 Jan 5;199(1):91-8.

[131] Belnoue E, Pihlgren M, McGaha TL, Tougne C, Rochat AF, Bossen C, et al. APRIL is critical for plasmablast survival in the bone marrow and poorly expressed by early-life bone marrow stromal cells. Blood. 2008 Mar 1;111(5):2755-64.

[132] Chu VT, Enghard P, Schurer S, Steinhauser G, Rudolph B, Riemekasten G, et al. Systemic activation of the immune system induces aberrant BAFF and APRIL expression in B cells in patients with systemic lupus erythematosus. Arthritis Rheum. 2009 Jul; 60(7):2083-93.

[133] Yokota T, Meka CS, Medina KL, Igarashi H, Comp PC, Takahashi M, et al. Paracrine regulation of fat cell formation in bone marrow cultures via adiponectin and prostaglandins. J Clin Invest. 2002 May;109(10):1303-10.

[134] Chitteti BR, Cheng YH, Poteat B, Rodriguez-Rodriguez S, Goebel WS, Carlesso N, et al. Impact of interactions of cellular components of the bone marrow microenvironment on hematopoietic stem and progenitor cell function. Blood. 2010 Apr 22;115(16): 3239-48.

[135] Naveiras O, Nardi V, Wenzel PL, Hauschka PV, Fahey F, Daley GQ. Bone-marrow adipocytes as negative regulators of the haematopoietic microenvironment. Nature. 2009 Jul 9;460(7252):259-63.

[136] Kincade PW. Plasticity of supporting cells in a stem cell factory. Immunity. 2010 Sep 24;33(3):291-3.

[137] Curtis SK, Cowden RR, Knagel JW. Ultrastructure of the bone marrow of the salamander *Plethodon glutinosus* (Caudata Plethodontidae). J Morphol. 1979;123(1):405-40.

[138] Zapata A, Cooper EL. The immune system: comparative histophysiology. In John Willey and Sons (Ed) Chichester (England). 1990.

[139] Chiba A, Torroba M, Honma Y, Zapata AG. Occurrence of lymphohaemopoietic tissue in the meninges of the stingray Dasyatis akajei (Elasmobranchii, chondricthyes). Am J Anat. 1988 Nov;183(3):268-76.

[140] Mattisson A, Fange R. The cellular structure of lymphomyeloid tissues in *Chimaera monstrosa* (Pisces, Holocephali). Biol Bull. 1986;171(1):660-71.

[141] Mattisson A, Fange R, Zapata AG. Histology and ultrastructure of the cranial lympho-haemopoietic tissue in *Chimaera monstrosa* (Pisces, Holocephali). Acta Zool Stockh. 1990;71(1):97-106.

[142] Scharrer E. The histology of the meningeal myeloid tissue in the ganoids Amia and Lepisosteus. Anat Rec. 1944;88(1):291-310.

[143] Dempster WT. The morphology of the amphibian endolymphatic organ. J Morphol. 1930;50(1):71-126.

[144] Sano Y, Imai C. Über ein blutbildendes Organ in der sella Turcica des Reisensalamanders (*Megalobatrachus japonicus*). Z Zellfors. 1961;53(1):471-80.

[145] Horton JD. Ontogeny of the immune system in amphibians. Am Zool. 1971;11(1): 219-28.

[146] Hadji-Azimi I, Schwager J, Thiebard CH, Perrenot N. B-lymphocyte differentitation in *Xenopus laevis larvae*. Dev Biol. 1982;90(1):253-8.

[147] Ardavin CF, Zapata A. Ultrastructure and changes during metamorphosis of the lympho-hemopoietic tissue of the larval anadromous sea lamprey Petromyzon marinus. Dev Comp Immunol. 1987 Winter;11(1):79-93.

[148] O'Brien LE, Bilder D. Beyond the niche: tissue-level coordination of stem cell dynamics. Annu Rev Cell Dev Biol. 2013;29:107-36.

Chapter 3

CXCR4 in Central and Peripheral Lymphoid Niches – Physiology, Pathology and Therapeutic Perspectives in Immune Deficiencies and Malignancies

Christelle Freitas, Alexandre Bignon,
Karl Balabanian and Ali Dalloul

Additional information is available at the end of the chapter

http://dx.doi.org/10.5772/58699

1. Introduction

The discovery of the cell-attracting chemokine Stromal cell Derived Factor 1 (SDF-1)/CXCL12 some twenty years ago and its ligand CXCR4 spurred tremendous research interest and generated an abundant literature reflecting the role of CXCR4 in various aspects of physiology and pathology, extending much beyond the role of CXCR4 as a co-receptor for Human Immunodeficiency Virus (HIV) on human T-cells [1-3].

Indeed, CXCL12 and CXCR4 are expressed in a complementary pattern during embryogenesis and are needed for endodermal migration [4, 5]. In addition, CXCL12 is induced by hypoxia through the transcription factor hypoxia-inducible factor 1 (HIF1) and accordingly stimulates CXCR4-positive stem cells that locate within cellular hypoxic niches [6, 7].

These features explain the lethal phenotype of *Cxcr4* null-mutated mice, hence the need for cell-specific conditional deletion models to be able to decipher its physiological functions [8]. This sharply contrasts with knockouts (KO) for most other chemokine/chemokine receptor pairs, which are almost viable throughout the adult life and point to the uniqueness of the CXCL12/CXCR4 pair in physiology. Note however that another CXCL12 receptor, ACKR3 (CXCR7) has been described, it may therefore overlap with the functions of CXCR4 [9-11]. As the physiological role of ACKR3 is not yet fully understood, its description is beyond the scope of this review.

Although CXCR4 signaling governs many aspects of stem cell development during embryonic and adult life, it also plays a role in the proper homing and migration of mature lymphocytes within the bone marrow and peripheral lymphoid organs. It also regulates B-cell differentiation as discussed later.

From a paradigmatic point of view, upon engagement with its ligand, three processes namely desensitization (homologous and heterologous), internalization, and degradation regulate CXCR4. The homologous desensitization, or becoming refractory to continued stimulation, involves the recruitment of G-protein-coupled receptor kinases (GRK) that promote the recruitment of β-Arrestin which prevents coupling of receptor to new G protein and initiate endocytosis in clathrin-coated pits [12]. The receptor may then be recycled or degraded following ubiquitination by the E3 ubiquitin ligase AIP4 [13]. Another mechanism of heterologous desensitization is mediated by the activation of second messenger dependent protein kinases such as protein kinase A and C (PKA and PKC) [14].

Abnormal expression and/or activity of CXCR4 due to an over stimulation or inhibition of transcription or internalization may therefore impact on several aspects of haematopoiesis and adaptive immune response, hence immune deficits and lymphoid neoplasms. A paradigmatic disease in this respect is the WHIM syndrome in which *CXCR4* is mutated and displays anomalous internalization and signaling properties [15]. More recently, one third of patients with Waldenström's macroglobilinemia, a disorder induced by an IgM-producing plasmacytoma, were shown to harbour the WHIM syndrome mutation [16].

Altogether understanding the molecules that regulate CXCR4 expression and/or activity in leukocytes is a very active area of investigation inasmuch as CXCR4 expression can be regulated by the processing of CXCL12 either by cleavage or binding to glycosaminoglycans (GAGs) within extracellular matrix [17, 18]. Engagement of either the T-cell or the B-cell receptor (TCR or BCR) on lymphocytes can down regulate cell surface CXCR4 expression pointing to cross talks between CXCL12/CXCR4 axis and other chemokine independent pathways [19, 20].

These findings already translated into clinical applications with the invention of the first CXCR4 antagonist, AMD3100 and as our knowledge widens will be followed by other drugs aimed at directly or indirectly regulating CXCR4 expression and signaling, for multiple medical purposes [21].

The physiological role of chemokine receptors is still a matter of intense research and attempts to generate pharmacological antagonists of these receptors are numerous [205]. Among these receptors, CXCR4 is the one with the most successful implications in medicine, especially in the fields of immunology and haematology. Our goals were therefore to summarize or current knowledge on CXCR4 in these fields. This review describes our current knowledge on CXCR4 involvement in the development of lymphoid progenitors (chapter 2), B-cells (chapter 3), T-cells (chapter 4) and as recently described, in NK-cells (chapter 5). We describedon one hand, pathological lymphoid disorders with quantitative abnormalities of CXCR4 including Idiopathic T-cell lymphocytopenia (chapter 7), chronic lymphocytic leukaemia (chapter 8), and solid tumours (chapter 9); and on the other hand, pathologies with qualitative mutations of CXCR4 including the WHIM Syndrome (chapter 6) and Waldenström's Macroglobulinemia (chapter 8).

2. CXCR4 and haematopoiesis

The CXCL12/CXCR4 axis is pivotal in the survival of Hematopoietic Stem Cells (HSC) and early-committed progenitors, in their mobilization from the bone marrow to the peripheral blood and migration back from the blood to the bone marrow. This later process continuously seeds suitable niches from the bone marrow with functional HSC thereby allowing the maintenance of haematopoiesis. Hematopoietic stem and progenitors expressing CXCR4 interact with CXCL12-expressing stromal cells within niches thus delivering survival, and mobilization signals to multipotent and committed progenitors.

The sequential appearance and disappearance of hematopoietic activity is notably governed by the CXCL12/CXCR4 axis and Very Late Antigen-4/Vascular cell adhesion molecule-1 (VLA4/VCAM-1) interactions [8, 22]. The role of CXCL12/CXCR4 interaction is needed for proper haematopoiesis throughout life as shown by induced deletion of *Cxcr4* in adult mice [23]. As mentioned above, the mechanisms of stem cell mobilization from the bone marrow are complex and regulated by factors of homing and adhesion to and within the niche. Adhesion is a multistep process that involves selectin-mediated tethering and rolling on stromal/endothelial cells followed by activation and ultimately firm adhesion mediated by integrins [24]. CD34$^+$cells express both CD162 a sialomucin that binds all selectins and VLA4 [25, 26]. Of interest in this respect, CXCL12 enhances integrin activation and at least *in vitro*, E-selectin thereby promoting adhesion of CD34$^+$cells to the niche under flow, which allows their subsequent transmigration through endothelium [27, 28]. CXCL12 synergizes with Granulocyte colony-stimulating factor (GCSF) which activate the VLA4-ligand VCAM1and Vascular endothelial growth factor (VEGF) which activate both VCAM1 and E-selectin [28-30]. These findings were also supported by injection of antibodies and several KO mice models, and most convincingly by grafting human CD34$^+$cells into severe combined immunodeficiency mice [31]; this work demonstrated the dependency of CD34$^+$cell engraftment on CXCR4, the expression of which is unregulated by Stem cell Factor (SCF) [32]. Altogether, these findings demonstrated that CXCR4 signaling is required for stem cells to migrate, survive and differentiate into dedicated bone marrow niches. Studies dealing with effects of CXCR4 on the cell cycle on HSCs are contradictory. Some indicate that CXCR4 is required for the quiescence of primitive HSCs whereas others pointed a synergy between SCF/Kit-ligand and CXCL12 for the proliferation of CXCR4$^+$progenitors [23, 33, 34]. Indeed, CXCR4 expression is higher in the most primitive CD34$^+$CD38$^-$progenitors and CXCL12 induced CD34$^+$cell proliferation in synergy with SCF or thrombopoietin TPO *in vitro*, further, cycling CD34$^+$cells may produce CXCL12 which protects them from apoptosis in an autocrine /paracrine manner [35, 36].

SCF and CXCL12 synergize for the maintenance of Kitand CXCR4 expressing HSC; this is highlighted by the finding that perivascular SCF expression pattern is very similar to that of

CXCL12 in bone marrow stromal cells especially in the perivascular niche [37]. This also is in agreement with the reported localization of the most primitive HSC adjacent to sinusoidal blood vessels [23, 38-40]. Most HSC are in contact with CXCL12-abundant reticular (CAR) cells mostly located around sinusoidal endothelial cells or near the endosteum [23]. Recent outstanding data from the groups of Morrison and Nagasawa using conditional deletion of *Cxcl12* in various stromal cells from the bone marrow redefined the HSC niches [37, 41]. These data demonstrate that the most primitive Cxcr4^{+}HSC reside in a perivascular region in close contact with endothelial cells and perivascular cells whereas early lymphoid progenitors occupy an endosteal niche in close contact with osteoblasts. This also implies that bone marrow progenitors directly interact with Cxcl12-expressing stromal cells and do not respond to a Cxcl12 gradient, but that Cxcr4 expression and responsiveness of various progenitors to Cxcl12 dictate their fate within the bone marrow. It is indeed striking that CXCR4 signaling mediates quiescence in primitive HSC whereas it induces the proliferation of early B-cell progenitors [8, 33]. Thus genetic ablation of the *Cxcr4* transcript relieves HSC from quiescence and allows them to differentiate and migrate to the blood, as they fail to home back into their bone marrow niches, these cells ultimately loose their self renewal capacity and ability to restore blood lineages upon grafting into syngeneic recipient.

Interference with the CXCL12/CXCR4 axis, was initially thought to prevent the binding and entry of HIV in T-cells using CXCR4 antagonist AMD3100, but this approach was not used in clinics until now [42]. However, a clinical trial is underway with the CXCR4 allosteric antagonist AMD11070 [43].

Nevertheless, interfering with CXCL12/CXCR4 axis by means of a pharmacological and reversible ligand of CXCR4, AMD3100, allows human HSC to leave their niche and migrate to the blood were they can easily be collected by apheresis in order to be grafted to a recipient (allograft) or to the donor himself (auto graft) following chemotherapy. AMD3100 is currently approved by the Food and Drug Administration (FDA) to induce in combination with GCSF the mobilization of human HSC [44]. Of interest, injection of GCSF to induce stem cell mobilization has long been used for clinical transplantation although its mechanism remained poorly understood [45]. Among the mechanisms evoked for GCSF action, is the degradation of CXCL12 by neutrophil elastase in parallel with up regulation of CXCR4 [46]. As CXCL12 has a very short half-life time (1 min), it is very sensitive to degradation by cleavage of its N-terminal domain [47].

In brief, CXCR4 is expressed on the most primitive and more committed hematopoietic progenitors and signals through interactions with CXCL12-expressing stromal cells in various bone marrow niches. CXCR4-mediated signaling results in distinct responses depending on the cell type (for instance quiescence *vs* proliferation), as also shown by predominant HSC and B-cell defects in *Cxcr4*-null mice, and neutrophil abnormalities in WHIM syndrome. It therefore remains to understand how CXCR4 is modulated and why progenitors behave differently following CXCR4 engagement. It is also unclear how progenitors move from one niche to another following CXCR4 engagement. Solving these challenges in the next years will undoubtedly teach us more in the fields of haematopoiesis and immunology.

3. CXCR4 and B-cell development

3.1. Early steps of B-cell development

Contrary to HSC, where CXCR4 signaling promotes survival and quiescence (see above), CXCR4 signaling in B-cell precursors promotes cell growth; this property led to the discovery of SDF1/CXCL12 [1]. The *Cxcl12*-null mice model established that Cxcl12 was necessary for B-cell lymphopoiesis in foetal liver and bone marrow, in addition to be necessary for bone marrow myelopoiesis [8]. This phenotype was close to that of *Cxcr4* null-mutated mice and is likely due to the inability of B-cells from the blood to home into bone marrow or foetal liver [48, 49]. Homing of B-cell precursors into specific bone marrow niches is indeed dependent on Cxcl12/Cxcr4 interactions [50]. Further experiments using adoptive transfer of cells from *Cxcr4*$^{-/-}$animals, established the need for Cxcr4 expression by B-cell progenitors throughout adult life in order to home into the bone marrow [51]. Although myeloid lineages were also affected, the defect was predominant on the B-cell lineage further pointing to the dependency of B-cells towards the CXCL12/CXCR4 axis. As for HSC, B-cell precursors depend on VLA4/VCAM1, and CXCR4/CXCL12 interactions for adhesion to stromal cells [52, 53].

Cxcl12-abundant reticular cells (CAR) are scattered throughout the bone marrow and in close contact with HSC and B-cell precursors [23, 54]. Recently, the redefinition of bone marrow niches established that early lymphoid precursors are in close contact with osteoblasts while committed B-lineage progenitors are in contact with the perivascular niche [37, 41]. Of interest, the ability of osteoblasts to support the differentiation of primitive HSC towards all stages of B-cells has been reported [55]. Furthermore, this differentiation appeared dependent on VCAM-1, CXCL12 and G protein αi subunit expression by osteoblasts [55, 56]. It should be mentioned that the expression of CXCR4 is sinusoidal in the B-cell lineage being highest in pre-B-cells and decreasing as cells develop in immature B-cells, and finally increases in mature B-cells; however the response of mature B-cells to CXCL12 remains poor despite their high CXCR4 expression [57]. Thus, as for HSC, B-cell precursors are highly sensitive to CXCL12; this is possibly linked to intrinsic properties of the cells such as sustained activation of the focal adhesion kinase pathway contrary to mature B-cells [53]. Altogether, the fine-tuning of CXCR4 response during B-cell maturation is complex and not only mediated by the regulation of CXCL12 production and proteolysis or by CXCR4 transcription and trafficking, but also by exogenous factors such as local production of CCR5 ligands that induce heterologous desensitization of CXCR4 to CXCL12 [58].

As discussed below, mutant mice harbouring the WHIM Syndrome linked *Cxcr4* mutation display abnormal B-cell differentiation at the early B-cell committed progenitors stages combined to anomalous B-cell positioning within secondary lymphoid organs [59]. This points to the role of CXCR4 in mature B-cells peripheral compartmentalization.

3.2. B-cell homing and positioning within secondary lymphoid organs and regulation of CXCR4 expression by BCR signaling

Following B-cell development in the bone marrow, transitional B-cells home into secondary lymphoid organs to become follicular naive B-cells that constitute a B-cell zone, called primary follicle, distinct from the T-cell one. B-cells use CCR7, CXCR4 and CXCR5 receptors to migrate from the blood into SLO [60, 61]. $CXCR5^{+}$B-cells are attracted and organize themselves around CXCL13-expressing follicular dendritic cells and marginal reticular cells at the periphery, whereas T-cells express CCR7 and are attracted by CCL19-and CCL21-producing reticular cells [62]. Following antigen-specific T-cell activation, T-cells down regulate CCR7 and move towards B-cell follicles. In parallel antigen-activated B-cells keep their CXCR5 expression unaltered while up regulating that of CCR7 thereby migrating towards the T-cell zone. Upon T-B encounter, antigen-specific B-cells then proliferate to constitute the dark zone of the germinal centre. These large cells so called centroblasts undergo Immunoglobulin (Ig) gene class-switch recombination, due to CD40/CD154 interactions on B-and T-cells respectively [63]. Class-switch recombination is also a prerequisite for B-cells to undergo somatic hyper mutation of the hyper variable regions of Ig genes, these random mutation process leads to antibody diversity [64]. Centroblasts constitute the dark zone and are $CXCR4^{+/hi}$. As centroblasts are promoted they down regulate CXCR4, enter a quiescent state and constitute the light zone made of $CD23^{+}CD83^{+}CXCR4^{-/lo}$ differentiated B-cells called centrocytes and are selected by antigen-presenting cells, so that B-cells with high affinity against antigen are positively selected for survival whereas low-affinity cells die in situ [65, 66]. Centrocytes ultimately give rise to plasma cells (PC) or to memory B-cells. B-cell differentiation in PC precursors also called plasmablasts, requires a synchronous up regulation of CXCR4 and down regulation of CXCR5, however migration of plasmablasts into splenic red pulp or lymph node medullary sinuses may not depend on CXCR4 expression inasmuch as mice with invalid *Cxcr4* in adult B-cells display PC in these respective anatomic sites [67-69]. A likely hypothesis is that CXCR4 is instead needed for long-lived PC survival in bone marrow niches and thatVLA4 interaction with VCAM1 is necessary in this respect [70, 71] whereas migration of antibody-producing cells into the bone marrow depends on the sphingosine-1 phosphate and its receptor S1P1 [68]. Finally, an unanswered and interesting question is whether long-lived PC and B-cell progenitors compete for the same bone marrow niches.

Thus CXCR4 expression is sinusoidal in mature B-cells and tightly regulated within the secondary follicle intuitively suggesting that anomalous expression or signaling of CXCR4 may lead to B-cell defects or malignancies.

It is well known that CXCR4 expression is regulated directly by interactions with CXCL12 and indirectly by BCR signaling. CXCL12 interaction with CXCR4 induces CXCR4 inactivation on T-and B-cells and myeloid cells as well, so it is a general mechanism at least on mature cells that lead to CXCR4 phosphorylation on its intracytoplasmic C-terminal tail, leading to G protein uncoupling and CXCR4 internalization [20, 72, 73]. As for most chemokine receptor which are G-protein-coupled receptors, binding of CXCR4 with its ligand induces conformational modifications and secondary activation of G-proteins and further intracellular recruitment of G-protein-coupled-receptor kinases (GRK) that phosphorylate CXCR4 on Ser/Thr residues and recruit β-arrestins resulting in CXCR4 desensitization and finally its internalization [14]. CXCR4 is internalized and its surface expression down regulated following BCR activation or CXCL12 binding. Both signaling pathways induce CXCR4 phosphorylation on Ser residues albeit on distinct residues [20]. In fact, the migratory B-cell response to CXCL12 depends on both intrinsic factors and CXCR4 expression; indeed naive and memory B-cells are more sensitive to CXCL12 than germinal center B-cells in this respect [74]. The inhibition of CXCL12-mediated chemotaxis following BCR engagement depends on PKC but not on Ca^{2+}due to CXCR4 internalization [20).

In conclusion, CXCR4 expression in B-cell lineage cells depends on CXCL12 expression at the pro/pre B-cell stages and PC that lack functional BCR, whereas it also depends on the BCR signal strength at more mature stages.

4. CXCR4 and T-cell development

Early thymic progenitors $cKit^{+}$which reside in the thymic double negative DN1 fraction differentiate from common lymphoid progenitor and migrate from the bone marrow to the thymus through the blood and invade the epithelial rudiment at the cortico-medullary junction through post capillary venules where they loose progressively non-T-cell differentiated capacity and become fully T-cell committed as also described in human thymocytes [75-79]. These DN thymocytes make some 2% of thymocytes and are further divided in differentiation stages based on CD44 and CD25 expression [80]. $CD44^{+}CD25^{-}$DN1 cells differentiate in $CD44^{+}CD25^{+}$DN2 cells following migration into the subcapsular zone, DN2 differentiate in $CD44^{-}CD25^{+}$DN3 cells and the later in $CD44^{-}CD25^{-}$cortical DN4 cells. In turn DN4 differentiate in $CD4^{+}CD8^{+}$double positive DP thymocytes that acquire surface CD3 and move toward the inner cortex [81]. Only two chemokines, CXCL12 and CCL25/thymus-expressed chemokine TECK are expressed in the thymic cortex namely by cortical epithelial thymic cells [82-84]. The CXCL12/CXCR4 axis is mandatory for retaining human double positive (DP) thymocytes in the cortex. In addition, in mice lacking Cxcr4 on thymocytes displayed defective DN migration to the cortex and defective DN to DP transition [82, 85-87]. DN migration can indeed be inhibited by AMD3100 [85, 88]. In the medulla, CCR7 promotes migration of mature thymocytes from the cortex. Thus, two opposing chemokine gradients regulate thymocyte migration from the cortex to the medulla.

TCR βselection starts at the DN stage and DN3 cells express pre-TCR made of a pre-TCR α chain associated to TCR-β chain. These cells are selected positively by interactions with stromal cells in the cortex and CXCR4 is crucial for this process. Indeed, it is functionally associated with the pre-TCR and needed to activate phosphatidyl inositol 3-kinase (PI3K) and Notch pathway, the later which is mandatory for T-cell differentiation [86, 87, 89, 90]. In brief, DN thymocytes expand and differentiate in the cortex due to migration retention and survival signals dependent on CXCR4, and as for HSC, CXCR4 activation up regulate adhesion molecules integrin-$\alpha 4\beta 1$/VLA4 [91, 92].

Thus CXCR4 plays a crucial and non-redundant role in thymic progenitors positioning in the cortex and in pre-TCR-mediated survival signals, the failure of which resulted in developmental arrest at the DN stage. Of interest, mice carrying a WHIM Syndrome heterozygous *Cxcr4* mutation display abnormal thymic maturation [59]. Once positive selection of DP thymocytes and negative selection of $CD4^{+}$and $CD8^{+}$single positive thymocytes has been achieved, these cells leave the thymus to seed secondary lymphoid organs. Naive T-cells do not rely on CXCR4 for thymic egress and to home into T-cell zones. However CXCR4 is needed for T-cell extravasation through high endothelial venules and entry into the lymph nodes [93].

As for the BCR, TCR engagement or signals that mimic PKC activation such as phorbol esters result in the phosphorylation and internalization of CXCR4 in T-cells, whereas infections with human herpes viruses HHV-6 or-7 down regulate*CXCR4* transcription [72, 94]. TCR-induced CXCR4 down modulation may therefore help T-cells stay at the site of self-antigen encounter by dendritic cells (DC) in the thymus in order to undergo negative selection in the medulla. Similar mechanism may operate in secondary lymphoid organs during presentation of antigens by DC. Of interest CXCR4 may increase the stability of the T-antigen-presenting cell (APC) immunologic synapses (IS) necessary for successful T-cell activation [95]. Interestingly, upon CXCL12 stimulation, a physical association between CXCR4 and TCR was described and promote TCR signal transduction. This association is responsible for prolonged extracellular signal regulated kinase activity, increased intracellular calcium ion concentrations, robust *activator protein-1* transcriptional activity, and SDF-1α costimulation of cytokine secretion. These pathways mediate costimulation of cytokine secretion by activated T-cells [96].

Following antigen encounter, naive T-cells differentiate in $CD4^{+}$or $CD8^{+}$memory T-cells that home into the bone marrow which is a survival site for memory T-cells [97, 98]. However CXCR4 does not seem to be necessary for the retention and survival of memory T-cells, as survival is mostly dependent on IL7 [99, 100].

In conclusion the physiological role of CXCR4 in T-cell development and functions predominates on thymic progenitors although impaired peripheral T-cell responses in WHIM syndrome and Idiopathic $CD4^{+}$T-cell Lymphocytopenia suggest a role of CXCR4 on peripheral T-cells.

5. CXCR4 and NK-cell development

Natural killer (NK) cells are lymphocytes of the innate immune system that are involved in the early control of infections by viruses and other intracellular pathogens. NK cells can be identified by expression of the activating NK receptor NK1.1 associated with the absence of T-cell CD3 receptor complex. Based on the membrane expression of the tumour necrosis factor superfamily member CD27 and the integrin CD11b, four maturation stages can be identified: $CD11b^{low}CD27^{low}$ ("double negative"), which likely encompass precursors, $CD11b^{low}CD27^{high}$ ("$CD11b^{low}$"), $CD11b^{high}CD27^{high}$ ("double positive"), and $CD11b^{high}CD27^{low}$ ("$CD27^{low}$") [101, 102]. These NK-cell subsets display heterogeneous distribution in lymphoid organs. The bone marrow plays a pivotal role in NK-cell development. NK-cell differentiation in the bone marrow is associated with coordinated and tightly regulated changes of both expression and activity of distinct GPCRs including CXCR4. Indeed, CXCR4 is highly expressed on NK precursors but it gradually declines during maturation [103]. Conversely, expression of CXCR3, CCR1, CX_3CR1 and S1P5 (Sphingosine-1-phosphate receptor 5) progressively increase with NK-cell maturation [104, 105]. CXCR4 and S1P5 appear as master regulators of NK-cell retention in, and egress from, the bone marrow. The exit process is obviously required for ensuring immunosurveillance. CXCR4 retains NK cells in the bone marrow parenchyma, whereas S1P5 promotes their exit from this organ through sinusoids. Using *S1p5*-null mice and knockin (KI) mouse model in which Cxcr4 cannot be desensitized, Mayol et al have recently showed that NK-cell exit from the bone marrow requires both Cxcr4 desensitization and S1p5 engagement by their corresponding ligands namely Cxcl12 and S1p, which are produced in the bone marrow and the bloodstream respectively [103]. In the bone marrow, CXCL12 is detected in different niches including the endosteal one and CAR cells as well [46, 106]. A recent study indicates that NK cells are found in close contact to CAR cells that also produce IL-15, another master regulator of NK-cell homeostasis [107]. Another work conducted by Sciumè and collaborators has involved the integrin chain $\alpha4$ and the Fractalkine/CX_3CL1 receptor CX_3CR1 in the positioning of mature NK cells in the sinusoidal compartment [104]. Once in the blood, the S1P concentration increases and S1P5 responsiveness decreases [108]. This responsiveness is recovered in the lymph nodes to allow NK-cell exit via lymphatics in a CXCR4-independent manner. The mechanism controlling NK-cell exit from the human bone marrow is likely to be similar to the one reported in the mouse counterpart. Several lines of evidence support this assertion. First, the absolute number of NK cells is deeply decreased in the peripheral blood of some WHIM syndrome patients that harbour a gain-of-*CXCR4*-function mutation [109]. Second, CXCR4 has been shown to retain human NK cells in the bone marrow and spleen of immunodeficient mice reconstituted with human immune system [110]. Finally, S1P5 was reported to be unregulated during human NK-cell differentiation [105].

6. CXCR4 in the WHIM syndrome

The WHIM syndrome (WS) is a rare immunological disorder characterized by the presence of warts (W), hypogammaglobulinemia (H), bacterial infections (I) and myelokathexis (M) meaning an abnormal retention of pro-apoptotic neutrophils in the bone marrow [111]. WS is an inherited pathological disorder with an autosomal dominant transmission. The WS estimate incidence was of 0.23 per million births but the prevalence is < 1/1 000 000. In fact, there are less than 60 documented cases in the world [112].

Inherited heterozygous autosomal dominant mutations of the *CXCR4* gene, which result in the truncation of the carboxyl-terminus of the receptor leading to a defect of CXCR4 inactivation, were found to be associated with the WS [113]. The disorder is clinically and genetically heterogeneous, since hypogammaglobulinemia and verrucosis were absent in some cases, and individuals with isolated myelokathexis were found to be "wild type" for the *CXCR4* gene [114]. As described previously, CXCR4 is expressed in hematopoietic cells. Consequently, the lack of CXCR4 inactivationis expected to generate significant immune and haematological disturbances [14].

Patients with WS exhibit a marked lymphopenia suggesting either a central defect of leukocyte differentiation in the bone marrow and the thymus or a peripheral defect such as an increase in apoptosis. Morphological analysis of bone marrow was performed in some patients; it evidenced an abnormal morphology of neutrophils called myelokathexis and an increase in neutrophil counts [111]. Furthermore the analysis of a bone marrow sample from one patient failed to detect abnormalities in lymphoid precursors and in immature and mature B-lymphocytes [115].

Biological features of WS include hypogammaglobulinemia, involving IgG. As expected, hypogammaglobulinemia combined to neutropenia in WS patients results in infections especially of encapsulated bacterias. Interestingly, infections transiently increase the number of neutrophil counts in the blood suggesting that neutropenia is caused by lack of neutrophils egress rather than to a defect in production. Patients also present recurrent pneumonias, sinusitis, urinary tract infection and skin infections among others. Strikingly, WS patients display high susceptibility to human papilloma virus (HPV) leading to skin lesions such as warts on hands, feet and trunk, genital and anal condylomas and mucosal lesions which often progress to carcinomas [116]. In two patients carrying the heterozygous *CXCR4* 1013 mutation, CXCR4 failed to internalize on lymphocytes upon stimulation. In PBMC, refractoriness of CXCR4 for desensitization and internalization led to an enhanced CXCL12-promoted chemotaxis [114]. Thus the lack of CXCR4 inactivation is associated with a gain of function of the receptor at least based on migration capacity criteria.

Despite lymphopenia and hypogammaglobulinemia, WS patients immunized with tetanus-toxoid produce normal amounts of antibodies against tetanus-toxoid 10 weeks after immunization. However, no specific antibodies were found one year after immunization suggesting a defect in long-lived PC and/or in memory B-cell response in WS [115].

There are currently no specific treatments for WS patients, albeit symptomatic use of GCSF and intravenous immunoglobulins (Ig) combined to anti-infectious agents is of some help [116], however CXCR4-targeted therapy is promising as described below.

In order to better understand this pathology, to characterise lymphoid differentiation and haematopoiesis defects in WS, we sought to generate an animal model. Balabanian et *al* generated a KI mouse strain that harbours a WS-associated heterozygous mutation of the *Cxcr4* gene (*i.e.,Cxcr4*$^{+/1013}$) to analyse the impact of Cxcr4 desensitization on leukocyte homeostasis [59]. These mice display a severe lymphoneutropenia. As in patients, CXCR4 failure to internalise doesn't lead to an increase in receptor expression in leukocytes in eitherbone marrow, thymus, spleen and blood. Moreover, all tested leukocyte subsets from *Cxcr4*$^{+/1013}$ mice, displayed increased sensitivity to Cxcl12-promoted chemotaxis compared to wild type mice as shown in blood T-and B-cells, and in DP, and SP thymocytes. Both frequency and absolute numbers of CD19^{+}B-cells were slightly, but significantly lower in the bone marrow from *Cxcr4*$^{+/1013}$ mice. In contrast to the myeloid series, leukocyte differentiation was altered in *Cxcr4*$^{+/1013}$ mice. Indeed, the absolute number of pro/pre-B cells, and to a lesser extent of immature B-cells, was reduced in *Cxcr4*$^{+/1013}$ mice. Therefore, the alteration of early B-cell development in *Cxcr4*$^{+/1013}$ mice was not associated with enhanced cell death. Regarding thymic differentiation, the absolute number of each thymic subset (from progenitors to mature SP thymocytes) was significantly decreased in *Cxcr4*$^{+/1013}$ mice. Together, these findings reveal that altered Cxcr4^{1013}-driven development of B-and T-cells leads to chronic circulating lymphopenia.

Regarding the periphery, the absolute number of each blood leukocyte subpopulation was significantly reduced in *Cxcr4*$^{+/1013}$ mice. Serum levels of natural IgA were similar in non-manipulated *Cxcr4*$^{+/1013}$ and WT mice, whereas IgM and IgG concentrations increased in *Cxcr4*$^{+/1013}$ mice contrary to patients as mentioned above. In parallel, mature T-and B-cells were abnormally compartmentalized in the periphery, with fewer primary follicles in the spleen and absence in lymph nodes, mirrored by an unfurling of the T-cell zone. These mice provide a model to decipher the role of CXCR4 desensitization in the homeostasis of B-and T-cells and to investigate which biological abnormalities of patients may be reversed by dampening the gain of CXCR4 function [59]. As mentioned above (chapter 4) CXCR4 may increase the stability of the T-APC interactions and is necessary for optimal T-cell activation [95]. Indeed, recent data demonstrated that WHIM-CXCR4–expressing retrogenic T-cells inhibits the formation of long-lasting T-APC interactions confirming the role of CXCR4 in the stability of T-APC Immune Synapse. Such anomalous synapse formation likely results in the failure to respond to vaccinations with the absence of specific antibody one year after vaccination observed in WS patient [117]. Thus, it will be interesting to investigate if there is an optimal adaptive immune response in these mice. In addition, the discovery of a mouse papillomavirus (MusPV) rendered possible to analyse HPV-infected *Cxcr4*$^{+/1013}$mice [118], as done in nude mice [119].Finally our mice will also be useful to test new drugs including molecules that target CXCR4.

AMD3100 (plerixafor) a selective and competitive antagonist of CXCR4 appears to be a potential treatment for patients with WS, and clinical studies have been conducted accordingly. One study reported the effect of daily injections with increased concentrations of AMD3100 (0.02-0.24 mg/kg) in 3 adults with WS; an increase of white blood cells and of absolute lymphocyte, neutrophil and monocyte counts was observed in all patients. Furthermore white blood cells mobilization is higher with AMD3100 than with GCSF [120]. Another study showed that treatment with AMD3100 every 2-4 days in 6 patients resulted in prompt leukocytosis [121]. Mice carrying the heterozygous *Cxcr4*$^{+/1013}$mutation, were also treated with AMD3100 or chalcone 4 that binds Cxcl12 and prevents signaling through Cxcr4. After single intraperitoneal injection of either AMD3100 or chalcone 4, the absolute numbers of total leukocytes in the blood of WT mice, including neutrophils, B and T cells was increased within 3 hours [59].Therefore, AMD3100 is able to reverse the pan leukopenia in WS mice and patients albeit transiently because of its short half-life. A more recent trial involving 3 WHIM patients treated with plerixafor subcutaneously twice daily for 6 months in combination with the Interferon-αinducer imiquimod, showed improvement in warts and in infections, albeit with partial restoration of Ig levels and vaccine responses [103].

To conclude, patients with WS have a gain of CXCR4 function, which leads to severe disorders of lymphocytes and neutrophils and hence of immune responses. How it affects mature T-and B-cells is still unclear. Undoubtedly, studies on homozygous and heterozygous mice and on the hematopoietic compartment will help understand the pathologies associated with CXCR4 dysfunction.

7. CXCR4 in the Idiopathic CD4⁺Lymphocytopenia syndrome

Idiopathic $CD4^+$T-cell Lymphocytopenia (ICL) is a rare haematological disorder characterized by a profound and persistent $CD4^+$T-cell defect, defined by an absolute $CD4^+$T-cell count <300 cells/$mm^3$with lack of HIV infection or other known immune defect or therapy associated with lymphopenia [122]. This clinical entity was defined in 1992 by the Centers for Disease Control and then some 258 cases have been reviewed in the literature [122, 123]. ICL patients often present life threatening opportunistic infections similar to those observed in acquired immunodeficiency (AIDS) syndrome. The most common infections in ICL are cryptococcal (*Cryptococcus neoformans*), genital HPV, non-tuberculous mycobacterial infections (*Mycobacterium avium*),in addition to progressive multifocal leukoencephalopathy [122, 124]. Malignancies are also common in ICL, including EBV-related B-cell lymphomas, Kaposi's sarcoma and as seen in the WHIM Syndrome, cervical or perineal neoplasias in the setting of long-term HPV infections [125-129]. In addition, autoimmune diseases (e.g. Sjögren syndrome) were also frequently reported [122].

The aetiology of ICL is unlikely to involve an infectious or environmental agent but may have a genetic basis as loss-of-function mutations have been reported in genes encoding regulators of the TCR diversity and signaling (*i.e.Unc119*, *MAGT1* and *RAG1*) [130-132]. ICL is considered as a heterogeneous syndrome possibly encompassing different disorders sharing the common feature of reduced circulating $CD4^+$T-cell counts. One issue is to determine whether the ICL results from a defective bone marrow production or thymic output, or an exacerbated peripheral consumption of $CD4^+$T-cells, or any combination of central or peripheral defect.

The defective production of $CD4^+$T-cell hypothesis is supported by a study reporting a regenerative failure of HSCs and lymphoid precursors in the bone marrow of ICL patients [133]. A work dealing with T-cell maturation has shown apparent restriction of the α/β and γ/δ TCR in ICL suggesting a disturbed thymic T-cell maturation in ICL [134]. In support of the central defect hypothesis, $CD34^+$HSCs derived from one ICL patient carrying a hypomorphic missense mutation in the *Recombination Activating Gene 1* (*RAG1*) were not able to repopulate humanized *$Rag2^{-/-}Il2Rgc^{-/-}$*mice, which is in favour for the lymphoid origin of the lymphopenia [132]. Finally, non-myeloablative allogeneic bone marrow transplantation in one patient was able to restore $CD4^+$T-cell numbers and functions [135].

In addition various works argue for an exacerbated peripheral consumption of $CD4^+$T-cells. Decreased T-cell responses as well as increased T-cell activation and apoptosis have been reported in ICL [136-138]. In terms of $CD4^+$T-cells activation, studies have reported enhanced propensity of ICL T-cells to undergo apoptosis, a process partially dependent on Fas and Fas ligand overexpression [137]. Markers for cell activation and turn-over, as indicated by HLA-DR, Ki-67 expression and BrdU labelling are also increased in $CD4^{+}$T-cells and this was inversely correlated with blood $CD4^{+}$T-cell counts [138]. In another report, abnormal $CD4^{+}$T-cells cycling was associated with levels of plasma lipopolysaccharide resulting from microbial translocation [139]. Therefore, immune activation and preferential loss of naive T-cells in ICL could result from chronic and persistent stimulation by an unidentified pathogen [122]. Mechanistic studies have also pointed out a defective TCR signal transduction in ICL raising the possibility that persistent T-cell activation leads to defective TCR signaling and may contribute to T-cell depletion [130-132]. Increased levels of the homeostatic cytokineIL7 in the sera of ICL patients inversely correlate with those of blood naive $CD4^{+}$T-cells and may reflect the triggering of a homeostatic response in order to restore normal $CD4^{+}$T-cell counts [133, 137, 140]. This accumulation of IL7 likely results from both a diminished consumption of the cytokine and a decrease of IL7 receptor α chain expression (CD127) on the reduced T-cell pool. Moreover the induction of phospho-STAT5 after IL7 stimulation was decreased in residual memory $CD4^{+}$T-cells of some ICL patients. These data suggest that high serum IL7 levels do not necessarily represent a compensatory response but may be further accentuating T-cell apoptosis and lymphopenia [141, 142]. In addition, while one study reported a successful therapy with IL2, decreased IL2 responses correlated with impaired IL7 responses, which may account for the loss of $CD4^{+}$T-cells homeostasis in ICL [142-144].

Regarding the CXCR4/CXCL12 axis, some studies demonstrated it role for T-cell production, homing, positioning and activation within secondary lymphoid tissues. Furthermore, alterations of CXCR4 expression or activity are likely to severely impact T-cell differentiation and trafficking. Thus, we hypothesized that expression or function of CXCR4 could be altered in ICL causing $CD4^{+}$T-cells alteration. Thus,ICL may mirror WHIM Syndrome as we observed decreased CXCR4 expression on $CD4^{+}$T-cells in ICL. We contributed to the identification of a defect in CXCR4 expression at the surface of $CD4^{+}$T-cells from 6 ICL patients with concomitant intracellular accumulation of CXCR4 and CXCL12. This suggested a defective intracellular routing of the chemokines/chemokines receptors complex. Analyses of CXCR4 fate following CXCL12 stimulation indicated that CXCR4 preserved its ability to internalize, but thereafter poorly recycled back to the plasma membrane of ICL T-cells. Altered membrane CXCR4 recovery resulted in a loss of CXCR4 function, as illustrated by the impaired CXCL12-promoted chemotaxis of ICL $CD4^{+}$T-cells [144]. These results were recently extended to 20 newly ICL patients and suggest that impaired membrane CXCR4 expression may contribute to the defective $CD4^{+}$T-cell homeostasis in the periphery but also in the thymus.

Altogether, ICL is a complex disorder with impaired membrane CXCR4 expression in peripheral lymphoid populations and in progenitors, possibly explaining the downstream lymphocytopenia. Deciphering the reasons for such dysfunction will help us discover new molecular targets for immune cell therapy. In this context, we derived induced pluripotent stem cells iPSCs, from $CD4^{+}$T-cells of an ICL patient; this will offer unprecedented opportunities to study T-cell differentiation in this disease.

8. CXCR4 in B-cell malignancies

Chronic lymphocytic leukaemia (CLL) is the most common leukaemia accounting for some 30% of adult leukaemia's in western countries It is due to an accumulation of small B-cells in the blood, the bone marrow and secondary lymphoid organs. Although this accumulation reflects lack of apoptosis, CLL cells do proliferate and their mitotic index and telomere length correlate with the degree of malignancy as reflected by the Binet [145] or Rai [146] clinical staging or the mutational status of Ig hyper variable regions, currently the gold biological standard for prognosis in this disease [147-149]. Somatically Ig hyper mutated (M-CLL) is typically indolent whereas Ig unmutated (U-CLL) is more aggressive [150, 151]. The cellular origin of CLL has been intensively debated, however contrary to the prevailing view that CLL originate from memory/activated B2 B-cells, we favour the recent hypothesis that CLL B-cells originate from malignant transformation of $CD5^{+}$B-cells [152]. Moreover both indolent and aggressive CLL originate from B-cells endowed with an auto reactive surface Ig which supports the hypothesis of a common molecular mechanism for both CLL types albeit occurring at two cellular differentiation stages [153].

CLL B-cells overexpress functional CXCR4 that may help B-cell to survive/proliferate and is associated with increased response to CXCL12 [154]. Indeed it has been noticed that CLL cells survive longer *in vitro* when cocultured with bone marrow stromal cells [155-157]. Interestingly survival mechanisms were linked to stimulation of CXCL12/CXCR4 and VLA4/VCAM axis [158-160]. This suggested that CLL cells are stimulated in protective microenvironments in the bone marrow or in secondary lymphoid organs and indeed proliferation centres with stromal cells and T-cells associated with CLL cells so called pseudo follicles were observed in these tissues [147, 161-164]. CLL cells in the blood are likely more sensitive to drugs than they are in lymphoid tissues.

Thus a new therapeutic strategy in CLL would be to down regulate or desensitize CXCR4 or disrupt CXCL12/CXCR4 interactions by AMD3100 on malignant B-cells in order to force them to leave their protective environments and undergo apoptosis in the bloodstream [165, 166].

As CLL are endowed with auto reactive surface Ig, they may be triggered repeatedly *in vivo* by auto antigens and BCR signaling is expected to down regulate CXCR4. Thus increased CXCR4 expression likely reflects the poor BCR signaling *in vivo* compared to that in normal B-cells; in addition CLL cells are surface Ig^{lo}. Indeed, CLL cells are anergic to anti-IgM stimulation, although U-CLL respond better in this respect than M-CLL in terms of proliferation and BCR signaling, and interestingly, U-CLL down regulate CXCR4 more efficiently than M-CLL upon sIg cross linking *in vitro* [167-169]. AMD3100 has been shown to potentiate Chemo/Immunotherapy in CLL *in vitro* [170]. This can be interpreted as either a blockade of the CXCR4 survival pathway or a help of CLL cells to detach from $CXCL12^{+}$stromal cells in SLO and move away from their protective microenvironment or both. These results need however to be reconciled with the finding that CXCR4 expression on CLL cells is lower in LO than it is in the blood [171].

As mentioned above (chapter 3), BCR-induced CXCR4 phosphorylation is PKC-but not Ca^{2+}-dependent [20]. Either CXCL12 interaction with CXCR4, or BCR engagement activates

PI3kinase-δ and CAL101 an inhibitor of this kinase was efficient against CLL [172]. Thus activating these pathways by molecules less toxic than phorbolmyristate acetate might prove useful to down regulate CXCR4, alternatively, pharmacological inhibitors of regulators, such as phosphatases of the PKC pathway might be useful as well.

In Waldenstom's Macroglobulinemia (WM), macroglobulinemia designs an increase in serum concentrations of IgM and causes much of the morbidity associated with the disease. WM is an indolent B-cell malignancy with a monoclonal proliferation of IgM-producing PCs that fail to undergo Ig isotype switching. WM is uncommon relative to plasma cells (PC) myeloma [173]. Although the pathogenesis of WM remained undetermined several data suggested that genetic factors contributed to the disease [174]. PCs from WM are $CXCR4^{hi}VLA4^{hi}$ and as for normal PC, malignant PC continuously home into the bone marrow and trans endothelial migration of PC from WM depends on CXCL12/CXCR4 [175]. It is now clear that WM results from somatic mutations in PC precursors.

A recent study evidenced somatic*CXCR4* mutation in 28% of the patients PC [16]. Five distinct somatic mutations were located in the *CXCR4* C-terminal tail, each of which were identical or functionally similar to mutations associated with WHIM syndrome (WS), resulting in the loss of regulatory Ser/Thr residues likely leading to impaired inactivation. CXCR4 is the second most frequent somatic mutation in WM next to L265P-MYD88 mutation that was detected in 90% of cases. Most interestingly, 98% of patients with CXCR4 mutation harboured the MYD88 mutation. These results imply that a gain of function CXCR4 is involved in the pathogeny of WM. It remains to understand how this helps malignant transformation to occur, and how to reconcile this observation with hypogammaglobulinemia in the WS. An interesting perspective here is the potential useful treatment of WM patients mutated or not for CXCR4, with AMD3100. It will undoubtedly represent the second therapeutic indication of plerixafor after WS.

9. CXCR4 and solid tumours

CXCR4 plays a critical role in the promotion of several solid tumours. A recent study showed that non-small lung cancer cell lines did express high CXCR4 despite secreting CXCL12 [176]; this questions the inability of CXCL12 to down regulate CXCR4 in these cells. Moreover these cells had a high self-renewal and tumour promoting activity *in vivo*. Thus CXCR4 may behave as a growth stimulator and help cancer stem cells home into protective niches within the solid tumour or metastasize within the bone marrow where they may compete with normal HSC, hence adverse effects on immunity and haematopoiesis.

A common feature of WS and ICL is the occurrence of cancers that are often linked to HPV infections. In ICL, some 15% of the patients experienced HPV infections and / or squamous and basal cell carcinomas of the skin, Bowen's disease, vulvar, cervical or bladder carcinomas [122]. Interestingly, HPV has been detected in these pathologies [177-181]. In WS, clinical manifestations at diagnosis include cervical papillomatosis that can lead to invasive cancer [112]. This highlights the potential and still undetermined role of CXCR4 (over or impaired membrane expression) in the control of immunity to HPV. It also points to the fact that the term "gain of function" for the *CXCR4*^{1013}mutation may be a misnomer as it points for an augmented capacity of the cells to migrate *in vitro* and most likely to stick to their niches *in vivo*. Of note, there is no correlation between HPV-associated cancers and Waldenstom's Macroglobulinemia, which only affects PC and leaves intact most immune cells with normal CXCR4. The germ line CXCR4^{1013}-mutation is clearly associated with a polyclonal loss of function of B-, T-and NK-cells *in vivo* and possibly of neutrophils as well and the respective contribution of these leukocytes to the adaptive immune response against HPV remains to be understood.

Within solid tumours or among malignant cells from leukaemia's, several lines of evidence suggest that there are cells called cancer stem cells or tumour initiating cells. They have been identified in many types of cancers including human gastric tumour, mammary gland, brain, prostate gland, colon pancreas, head and neck and liver [182-190]. On primary cultures of cells from a human gastric tumour, those cells are able to grow on extracellular matrix and to form spheroid, structures specific of stem cells. Furthermore those cells are more resistant to chemotherapy and capable of self-renewal [182]. This feature could explain the failure in the treatment of cancers. In a Gefitinib-resistant non-small cell lung cancer cells line (A549/GR), experiments show that these cells possess some features of stem cells such as the acquisition of epithelial mesenchymal transition property. More interestingly, there is a high proportion of CXCR4^{+}cells in A549/GR, with high self-renewal activity *in vitro,* high tumorigenic potential *in vivo,* a strong sphere-forming activity and a resistance to radiation. The use of si-RNA specific of CXCR4 or the AMD3100 are both able to suppressed sphere forming activity in those cell lines, demonstrating an important role of CXCR4 in maintaining cancer stem cells features in A549/GR [176]. Cancer stem cells are potential targets for therapies in order to eliminate malignant cells in solid tumours resistant to chemotherapy.

The CXCL12/CXCR4 axis is known to be involved in tumour growth and metastatic process. The microenvironment is essential for tumour development and stromal and malignant cells communicate *via* growth factors and cytokines [191, 192]. In order to understand the impact of Cxcr4 in the microenvironment on metastasis of tumour cells, D'Alterio et al injected murine melanoma B16 cells on C57BL/6 *Cxcr4*$^{+/+}$or *Cxcr4*$^{+/-}$mice in the presence of AMD3100. In *Cxcr4*$^{+/-}$ mice nodule size were significantly smaller and bone marrow-derived cells recruitment was lower compared to *Cxcr4*$^{+/+}$. Furthermore, the Cxcr4 inhibitor AMD3100 preserves the pulmonary architecture in Cxcr4$^{+/-}$mice by reducing lung metastases. *Cxcr4*$^{+/-}$mice also show a decrease in LY6G-positive myeloid/granulocytic cells and in p38 MAPK activation in lungs compared to *Cxcr4*$^{+/+}$mice [193]. These experiments demonstrate the importance CXCL12/ CXCR4 axis in the microenvironment regarding metastases and tumour growth. Since CXCL12/CXCR4 axis may promote cancer cell survival, invasion, and tumour-initiating cell phenotype. Therefore blocking this axis may be a potential approach to target various components in solid tumours.

The CXCL12/CXCR4 axis is also involved in the recruitment of immune cells into the tumour as described for regulatory T-cells (Tregs). Indeed Tregs expressing CXCR4 are recruited in a number of tumours including ovarian cancer, adenocarcinoma of the lung, malignant mesothelioma and the myelodysplastic syndromes [194-197]. In cancer, the balance between Tregs- and T-effector cells is often deregulated and Tregs are recruited to the tumour, a process that suppress the anti-tumour immune response leading to tumour growth. In basal-like breast cancer, it was shown that there is a positive correlation between CXCL12 expression in the tumour and Tregs recruitment correlated with a poor survival in patients [198]. This was also shown in lung adenocarcinomas and malignant mesothelioma [195, 196]. In addition Yan et *al* showed that the recruitment of Tregs is correlated with hypoxia-induced CXCR4 expression in basal-like breast cancers. Furthermore, in lung adenocarcinoma, CXCR4 positive Tregs cells are able to regulate the immune response and to secrete tumour growth factor-β which up-regulate CXCR4 on naive T-cells and contribute to their migration and retention in the tumour microenvironment and that contribute to increase pathogenesis [195]. Overall, these observations suggest that CXCL12 expression may influence tumour progression by shaping the immune cell population infiltrating lung adenocarcinoma tumours. In human ovarian carcinoma, Tregs preferentially move to tumours and ascites and suppress tumour-specific T-cell immunity and contribute to growth of human tumours *in vivo.* The recruitment of Tregs represents a mechanism by which tumours may promote immune advantage. Altogether, these results suggest that AMD3100 or other drugs that target the CXCL12/CXCR4 axis may be useful adjuncts for immune-chemotherapy in some cancers [199].

10. Conclusion/therapeutic perspectives

The contribution of the CXCL12/CXCR4 signaling axis to many aspects of physiology and pathology has been increasingly appreciated (Figure 1 & 2). Deregulations of CXCR4 signaling and/or expression are associated with several disorders including lymphoid and autoimmune diseases, solid tumours and immune defects [15, 200-203]. Therefore, interference with the CXCL12/CXCR4 interaction or modulation of CXCR4 expression and/or activity is potentially interesting in the treatment of diseases with loss or gain of function of CXCR4. In this context and for the last two decades, many pharmaceutical companies have been trying to develop specific drugs targeting this axis [204, 205]. In light of recent clinical investigations, the bicyclamplerixafor (AMD3100), has already been approved by the FDA for HSC mobilization in patients with non-Hodgkin's lymphoma and multiple myeloma [206]. In addition, Plerixafor is already under Phase I of clinical investigation in Glioma, Acute Myeloid leukaemia, CLL and WS [204, 207]. Furthermore, regarding the low half-life, the constraining frequency of administration of AMD3100, several promising CXCR4 antagonists have been developed to block CXCL12/CXCR4 interactions that are currently under different stages of clinical trials mainly for HSC transplantation in patients with multiple myeloma and non-Hodgkin's lymphoma [205]. Most of them neutralize the interaction of CXCL12 with CXCR4 by binding to the receptor. Another way is to directly target CXCL12 by developing molecules that interact directly with it thus diverting the chemokine from its receptor. For example blocking interactions of CXCL12 with the extracellular matrix and cell surface glycosaminoglycans may be an approach by developing CXCL12 binding heparansulphates. In addition other strategies will be to generate molecules interfering with intracellular trafficking and expression of CXCR4 which are two different ways to normalize CXCR4 expression and functioning in pathologies associated with CXCR4 anomalies as seen both in ICL and WS. As mentioned in the introduction, ACKR3 may share overlapping functions with CXCR4 and is possibly involved in the same pathologies and need to be investigated in this respect [208, 209].

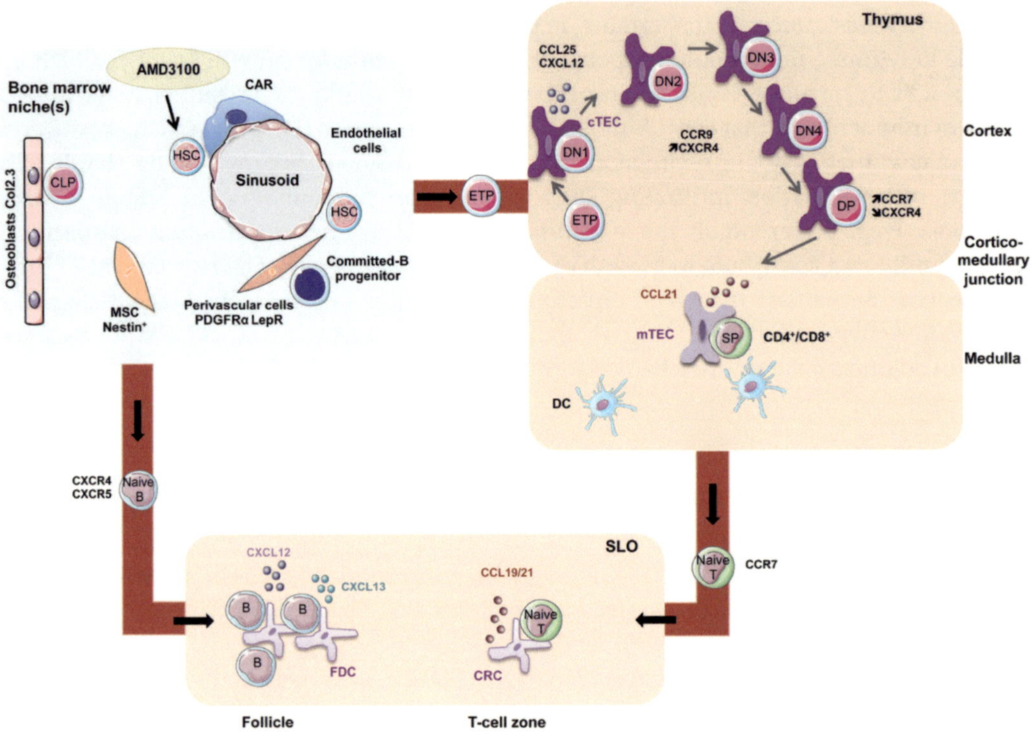

Figure 1. CXCR4 in lymphopoiesis. Most primitive hematopoietic stem cells (HSC) and downstream lymphoid progenitors, including common lymphoid (CLP) and committed B-cell precursors express CXCR4 and the differentiation of HSC and B-cell precursors is CXCR4-dependent. These cells interact with CXCL12-expressing stromal cells. These stromal cells comprise Nestin⁺Mesenchymal stem cells MSC, CXCL12-producing reticular cells (CAR), perivascular cells expressing PDGFR and LeptinR, and osteoblasts. AMD3100 is already used to allow the migration of HSC in the peripheral blood of patients in order to be collected for grafting. Recent data suggest that B-cell progenitors locate near the perivascular niche. CLP locate in the osteoblastic niche. Downstream progenitors migrate through the blood and enter the thymus at the cortico-medullary junction where they constitute early thymic progenitors (ETP). ETP further move inside the cortex and differentiate in several double negative $CD4^-CD8^-$ stages DN1 to DN4. This process depends on the production of CXCL12 by cortical thymic epithelial cells (cTEC). At the DN4 stage, thymocytes downregulate CXCR4, upregulate CCR7 and become $CD4^+CD8^+(CD3^+)$ cells double positive DP cells. DP thymocytes are attracted by CCL21-producing thymic medullary epithelial cells (mTEC). At the end of differentiation, and negative selection due to interactions with dendritic cells (DC) in the medulla, naive single positive (SP) $CD4^+$or $CD8^+$thymocytes egress from the thymus to the bloodstream and further migrate to the T-cell zones of secondary lymphoid organs (SLO). In parallel B-cell precursors differentiate within the bone marrow in naive B-cells that express CXCR4/5 and migrate to the SLO, where they organize themselves around CXCL12/13-producing follicular dendritic cells (FDC) and marginal central reticular cells (CRC). Following T-dependent response to antigen the germinal centre (GC)reaction is initiated (figure 2).

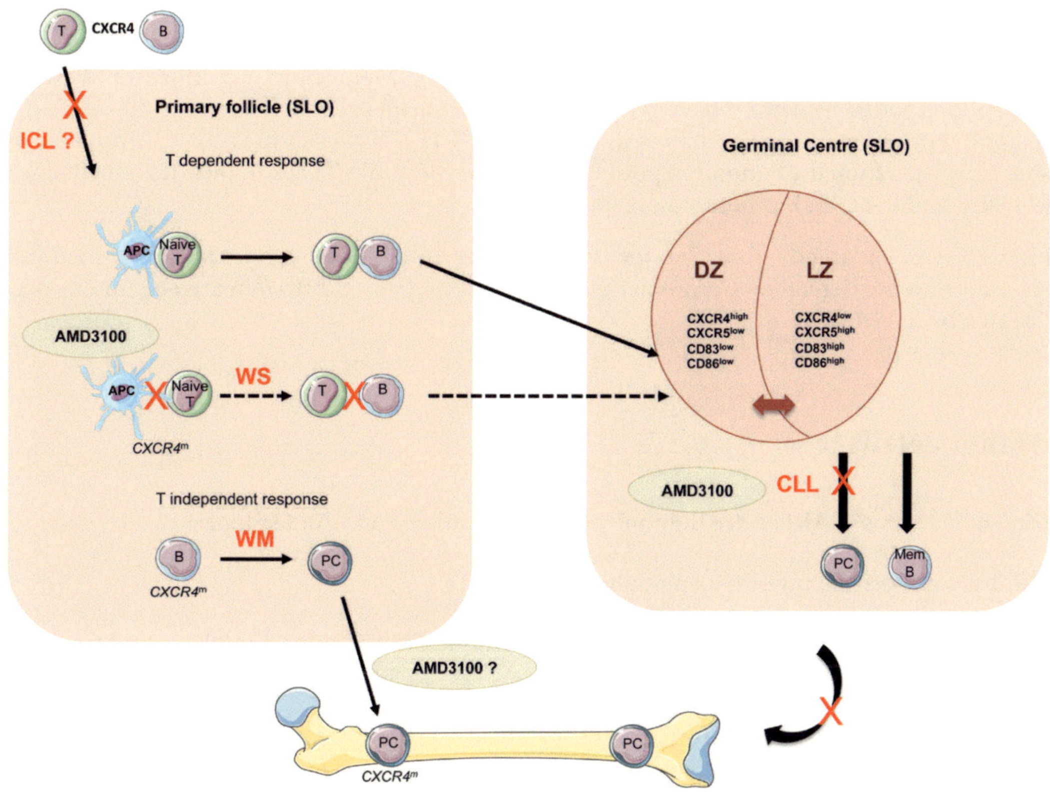

Figure 2. CXCR4 in T-dependent antigen response, and in pathology. Following antigen capture and processing, dendritic antigen-presenting cells (APC) move to the T-cell zone in the SLO to prime antigen-specific T-cells whereas follicular B-cells are stimulated by soluble antigen. In turn T-and B-cells move towards each other and antigen-specific B-cells expand to generate the dark zone DZ made mostly of centroblasts. These cells differentiate in centrocytes that constitute the light zone and downregulate CXCR4. Ultimately, centrocytes differentiate in memory B-cells or plasma cells (PC), the laterupregulate CXCR4 and migrate to the bone marrow. In the WS, one hypothesis is that the germ-line*CXCR4^m* mutation impairs T-APC and T-B interactions leading to deficient GC formation. In Waldenstom's Macro-globulinemiaWM, the somatic *CXCR4^m* mutation in B-cells confers a survival advantage and drives PC to migrate and home into the bone marrow were they become malignant due to additional gain of function mutations such as in the *MYD88* gene. In the Idiopathic CD4$^+$T-cell LymphocytopenialCL, downregulation of CXCR4 expression in T-cells may concur to the deficient adaptive immune response. In chronic lymphocytic leukaemia, PC differentiation is impaired and leukemic B-cells express CXCR4, which keeps them in a protective environment in contact with CXCL12-producing stromal cells thereby protecting them from immune/chemotherapeutic drugs. In this context, AMD3100 plerixafor may help sensitize chronic lymphocytic leukaemia CLL cells to drugs. Plerixafor is already used for the treatment of WHIM Syndrome WS, and may prove useful in the treatment of WM.

A new therapeutic era has begun, next years will witness the use of anti-chemokines receptors to prevent malignant cells from interacting with their protective stromal cells and of inflammatory cells to migrate into tissues, these new families of drugs should clearly improve the efficacy of current anti-cancer, anti-autoimmune and anti-inflammatory molecules [182].

Acknowledgements

This work was supported by the Assistance Publique-Hôpitaux de Paris (Grant 07018) and the Agence Nationale de la Recherche (Grant 2010 JCJC 1104 01). C.F., A.B., K.B., and A.D. are membres of the Laboratory of Excellence in Research on Medication and Innovative Therapeutics, supported by a grant from the Agence Nationale de la Recherche (ANR-10-LABX-33) under the program "Investissements d'Avenir" ANR-11-IDEX-0003-01. A.B., C.F. are fellowship recipients from the Fondation pour la Recherche Médicale (FDT20130928127), the French Ministry and the DIM Biothérapies, respectively.

Figures were adapted from Servier Medical Art http://smart.servier.fr/servier-medical-artunder licence Creative Commons Attribution 3.0 France http://creativecommons.org/licenses/by/3.0/fr/.

Author details

Christelle Freitas[1,2], Alexandre Bignon[1,2], Karl Balabanian[1,2] and Ali Dalloul[1,2*]

*Address all correspondence to: ali.dalloul@u-psud.fr

1 Université Paris-Sud, Laboratoire "Cytokines, Chimiokines et Immunopathologie," Clamart, France

2 INSERM, Laboratoire d'Excellence en Recherche sur le Médicament et l'Innovation Thérapeutique, Clamart, France

Christelle Freitas and Alexandre Bignon contributed equally to this work.

References

[1] Nagasawa T, Kikutani H, Kishimoto T. Molecular cloning and structure of a pre-B-cell growth-stimulating factor. Proc Natl Acad Sci U S A. 1994 Mar 15;91 (6):2305-9.

[2] Oberlin E, Amara A, Bachelerie F, Bessia C, Virelizier JL, Arenzana-Seisdedos F, et al. The CXC chemokine SDF-1 is the ligand for LESTR/fusin and prevents infection by T-cell-line-adapted HIV-1. Nature. 1996 Aug 29;382 (6594):833-5.

[3] Berger EA, Murphy PM, Farber JM. Chemokine receptors as HIV-1 coreceptors: roles in viral entry, tropism, and disease. Annu Rev Immunol. 1999;17:657-700.

[4] Katsumoto K, Kume S. Endoderm and mesoderm reciprocal signaling mediated by CXCL12 and CXCR4 regulates the migration of angioblasts and establishes the pancreatic fate. Development. 2011 May;138 (10):1947-55.

[5] Nair S, Schilling TF. Chemokine signaling controls endodermal migration during zebrafish gastrulation. Science. 2008 Oct 3;322 (5898):89-92.

[6] Ceradini DJ, Kulkarni AR, Callaghan MJ, Tepper OM, Bastidas N, Kleinman ME, et al. Progenitor cell trafficking is regulated by hypoxic gradients through HIF-1 induction of SDF-1. Nat Med. 2004 Aug;10 (8):858-64.

[7] Mohyeldin A, Garzon-Muvdi T, Quinones-Hinojosa A. Oxygen in stem cell biology: a critical component of the stem cell niche. Cell Stem Cell. 2010 Aug 6;7 (2):150-61.

[8] Nagasawa T, Hirota S, Tachibana K, Takakura N, Nishikawa S, Kitamura Y, et al. Defects of B-cell lymphopoiesis and bone-marrow myelopoiesis in mice lacking the CXC chemokine PBSF/SDF-1. Nature. 1996 Aug 15;382 (6592):635-8.

[9] Law NM, Rosenzweig SA. Characterization of the G-protein linked orphan receptor GPRN1/RDC1. Biochem Biophys Res Commun. 1994 May 30;201 (1):458-65.

[10] Balabanian K, Lagane B, Infantino S, Chow KY, Harriague J, Moepps B, et al. The chemokine SDF-1/CXCL12 binds to and signals through the orphan receptor RDC1 in T lymphocytes. J Biol Chem. 2005 Oct 21;280 (42):35760-6.

[11] Burns JM, Summers BC, Wang Y, Melikian A, Berahovich R, Miao Z, et al. A novel chemokine receptor for SDF-1 and I-TAC involved in cell survival, cell adhesion, and tumor development. J Exp Med. 2006 Sep 4;203 (9):2201-13.

[12] Marchese A, Benovic JL. Agonist-promoted ubiquitination of the G protein-coupled receptor CXCR4 mediates lysosomal sorting. J Biol Chem. 2001 Dec 7;276 (49): 45509-12.

[13] Marchese A, Raiborg C, Santini F, Keen JH, Stenmark H, Benovic JL. The E3 ubiquitin ligase AIP4 mediates ubiquitination and sorting of the G protein-coupled receptor CXCR4. Dev Cell. 2003 Nov;5 (5):709-22.

[14] Busillo JM, Benovic JL. Regulation of CXCR4 signaling. Biochim Biophys Acta. 2007 Apr;1768 (4):952-63.

[15] Bignon A, Biajoux V, Bouchet-Delbos L, Emilie D, Lortholary O, Balabanian K. [CXCR4, a therapeutic target in rare immunodeficiencies?]. Med Sci (Paris). 2011 Apr;27 (4):391-7.

[16] Hunter Z, Xu L, Yang G, Zhou Y, Liu X, Cao Y, et al. The genomic landscape of Waldenstom's Macroglobulinemia is characterized by highly recurring MYD88 and WHIM-like CXCR4 mutations, and small somatic deletions associated with B-cell lymphomagenesis. Blood. 2013 Dec 23.

[17] Van Damme J, Struyf S, Wuyts A, Van Coillie E, Menten P, Schols D, et al. The role of CD26/DPP IV in chemokine processing. Chem Immunol. 1999;72:42-56.

[18] Fermas S, Gonnet F, Sutton A, Charnaux N, Mulloy B, Du Y, et al. Sulfated oligosaccharides (heparin and fucoidan) binding and dimerization of stromal cell-derived

factor-1 (SDF-1/CXCL 12) are coupled as evidenced by affinity CE-MS analysis. Glycobiology. 2008 Dec;18 (12):1054-64.

[19] Peacock JW, Jirik FR. TCR activation inhibits chemotaxis toward stromal cell-derived factor-1: evidence for reciprocal regulation between CXCR4 and the TCR. J Immunol. 1999 Jan 1;162 (1):215-23.

[20] Guinamard R, Signoret N, Ishiai M, Marsh M, Kurosaki T, Ravetch JV. B cell antigen receptor engagement inhibits stromal cell-derived factor (SDF)-1alpha chemotaxis and promotes protein kinase C (PKC)-induced internalization of CXCR4. J Exp Med. 1999 May 3;189 (9):1461-6.

[21] Este JA, Cabrera C, De Clercq E, Struyf S, Van Damme J, Bridger G, et al. Activity of different bicyclam derivatives against human immunodeficiency virus depends on their interaction with the CXCR4 chemokine receptor. Mol Pharmacol. 1999 Jan;55 (1):67-73.

[22] Arroyo AG, Yang JT, Rayburn H, Hynes RO. Alpha4 integrins regulate the proliferation/differentiation balance of multilineage hematopoietic progenitors in vivo. Immunity. 1999 Nov;11 (5):555-66.

[23] Sugiyama T, Kohara H, Noda M, Nagasawa T. Maintenance of the hematopoietic stem cell pool by CXCL12-CXCR4 chemokine signaling in bone marrow stromal cell niches. Immunity. 2006 Dec;25 (6):977-88.

[24] Frenette PS, Subbarao S, Mazo IB, von Andrian UH, Wagner DD. Endothelial selectins and vascular cell adhesion molecule-1 promote hematopoietic progenitor homing to bone marrow. Proc Natl Acad Sci U S A. 1998 Nov 24;95 (24):14423-8.

[25] Spertini O, Cordey AS, Monai N, Giuffre L, Schapira M. P-selectin glycoprotein ligand 1 is a ligand for L-selectin on neutrophils, monocytes, and CD34+hematopoietic progenitor cells. J Cell Biol. 1996 Oct;135 (2):523-31.

[26] Oostendorp RA, Dormer P. VLA-4-mediated interactions between normal human hematopoietic progenitors and stromal cells. Leuk Lymphoma. 1997 Feb;24 (5-6): 423-35.

[27] Peled A, Grabovsky V, Habler L, Sandbank J, Arenzana-Seisdedos F, Petit I, et al. The chemokine SDF-1 stimulates integrin-mediated arrest of CD34 (+) cells on vascular endothelium under shear flow. J Clin Invest. 1999 Nov;104 (9):1199-211.

[28] Naiyer AJ, Jo DY, Ahn J, Mohle R, Peichev M, Lam G, et al. Stromal derived factor-1-induced chemokinesis of cord blood CD34 (+) cells (long-term culture-initiating cells) through endothelial cells is mediated by E-selectin. Blood. 1999 Dec 15;94 (12):4011-9.

[29] Fuste B, Escolar G, Marin P, Mazzara R, Ordinas A, Diaz-Ricart M. G-CSF increases the expression of VCAM-1 on stromal cells promoting the adhesion of CD34+hematopoietic cells: studies under flow conditions. Exp Hematol. 2004 Aug;32 (8):765-72.

[30] Jo DY, Rafii S, Hamada T, Moore MA. Chemotaxis of primitive hematopoietic cells in response to stromal cell-derived factor-1. J Clin Invest. 2000 Jan;105 (1):101-11.

[31] Peled A, Petit I, Kollet O, Magid M, Ponomaryov T, Byk T, et al. Dependence of human stem cell engraftment and repopulation of NOD/SCID mice on CXCR4. Science. 1999 Feb 5;283 (5403):845-8.

[32] Aiuti A, Webb IJ, Bleul C, Springer T, Gutierrez-Ramos JC. The chemokine SDF-1 is a chemoattractant for human CD34+hematopoietic progenitor cells and provides a new mechanism to explain the mobilization of CD34+progenitors to peripheral blood. J Exp Med. 1997 Jan 6;185 (1):111-20.

[33] Nie Y, Han YC, Zou YR. CXCR4 is required for the quiescence of primitive hematopoietic cells. J Exp Med. 2008 Apr 14;205 (4):777-83.

[34] Eaves CJ. SDF-1 tells stem cells to mind their P's and Z's. J Clin Invest. 2005 Jan;115 (1):27-9.

[35] Lataillade JJ, Clay D, Dupuy C, Rigal S, Jasmin C, Bourin P, et al. Chemokine SDF-1 enhances circulating CD34 (+) cell proliferation in synergy with cytokines: possible role in progenitor survival. Blood. 2000 Feb 1;95 (3):756-68.

[36] Lataillade JJ, Clay D, Bourin P, Herodin F, Dupuy C, Jasmin C, et al. Stromal cell-derived factor 1 regulates primitive hematopoiesis by suppressing apoptosis and by promoting G (0)/G (1) transition in CD34 (+) cells: evidence for an autocrine/paracrine mechanism. Blood. 2002 Feb 15;99 (4):1117-29.

[37] Ding L, Morrison SJ. Haematopoietic stem cells and early lymphoid progenitors occupy distinct bone marrow niches. Nature. 2013 Mar 14;495 (7440):231-5.

[38] Kiel MJ, Yilmaz OH, Iwashita T, Terhorst C, Morrison SJ. SLAM family receptors distinguish hematopoietic stem and progenitor cells and reveal endothelial niches for stem cells. Cell. 2005 Jul 1;121 (7):1109-21.

[39] Mendez-Ferrer S, Michurina TV, Ferraro F, Mazloom AR, Macarthur BD, Lira SA, et al. Mesenchymal and haematopoietic stem cells form a unique bone marrow niche. Nature. 2010 Aug 12;466 (7308):829-34.

[40] Ding L, Saunders TL, Enikolopov G, Morrison SJ. Endothelial and perivascular cells maintain haematopoietic stem cells. Nature. 2012 Jan 26;481 (7382):457-62.

[41] Greenbaum A, Hsu YM, Day RB, Schuettpelz LG, Christopher MJ, Borgerding JN, et al. CXCL12 in early mesenchymal progenitors is required for haematopoietic stem-cell maintenance. Nature. 2013 Mar 14;495 (7440):227-30.

[42] Donzella GA, Schols D, Lin SW, Este JA, Nagashima KA, Maddon PJ, et al. AMD3100, a small molecule inhibitor of HIV-1 entry via the CXCR4 co-receptor. Nat Med. 1998 Jan;4 (1):72-7.

[43] Mosi RM, Anastassova V, Cox J, Darkes MC, Idzan SR, Labrecque J, et al. The molecular pharmacology of AMD11070: an orally bioavailable CXCR4 HIV entry inhibitor. Biochem Pharmacol. 2012 Feb 15;83 (4):472-9.

[44] Pusic I, DiPersio JF. Update on clinical experience with AMD3100, an SDF-1/CXCL12-CXCR4 inhibitor, in mobilization of hematopoietic stem and progenitor cells. Curr Opin Hematol. 2010 Jul;17 (4):319-26.

[45] Publicover A, Richardson DS, Davies A, Hill KS, Hurlock C, Hutchins D, et al. Use of a biosimilar granulocyte colony-stimulating factor for peripheral blood stem cell mobilization: an analysis of mobilization and engraftment. Br J Haematol. 2013 Jul;162 (1):107-11.

[46] Petit I, Szyper-Kravitz M, Nagler A, Lahav M, Peled A, Habler L, et al. G-CSF induces stem cell mobilization by decreasing bone marrow SDF-1 and up-regulating CXCR4. Nat Immunol. 2002 Jul;3 (7):687-94.

[47] Lambeir AM, Proost P, Durinx C, Bal G, Senten K, Augustyns K, et al. Kinetic investigation of chemokine truncation by CD26/dipeptidyl peptidase IV reveals a striking selectivity within the chemokine family. J Biol Chem. 2001 Aug 10;276 (32):29839-45.

[48] Zou YR, Kottmann AH, Kuroda M, Taniuchi I, Littman DR. Function of the chemokine receptor CXCR4 in haematopoiesis and in cerebellar development. Nature. 1998 Jun 11;393 (6685):595-9.

[49] Ma Q, Jones D, Springer TA. The chemokine receptor CXCR4 is required for the retention of B lineage and granulocytic precursors within the bone marrow microenvironment. Immunity. 1999 Apr;10 (4):463-71.

[50] D'Apuzzo M, Rolink A, Loetscher M, Hoxie JA, Clark-Lewis I, Melchers F, et al. The chemokine SDF-1, stromal cell-derived factor 1, attracts early stage B cell precursors via the chemokine receptor CXCR4. Eur J Immunol. 1997 Jul;27 (7):1788-93.

[51] Kawabata K, Ujikawa M, Egawa T, Kawamoto H, Tachibana K, Iizasa H, et al. A cell-autonomous requirement for CXCR4 in long-term lymphoid and myeloid reconstitution. Proc Natl Acad Sci U S A. 1999 May 11;96 (10):5663-7.

[52] Dittel BN, McCarthy JB, Wayner EA, LeBien TW. Regulation of human B-cell precursor adhesion to bone marrow stromal cells by cytokines that exert opposing effects on the expression of vascular cell adhesion molecule-1 (VCAM-1). Blood. 1993 May 1;81 (9):2272-82.

[53] Glodek AM, Honczarenko M, Le Y, Campbell JJ, Silberstein LE. Sustained activation of cell adhesion is a differentially regulated process in B lymphopoiesis. J Exp Med. 2003 Feb 17;197 (4):461-73.

[54] Nagasawa T. The chemokine CXCL12 and regulation of HSC and B lymphocyte development in the bone marrow niche. Adv Exp Med Biol. 2007;602:69-75.

[55] Zhu J, Garrett R, Jung Y, Zhang Y, Kim N, Wang J, et al. Osteoblasts support B-lymphocyte commitment and differentiation from hematopoietic stem cells. Blood. 2007 May 1;109 (9):3706-12.

[56] Wu JY, Purton LE, Rodda SJ, Chen M, Weinstein LS, McMahon AP, et al. Osteoblastic regulation of B lymphopoiesis is mediated by Gs{alpha}-dependent signaling pathways. Proc Natl Acad Sci U S A. 2008 Nov 4;105 (44):16976-81.

[57] Honczarenko M, Douglas RS, Mathias C, Lee B, Ratajczak MZ, Silberstein LE. SDF-1 responsiveness does not correlate with CXCR4 expression levels of developing human bone marrow B cells. Blood. 1999 Nov 1;94 (9):2990-8.

[58] Honczarenko M, Le Y, Glodek AM, Majka M, Campbell JJ, Ratajczak MZ, et al. CCR5-binding chemokines modulate CXCL12 (SDF-1)-induced responses of progenitor B cells in human bone marrow through heterologous desensitization of the CXCR4 chemokine receptor. Blood. 2002 Oct 1;100 (7):2321-9.

[59] Balabanian K, Brotin E, Biajoux V, Bouchet-Delbos L, Lainey E, Fenneteau O, et al. Proper desensitization of CXCR4 is required for lymphocyte development and peripheral compartmentalization in mice. Blood. 2012 Jun 14;119 (24):5722-30.

[60] Cyster JG. B cell follicles and antigen encounters of the third kind. Nat Immunol. 2010 Nov;11 (11):989-96.

[61] Pereira JP, Kelly LM, Cyster JG. Finding the right niche: B-cell migration in the early phases of T-dependent antibody responses. Int Immunol. 2010 Jun;22 (6):413-9.

[62] Okada T, Ngo VN, Ekland EH, Forster R, Lipp M, Littman DR, et al. Chemokine requirements for B cell entry to lymph nodes and Peyer's patches. J Exp Med. 2002 Jul 1;196 (1):65-75.

[63] Revy P, Muto T, Levy Y, Geissmann F, Plebani A, Sanal O, et al. Activation-induced cytidine deaminase (AID) deficiency causes the autosomal recessive form of the Hyper-IgM syndrome (HIGM2). Cell. 2000 Sep 1;102 (5):565-75.

[64] Neuberger MS, Milstein C. Somatic hypermutation. Curr Opin Immunol. 1995 Apr;7 (2):248-54.

[65] Victora GD, Dominguez-Sola D, Holmes AB, Deroubaix S, Dalla-Favera R, Nussenzweig MC. Identification of human germinal center light and dark zone cells and their relationship to human B-cell lymphomas. Blood. 2012 Sep 13;120 (11):2240-8.

[66] Hennino A, Berard M, Krammer PH, Defrance T. FLICE-inhibitory protein is a key regulator of germinal center B cell apoptosis. J Exp Med. 2001 Feb 19;193 (4):447-58.

[67] Allen CD, Ansel KM, Low C, Lesley R, Tamamura H, Fujii N, et al. Germinal center dark and light zone organization is mediated by CXCR4 and CXCR5. Nat Immunol. 2004 Sep;5 (9):943-52.

[68] Kabashima K, Haynes NM, Xu Y, Nutt SL, Allende ML, Proia RL, et al. Plasma cell S1P1 expression determines secondary lymphoid organ retention versus bone marrow tropism. J Exp Med. 2006 Nov 27;203 (12):2683-90.

[69] Nie Y, Waite J, Brewer F, Sunshine MJ, Littman DR, Zou YR. The role of CXCR4 in maintaining peripheral B cell compartments and humoral immunity. J Exp Med. 2004 Nov 1;200 (9):1145-56.

[70] Slifka MK, Antia R, Whitmire JK, Ahmed R. Humoral immunity due to long-lived plasma cells. Immunity. 1998 Mar;8 (3):363-72.

[71] Koni PA, Joshi SK, Temann UA, Olson D, Burkly L, Flavell RA. Conditional vascular cell adhesion molecule 1 deletion in mice: impaired lymphocyte migration to bone marrow. J Exp Med. 2001 Mar 19;193 (6):741-54.

[72] Signoret N, Oldridge J, Pelchen-Matthews A, Klasse PJ, Tran T, Brass LF, et al. Phorbol esters and SDF-1 induce rapid endocytosis and down modulation of the chemokine receptor CXCR4. J Cell Biol. 1997 Nov 3;139 (3):651-64.

[73] Haribabu B, Richardson RM, Fisher I, Sozzani S, Peiper SC, Horuk R, et al. Regulation of human chemokine receptors CXCR4. Role of phosphorylation in desensitization and internalization. J Biol Chem. 1997 Nov 7;272 (45):28726-31.

[74] Bleul CC, Schultze JL, Springer TA. B lymphocyte chemotaxis regulated in association with microanatomic localization, differentiation state, and B cell receptor engagement. J Exp Med. 1998 Mar 2;187 (5):753-62.

[75] Godfrey DI, Zlotnik A. Control points in early T-cell development. Immunol Today. 1993 Nov;14 (11):547-53.

[76] Inlay MA, Bhattacharya D, Sahoo D, Serwold T, Seita J, Karsunky H, et al. Ly6d marks the earliest stage of B-cell specification and identifies the branchpoint between B-cell and T-cell development. Genes Dev. 2009 Oct 15;23 (20):2376-81.

[77] Petrie HT, Zuniga-Pflucker JC. Zoned out: functional mapping of stromal signaling microenvironments in the thymus. Annu Rev Immunol. 2007;25:649-79.

[78] Dalloul AH, Patry C, Salamero J, Canque B, Grassi F, Schmitt C. Functional and phenotypic analysis of thymic CD34+CD1a-progenitor-derived dendritic cells: predominance of CD1a+differentiation pathway. J Immunol. 1999 May 15;162 (10):5821-8.

[79] Blom B, Spits H. Development of human lymphoid cells. Annu Rev Immunol. 2006;24:287-320.

[80] Shortman K, Egerton M, Spangrude GJ, Scollay R. The generation and fate of thymocytes. Semin Immunol. 1990 Jan;2 (1):3-12.

[81] Ehrlich LI, Oh DY, Weissman IL, Lewis RS. Differential contribution of chemotaxis and substrate restriction to segregation of immature and mature thymocytes. Immunity. 2009 Dec 18;31 (6):986-98.

[82] Plotkin J, Prockop SE, Lepique A, Petrie HT. Critical role for CXCR4 signaling in progenitor localization and T cell differentiation in the postnatal thymus. J Immunol. 2003 Nov 1;171 (9):4521-7.

[83] Griffith AV, Fallahi M, Nakase H, Gosink M, Young B, Petrie HT. Spatial mapping of thymic stromal microenvironments reveals unique features influencing T lymphoid differentiation. Immunity. 2009 Dec 18;31 (6):999-1009.

[84] Love PE, Bhandoola A. Signal integration and crosstalk during thymocyte migration and emigration. Nat Rev Immunol. 2011 Jul;11 (7):469-77.

[85] Halkias J, Melichar HJ, Taylor KT, Ross JO, Yen B, Cooper SB, et al. Opposing chemokine gradients control human thymocyte migration in situ. J Clin Invest. 2013 May 1;123 (5):2131-42.

[86] Trampont PC, Tosello-Trampont AC, Shen Y, Duley AK, Sutherland AE, Bender TP, et al. CXCR4 acts as a costimulator during thymic beta-selection. Nat Immunol. 2010 Feb;11 (2):162-70.

[87] Janas ML, Varano G, Gudmundsson K, Noda M, Nagasawa T, Turner M. Thymic development beyond beta-selection requires phosphatidylinositol 3-kinase activation by CXCR4. J Exp Med. 2010 Jan 18;207 (1):247-61.

[88] Nakano H, Mori S, Yonekawa H, Nariuchi H, Matsuzawa A, Kakiuchi T. A novel mutant gene involved in T-lymphocyte-specific homing into peripheral lymphoid organs on mouse chromosome 4. Blood. 1998 Apr 15;91 (8):2886-95.

[89] Pui JC, Allman D, Xu L, DeRocco S, Karnell FG, Bakkour S, et al. Notch1 expression in early lymphopoiesis influences B versus T lineage determination. Immunity. 1999 Sep;11 (3):299-308.

[90] Radtke F, Wilson A, Stark G, Bauer M, van Meerwijk J, MacDonald HR, et al. Deficient T cell fate specification in mice with an induced inactivation of Notch1. Immunity. 1999 May;10 (5):547-58.

[91] Mojcik CF, Salomon DR, Chang AC, Shevach EM. Differential expression of integrins on human thymocyte subpopulations. Blood. 1995 Dec 1;86 (11):4206-17.

[92] Salomon DR, Mojcik CF, Chang AC, Wadsworth S, Adams DH, Coligan JE, et al. Constitutive activation of integrin alpha 4 beta 1 defines a unique stage of human thymocyte development. J Exp Med. 1994 May 1;179 (5):1573-84.

[93] Bromley SK, Mempel TR, Luster AD. Orchestrating the orchestrators: chemokines in control of T cell traffic. Nat Immunol. 2008 Sep;9 (9):970-80.

[94] Yasukawa M, Hasegawa A, Sakai I, Ohminami H, Arai J, Kaneko S, et al. Down-regulation of CXCR4 by human herpesvirus 6 (HHV-6) and HHV-7. J Immunol. 1999 May 1;162 (9):5417-22.

[95] Molon B, Gri G, Bettella M, Gomez-Mouton C, Lanzavecchia A, Martinez AC, et al. T cell costimulation by chemokine receptors. Nat Immunol. 2005 May;6 (5):465-71.

[96] Kumar A, Humphreys TD, Kremer KN, Bramati PS, Bradfield L, Edgar CE, et al. CXCR4 physically associates with the T cell receptor to signal in T cells. Immunity. 2006 Aug;25 (2):213-24.

[97] Becker TC, Coley SM, Wherry EJ, Ahmed R. Bone marrow is a preferred site for homeostatic proliferation of memory CD8 T cells. J Immunol. 2005 Feb 1;174 (3):1269-73.

[98] Tokoyoda K, Radbruch A. Signals controlling rest and reactivation of T helper memory lymphocytes in bone marrow. Cell Mol Life Sci. 2012 May;69 (10):1609-13.

[99] Tokoyoda K, Zehentmeier S, Hegazy AN, Albrecht I, Grun JR, Lohning M, et al. Professional memory CD4+T lymphocytes preferentially reside and rest in the bone marrow. Immunity. 2009 May;30 (5):721-30.

[100] Nemoto Y, Kanai T, Takahara M, Oshima S, Nakamura T, Okamoto R, et al. Bone marrow-mesenchymal stem cells are a major source of interleukin-7 and sustain colitis by forming the niche for colitogenic CD4 memory T cells. Gut. 2013 Aug;62 (8): 1142-52.

[101] Hayakawa Y, Smyth MJ. CD27 dissects mature NK cells into two subsets with distinct responsiveness and migratory capacity. J Immunol. 2006 Feb 1;176 (3):1517-24.

[102] Chiossone L, Chaix J, Fuseri N, Roth C, Vivier E, Walzer T. Maturation of mouse NK cells is a 4-stage developmental program. Blood. 2009 May 28;113 (22):5488-96.

[103] Mayol K, Biajoux V, Marvel J, Balabanian K, Walzer T. Sequential desensitization of CXCR4 and S1P5 controls natural killer cell trafficking. Blood. 2011 Nov 3;118 (18): 4863-71.

[104] Sciume G, De Angelis G, Benigni G, Ponzetta A, Morrone S, Santoni A, et al. CX3CR1 expression defines 2 KLRG1+mouse NK-cell subsets with distinct functional properties and positioning in the bone marrow. Blood. 2011 Apr 28;117 (17):4467-75.

[105] Walzer T, Chiossone L, Chaix J, Calver A, Carozzo C, Garrigue-Antar L, et al. Natural killer cell trafficking in vivo requires a dedicated sphingosine 1-phosphate receptor. Nat Immunol. 2007 Dec;8 (12):1337-44.

[106] Tokoyoda K, Egawa T, Sugiyama T, Choi BI, Nagasawa T. Cellular niches controlling B lymphocyte behavior within bone marrow during development. Immunity. 2004 Jun;20 (6):707-18.

[107] Noda M, Omatsu Y, Sugiyama T, Oishi S, Fujii N, Nagasawa T. CXCL12-CXCR4 chemokine signaling is essential for NK-cell development in adult mice. Blood. 2011 Jan 13;117 (2):451-8.

[108] Cyster JG, Schwab SR. Sphingosine-1-phosphate and lymphocyte egress from lymphoid organs. Annu Rev Immunol. 2012;30:69-94.

[109] Siedlar M, Rudzki Z, Strach M, Trzyna E, Pituch-Noworolska A, Blaut-Szlosarczyk A, et al. Familial occurrence of warts, hypogammaglobulinemia, infections, and myelokathexis (WHIM) syndrome. Arch Immunol Ther Exp (Warsz). 2008 Nov-Dec;56 (6):419-25.

[110] Beider K, Nagler A, Wald O, Franitza S, Dagan-Berger M, Wald H, et al. Involvement of CXCR4 and IL-2 in the homing and retention of human NK and NK T cells to the bone marrow and spleen of NOD/SCID mice. Blood. 2003 Sep 15;102 (6):1951-8.

[111] Gorlin RJ, Gelb B, Diaz GA, Lofsness KG, Pittelkow MR, Fenyk JR, Jr. WHIM syndrome, an autosomal dominant disorder: clinical, hematological, and molecular studies. Am J Med Genet. 2000 Apr 24;91 (5):368-76.

[112] Beaussant Cohen S, Fenneteau O, Plouvier E, Rohrlich PS, Daltroff G, Plantier I, et al. Description and outcome of a cohort of 8 patients with WHIM syndrome from the French Severe Chronic Neutropenia Registry. Orphanet J Rare Dis. 2012;7:71.

[113] Hernandez PA, Gorlin RJ, Lukens JN, Taniuchi S, Bohinjec J, Francois F, et al. Mutations in the chemokine receptor gene CXCR4 are associated with WHIM syndrome, a combined immunodeficiency disease. Nat Genet. 2003 May;34 (1):70-4.

[114] Balabanian K, Lagane B, Pablos JL, Laurent L, Planchenault T, Verola O, et al. WHIM syndromes with different genetic anomalies are accounted for by impaired CXCR4 desensitization to CXCL12. Blood. 2005 Mar 15;105 (6):2449-57.

[115] Gulino AV, Moratto D, Sozzani S, Cavadini P, Otero K, Tassone L, et al. Altered leukocyte response to CXCL12 in patients with warts hypogammaglobulinemia, infections, myelokathexis (WHIM) syndrome. Blood. 2004 Jul 15;104 (2):444-52.

[116] Kawai T, Malech HL. WHIM syndrome: congenital immune deficiency disease. Curr Opin Hematol. 2009 Jan;16 (1):20-6.

[117] Kallikourdis M, Trovato AE, Anselmi F, Sarukhan A, Roselli G, Tassone L, et al. The CXCR4 mutations in WHIM syndrome impair the stability of the T-cell immunologic synapse. Blood. 2013 Aug 1;122 (5):666-73.

[118] Ingle A, Ghim S, Joh J, Chepkoech I, Bennett Jenson A, Sundberg JP. Novel laboratory mouse papillomavirus (MusPV) infection. Vet Pathol. 2011 Mar;48 (2):500-5.

[119] Handisurya A, Day PM, Thompson CD, Buck CB, Pang YY, Lowy DR, et al. Characterization of Mus musculus papillomavirus 1 infection in situ reveals an unusual pattern of late gene expression and capsid protein localization. J Virol. 2013 Dec;87 (24): 13214-25.

[120] McDermott DH, Liu Q, Ulrick J, Kwatemaa N, Anaya-O'Brien S, Penzak SR, et al. The CXCR4 antagonist plerixafor corrects panleukopenia in patients with WHIM syndrome. Blood. 2011 Nov 3;118 (18):4957-62.

[121] Dale DC, Bolyard AA, Kelley ML, Westrup EC, Makaryan V, Aprikyan A, et al. The CXCR4 antagonist plerixafor is a potential therapy for myelokathexis, WHIM syndrome. Blood. 2011 Nov 3;118 (18):4963-6.

[122] Ahmad DS, Esmadi M, Steinmann WC. Idiopathic CD4 Lymphocytopenia: Spectrum of opportunistic infections, malignancies, and autoimmune diseases. Avicenna J Med. 2013 Apr;3 (2):37-47.

[123] Smith DK, Neal JJ, Holmberg SD. Unexplained opportunistic infections and CD4+T-lymphocytopenia without HIV infection. An investigation of cases in the United States. The Centers for Disease Control Idiopathic CD4+T-lymphocytopenia Task Force. N Engl J Med. 1993 Feb 11;328 (6):373-9.

[124] Walker UA, Warnatz K. Idiopathic CD4 lymphocytopenia. Curr Opin Rheumatol. 2006 Jul;18 (4):389-95.

[125] Kojima M, Sakurai S, Morita Y, Nakamura N, Sugihara S, Shimano S. EBV (+) B-cell lymphoproliferative disorder associated with subsequent development of Burkitt lymphoma in a patient with idiopathic CD4 (+) T-lymphocytopenia. J Clin Exp Hematop. 2008 Nov;48 (2):55-9.

[126] Fierro MT, Savoia P, Quaglino P, Novero D, Bernengo MG. Disseminated Kaposi's sarcoma associated with idiopathic CD4+lymphocytopenia and low dose steroid therapy. Clin Exp Dermatol. 2005 Jul;30 (4):395-7.

[127] Inhoff O, Doerries K, Doerries R, Scharf J, Groden C, Goerdt S, et al. Disseminated cutaneous Kaposi sarcoma and progressive multifocal leukoencephalopathy in a patient with idiopathic CD4+T lymphocytopenia. Arch Dermatol. 2007 May;143 (5): 673-5.

[128] Alisjahbana B, Dinata R, Sutedja E, Suryahudaya I, Soedjana H, Hidajat NN, et al. Disfiguring generalized verrucosis in an indonesian man with idiopathic CD4 lymphopenia. Arch Dermatol. 2010 Jan;146 (1):69-73.

[129] Stetson CL, Rapini RP, Tyring SK, Kimbrough RC. CD4+T lymphocytopenia with disseminated HPV. J Cutan Pathol. 2002 Sep;29 (8):502-5.

[130] Gorska MM, Alam R. A mutation in the human Uncoordinated 119 gene impairs TCR signaling and is associated with CD4 lymphopenia. Blood. 2012 Feb 9;119 (6): 1399-406.

[131] Li FY, Chaigne-Delalande B, Kanellopoulou C, Davis JC, Matthews HF, Douek DC, et al. Second messenger role for Mg2+revealed by human T-cell immunodeficiency. Nature. 2011 Jul 28;475 (7357):471-6.

[132] Kuijpers TW, Ijspeert H, van Leeuwen EM, Jansen MH, Hazenberg MD, Weijer KC, et al. Idiopathic CD4+T lymphopenia without autoimmunity or granulomatous disease in the slipstream of RAG mutations. Blood. 2011 Jun 2;117 (22):5892-6.

[133] Isgro A, Sirianni MC, Gramiccioni C, Mezzaroma I, Fantauzzi A, Aiuti F. Idiopathic CD4+lymphocytopenia may be due to decreased bone marrow clonogenic capability. Int Arch Allergy Immunol. 2005 Apr;136 (4):379-84.

[134] Fruhwirth M, Clodi K, Heitger A, Neu N. Lymphocyte diversity in a 9-year-old boy with idiopathic CD4+T cell lymphocytopenia. Int Arch Allergy Immunol. 2001 May; 125 (1):80-5.

[135] Cervera C, Fernandez-Aviles F, de la Calle-Martin O, Bosch X, Rovira M, Plana M, et al. Non-myeloablative hematopoietic stem cell transplantation in the treatment of severe idiopathic CD4+lymphocytopenia. Eur J Haematol. 2011 Jul;87 (1):87-91.

[136] Laurence J, Mitra D, Steiner M, Lynch DH, Siegal FP, Staiano-Coico L. Apoptotic depletion of CD4+T cells in idiopathic CD4+T lymphocytopenia. J Clin Invest. 1996 Feb 1;97 (3):672-80.

[137] Roger PM, Bernard-Pomier G, Counillon E, Breittmayer JP, Bernard A, Dellamonica P. Overexpression of Fas/CD95 and Fas-induced apoptosis in a patient with idiopathic CD4+T lymphocytopenia. Clin Infect Dis. 1999 May;28 (5):1012-6.

[138] Zonios DI, Falloon J, Bennett JE, Shaw PA, Chaitt D, Baseler MW, et al. Idiopathic CD4+lymphocytopenia: natural history and prognostic factors. Blood. 2008 Jul 15;112 (2):287-94.

[139] Lee PI, Ciccone EJ, Read SW, Asher A, Pitts R, Douek DC, et al. Evidence for translocation of microbial products in patients with idiopathic CD4 lymphocytopenia. J Infect Dis. 2009 Jun 1;199 (11):1664-70.

[140] Malaspina A, Moir S, Chaitt DG, Rehm CA, Kottilil S, Falloon J, et al. Idiopathic CD4+T lymphocytopenia is associated with increases in immature/transitional B cells and serum levels of IL-7. Blood. 2007 Mar 1;109 (5):2086-8.

[141] Puronen CE, Thompson WL, Imamichi H, Beq S, Hodge JN, Rehm C, et al. Decreased interleukin 7 responsiveness of T lymphocytes in patients with idiopathic CD4 lymphopenia. J Infect Dis. 2012 May 1;205 (9):1382-90.

[142] Bugault F, Benati D, Mouthon L, Landires I, Rohrlich P, Pestre V, et al. Altered responses to homeostatic cytokines in patients with idiopathic CD4 lymphocytopenia. PLoS One. 2013;8 (1):e55570.

[143] Cunningham-Rundles C, Murray HW, Smith JP. Treatment of idiopathic CD4 T lymphocytopenia with IL-2. Clin Exp Immunol. 1999 May;116 (2):322-5.

[144] Scott-Algara D, Balabanian K, Chakrabarti LA, Mouthon L, Dromer F, Didier C, et al. Idiopathic CD4+T-cell lymphocytopenia is associated with impaired membrane expression of the chemokine receptor CXCR4. Blood. 2010 May 6;115 (18):3708-17.

[145] Dighiero G, Follezou JY, Binet JL. Surface immunoglobulins in lymphocytic leukemias. Blood. 1977 Jul;50 (1):179.

[146] Rai KR, Sawitsky A, Cronkite EP, Chanana AD, Levy RN, Pasternack BS. Clinical staging of chronic lymphocytic leukemia. Blood. 1975 Aug;46 (2):219-34.

[147] Granziero L, Ghia P, Circosta P, Gottardi D, Strola G, Geuna M, et al. Survivin is expressed on CD40 stimulation and interfaces proliferation and apoptosis in B-cell chronic lymphocytic leukemia. Blood. 2001 May 1;97 (9):2777-83.

[148] Damle RN, Batliwalla FM, Ghiotto F, Valetto A, Albesiano E, Sison C, et al. Telomere length and telomerase activity delineate distinctive replicative features of the B-CLL subgroups defined by immunoglobulin V gene mutations. Blood. 2004 Jan 15;103 (2): 375-82.

[149] Messmer BT, Messmer D, Allen SL, Kolitz JE, Kudalkar P, Cesar D, et al. In vivo measurements document the dynamic cellular kinetics of chronic lymphocytic leukemia B cells. J Clin Invest. 2005 Mar;115 (3):755-64.

[150] Rosenwald A, Alizadeh AA, Widhopf G, Simon R, Davis RE, Yu X, et al. Relation of gene expression phenotype to immunoglobulin mutation genotype in B cell chronic lymphocytic leukemia. J Exp Med. 2001 Dec 3;194 (11):1639-47.

[151] Klein U, Tu Y, Stolovitzky GA, Mattioli M, Cattoretti G, Husson H, et al. Gene expression profiling of B cell chronic lymphocytic leukemia reveals a homogeneous phenotype related to memory B cells. J Exp Med. 2001 Dec 3;194 (11):1625-38.

[152] Seifert M, Sellmann L, Bloehdorn J, Wein F, Stilgenbauer S, Durig J, et al. Cellular origin and pathophysiology of chronic lymphocytic leukemia. J Exp Med. 2012 Nov 19;209 (12):2183-98.

[153] Herve M, Xu K, Ng YS, Wardemann H, Albesiano E, Messmer BT, et al. Unmutated and mutated chronic lymphocytic leukemias derive from self-reactive B cell precursors despite expressing different antibody reactivity. J Clin Invest. 2005 Jun;115 (6): 1636-43.

[154] Mohle R, Failenschmid C, Bautz F, Kanz L. Overexpression of the chemokine receptor CXCR4 in B cell chronic lymphocytic leukemia is associated with increased functional response to stromal cell-derived factor-1 (SDF-1). Leukemia. 1999 Dec;13 (12): 1954-9.

[155] Burger JA, Zvaifler NJ, Tsukada N, Firestein GS, Kipps TJ. Fibroblast-like synoviocytes support B-cell pseudoemperipolesis via a stromal cell-derived factor-1-and CD106 (VCAM-1)-dependent mechanism. J Clin Invest. 2001 Feb;107 (3):305-15.

[156] Lagneaux L, Delforge A, Bron D, De Bruyn C, Stryckmans P. Chronic lymphocytic leukemic B cells but not normal B cells are rescued from apoptosis by contact with normal bone marrow stromal cells. Blood. 1998 Apr 1;91 (7):2387-96.

[157] Burger JA, Gribben JG. The microenvironment in chronic lymphocytic leukemia (CLL) and other B cell malignancies: Insight into disease biology and new targeted therapies. Semin Cancer Biol. 2014 Feb;24:71-81.

[158] de la Fuente MT, Casanova B, Moyano JV, Garcia-Gila M, Sanz L, Garcia-Marco J, et al. Engagement of alpha4beta1 integrin by fibronectin induces in vitro resistance of B chronic lymphocytic leukemia cells to fludarabine. J Leukoc Biol. 2002 Mar;71 (3): 495-502.

[159] Tsukada N, Burger JA, Zvaifler NJ, Kipps TJ. Distinctive features of "nurselike" cells that differentiate in the context of chronic lymphocytic leukemia. Blood. 2002 Feb 1;99 (3):1030-7.

[160] Kumar CC. Signaling by integrin receptors. Oncogene. 1998 Sep 17;17 (11 Reviews): 1365-73.

[161] Bagnara D, Kaufman MS, Calissano C, Marsilio S, Patten PE, Simone R, et al. A novel adoptive transfer model of chronic lymphocytic leukemia suggests a key role for T lymphocytes in the disease. Blood. 2011 May 19;117 (20):5463-72.

[162] Furman RR, Asgary Z, Mascarenhas JO, Liou HC, Schattner EJ. Modulation of NF-kappa B activity and apoptosis in chronic lymphocytic leukemia B cells. J Immunol. 2000 Feb 15;164 (4):2200-6.

[163] Ghia P, Strola G, Granziero L, Geuna M, Guida G, Sallusto F, et al. Chronic lymphocytic leukemia B cells are endowed with the capacity to attract CD4+, CD40L+T cells by producing CCL22. Eur J Immunol. 2002 May;32 (5):1403-13.

[164] Chilosi M, Pizzolo G, Caligaris-Cappio F, Ambrosetti A, Vinante F, Morittu L, et al. Immunohistochemical demonstration of follicular dendritic cells in bone marrow involvement of B-cell chronic lymphocytic leukemia. Cancer. 1985 Jul 15;56 (2):328-32.

[165] Burger M, Hartmann T, Krome M, Rawluk J, Tamamura H, Fujii N, et al. Small peptide inhibitors of the CXCR4 chemokine receptor (CD184) antagonize the activation, migration, and antiapoptotic responses of CXCL12 in chronic lymphocytic leukemia B cells. Blood. 2005 Sep 1;106 (5):1824-30.

[166] Burger JA, Ghia P, Rosenwald A, Caligaris-Cappio F. The microenvironment in mature B-cell malignancies: a target for new treatment strategies. Blood. 2009 Oct 15;114 (16):3367-75.

[167] Lankester AC, van Schijndel GM, van der Schoot CE, van Oers MH, van Noesel CJ, van Lier RA. Antigen receptor nonresponsiveness in chronic lymphocytic leukemia B cells. Blood. 1995 Aug 1;86 (3):1090-7.

[168] Stevenson FK, Krysov S, Davies AJ, Steele AJ, Packham G. B-cell receptor signaling in chronic lymphocytic leukemia. Blood. 2011 Oct 20;118 (16):4313-20.

[169] Vlad A, Deglesne PA, Letestu R, Saint-Georges S, Chevallier N, Baran-Marszak F, et al. Down-regulation of CXCR4 and CD62L in chronic lymphocytic leukemia cells is triggered by B-cell receptor ligation and associated with progressive disease. Cancer Res. 2009 Aug 15;69 (16):6387-95.

[170] Stamatopoulos B, Meuleman N, De Bruyn C, Pieters K, Mineur P, Le Roy C, et al. AMD3100 disrupts the cross-talk between chronic lymphocytic leukemia cells and a mesenchymal stromal or nurse-like cell-based microenvironment: pre-clinical evidence for its association with chronic lymphocytic leukemia treatments. Haematologica. 2012 Apr;97 (4):608-15.

[171] Herishanu Y, Perez-Galan P, Liu D, Biancotto A, Pittaluga S, Vire B, et al. The lymph node microenvironment promotes B-cell receptor signaling, NF-kappaB activation, and tumor proliferation in chronic lymphocytic leukemia. Blood. 2011 Jan 13;117 (2): 563-74.

[172] Lannutti BJ, Meadows SA, Herman SE, Kashishian A, Steiner B, Johnson AJ, et al. CAL-101, a p110delta selective phosphatidylinositol-3-kinase inhibitor for the treatment of B-cell malignancies, inhibits PI3K signaling and cellular viability. Blood. 2011 Jan 13;117 (2):591-4.

[173] Owen RG, Treon SP, Al-Katib A, Fonseca R, Greipp PR, McMaster ML, et al. Clinicopathological definition of Waldenstrom's macroglobulinemia: consensus panel recommendations from the Second International Workshop on Waldenstrom's Macroglobulinemia. Semin Oncol. 2003 Apr;30 (2):110-5.

[174] Treon SP, Hunter ZR, Aggarwal A, Ewen EP, Masota S, Lee C, et al. Characterization of familial Waldenstrom's macroglobulinemia. Ann Oncol. 2006 Mar;17 (3):488-94.

[175] Ngo HT, Leleu X, Lee J, Jia X, Melhem M, Runnels J, et al. SDF-1/CXCR4 and VLA-4 interaction regulates homing in Waldenstrom macroglobulinemia. Blood. 2008 Jul 1;112 (1):150-8.

[176] Jung MJ, Rho JK, Kim YM, Jung JE, Jin YB, Ko YG, et al. Upregulation of CXCR4 is functionally crucial for maintenance of stemness in drug-resistant non-small cell lung cancer cells. Oncogene. 2013 Jan 10;32 (2):209-21.

[177] Ostrow RS, Bender M, Niimura M, Seki T, Kawashima M, Pass F, et al. Human papillomavirus DNA in cutaneous primary and metastasized squamous cell carcinomas from patients with epidermodysplasia verruciformis. Proc Natl Acad Sci U S A. 1982 Mar;79 (5):1634-8.

[178] Eliezri YD, Silverstein SJ, Nuovo GJ. Occurrence of human papillomavirus type 16 DNA in cutaneous squamous and basal cell neoplasms. J Am Acad Dermatol. 1990 Nov;23 (5 Pt 1):836-42.

[179] Hahn A, Loning T, Hoos A, Henke P. Immunohistochemistry (S 100, KL 1) and human papillomavirus DNA hybridization on morbus Bowen and bowenoid papulosis. Virchows Arch A Pathol Anat Histopathol. 1988;413 (2):113-22.

[180] Bryant P, Davies P, Wilson D. Detection of human papillomavirus DNA in cancer of the urinary bladder by in situ hybridisation. Br J Urol. 1991 Jul;68 (1):49-52.

[181] Schneider A, Oltersdorf T, Schneider V, Gissmann L. Distribution pattern of human papilloma virus 16 genome in cervical neoplasia by molecular in situ hybridization of tissue sections. Int J Cancer. 1987 Jun 15;39 (6):717-21.

[182] Fukamachi H, Seol HS, Shimada S, Funasaka C, Baba K, Kim JH, et al. CD49f (high) cells retain sphere-forming and tumor-initiating activities in human gastric tumors. PLoS One. 2013;8 (8):e72438.

[183] Al-Hajj M, Wicha MS, Benito-Hernandez A, Morrison SJ, Clarke MF. Prospective identification of tumorigenic breast cancer cells. Proc Natl Acad Sci U S A. 2003 Apr 1;100 (7):3983-8.

[184] Singh SK, Clarke ID, Terasaki M, Bonn VE, Hawkins C, Squire J, et al. Identification of a cancer stem cell in human brain tumors. Cancer Res. 2003 Sep 15;63 (18):5821-8.

[185] Collins AT, Berry PA, Hyde C, Stower MJ, Maitland NJ. Prospective identification of tumorigenic prostate cancer stem cells. Cancer Res. 2005 Dec 1;65 (23):10946-51.

[186] O'Brien CA, Pollett A, Gallinger S, Dick JE. A human colon cancer cell capable of initiating tumour growth in immunodeficient mice. Nature. 2007 Jan 4;445 (7123): 106-10.

[187] Ricci-Vitiani L, Lombardi DG, Pilozzi E, Biffoni M, Todaro M, Peschle C, et al. Identification and expansion of human colon-cancer-initiating cells. Nature. 2007 Jan 4;445 (7123):111-5.

[188] Li C, Heidt DG, Dalerba P, Burant CF, Zhang L, Adsay V, et al. Identification of pancreatic cancer stem cells. Cancer Res. 2007 Feb 1;67 (3):1030-7.

[189] Prince ME, Sivanandan R, Kaczorowski A, Wolf GT, Kaplan MJ, Dalerba P, et al. Identification of a subpopulation of cells with cancer stem cell properties in head and neck squamous cell carcinoma. Proc Natl Acad Sci U S A. 2007 Jan 16;104 (3):973-8.

[190] Yang W, Yan HX, Chen L, Liu Q, He YQ, Yu LX, et al. Wnt/beta-catenin signaling contributes to activation of normal and tumorigenic liver progenitor cells. Cancer Res. 2008 Jun 1;68 (11):4287-95.

[191] Vicari AP, Caux C. Chemokines in cancer. Cytokine Growth Factor Rev. 2002 Apr;13 (2):143-54.

[192] Balkwill F. Cancer and the chemokine network. Nat Rev Cancer. 2004 Jul;4 (7):540-50.

[193] D'Alterio C, Barbieri A, Portella L, Palma G, Polimeno M, Riccio A, et al. Inhibition of stromal CXCR4 impairs development of lung metastases. Cancer Immunol Immunother. 2012 Oct;61 (10):1713-20.

[194] Wei S, Kryczek I, Edwards RP, Zou L, Szeliga W, Banerjee M, et al. Interleukin-2 administration alters the CD4+FOXP3+T-cell pool and tumor trafficking in patients with ovarian carcinoma. Cancer Res. 2007 Aug 1;67 (15):7487-94.

[195] Wald O, Izhar U, Amir G, Avniel S, Bar-Shavit Y, Wald H, et al. CD4+CXCR4highCD69+T cells accumulate in lung adenocarcinoma. J Immunol. 2006 Nov 15;177 (10):6983-90.

[196] Shimizu Y, Dobashi K, Imai H, Sunaga N, Ono A, Sano T, et al. CXCR4+FOXP3+CD25+lymphocytes accumulate in CXCL12-expressing malignant pleural mesothelioma. Int J Immunopathol Pharmacol. 2009 Jan-Mar;22 (1):43-51.

[197] Kotsianidis I, Bouchliou I, Nakou E, Spanoudakis E, Margaritis D, Christophoridou AV, et al. Kinetics, function and bone marrow trafficking of CD4+CD25+FOXP3+regulatory T cells in myelodysplastic syndromes (MDS). Leukemia. 2009 Mar;23 (3): 510-8.

[198] Yan M, Jene N, Byrne D, Millar EK, O'Toole SA, McNeil CM, et al. Recruitment of regulatory T cells is correlated with hypoxia-induced CXCR4 expression, and is associated with poor prognosis in basal-like breast cancers. Breast Cancer Res. 2011;13 (2):R47.

[199] Curiel TJ, Coukos G, Zou L, Alvarez X, Cheng P, Mottram P, et al. Specific recruitment of regulatory T cells in ovarian carcinoma fosters immune privilege and predicts reduced survival. Nat Med. 2004 Sep;10 (9):942-9.

[200] Pistoia V, Corcione A, Dallegri F, Ottonello L. Lymphoproliferative disorders and chemokines. Curr Drug Targets. 2006 Jan;7 (1):81-90.

[201] Chong BF, Mohan C. Targeting the CXCR4/CXCL12 axis in systemic lupus erythematosus. Expert Opin Ther Targets. 2009 Oct;13 (10):1147-53.

[202] Biajoux V, Bignon A, Freitas C, Martinez V, Thelen M, Lima G, et al. Expression of CXCL12 receptors in B cells from Mexican Mestizos patients with systemic Lupus erythematosus. J Transl Med. 2012;10:251.

[203] Duda DG, Kozin SV, Kirkpatrick ND, Xu L, Fukumura D, Jain RK. CXCL12 (SDF1alpha)-CXCR4/CXCR7 pathway inhibition: an emerging sensitizer for anticancer therapies? Clin Cancer Res. 2011 Apr 15;17 (8):2074-80.

[204] Debnath B, Xu S, Grande F, Garofalo A, Neamati N. Small molecule inhibitors of CXCR4. Theranostics. 2013;3 (1):47-75.

[205] Bachelerie F, Ben-Baruch A, Burkhardt AM, Combadiere C, Farber JM, Graham GJ, et al. International Union of Pharmacology. LXXXIX. Update on the extended family of chemokine receptors and introducing a new nomenclature for atypical chemokine receptors. Pharmacol Rev. 2014 Jan;66 (1):1-79.

[206] Calandra G, Bridger G, Fricker S. CXCR4 in clinical hematology. Curr Top Microbiol Immunol. 2010;341:173-91.

[207] McDermott DH, Liu Q, Velez D, Lopez L, Anaya-O'Brien S, Ulrick J, et al. A phase 1 clinical trial of long-term, low-dose treatment of WHIM syndrome with the CXCR4 antagonist plerixafor. Blood. 2014 Feb 12.

[208] Melo Rde C, Longhini AL, Bigarella CL, Baratti MO, Traina F, Favaro P, et al. CXCR7 Is Highly Expressed in Acute Lymphoblastic Leukemia and Potentiates CXCR4 Response to CXCL12. PLoS One. 2014;9 (1):e85926.

[209] Torossian F, Anginot A, Chabanon A, Clay D, Guerton B, Desterke C, et al. CXCR7 participates in CXCL12-induced CD34+cell cycling through beta-arrestin-dependent Akt activation. Blood. 2014 Jan 9;123 (2):191-202.

Chapter 4

Hematopoietic Stem Cells, Tumor Cells and Lymphocytes — Party in the Bone Marrow

Adriana Bonomo, Ana Carolina Monteiro and Alex Balduíno

Additional information is available at the end of the chapter

http://dx.doi.org/10.5772/58843

1. Introduction

1.1. The hematopoietic stem cell niche

1.1.1. Hematopoietic system development: distinct niches activities

During vertebrate embryogenesis, different anatomical sites are responsible for creating specific conditions to promote hematopoietic stem cells self-renewal, expansion, commitment, and differentiation of the hematopoietic stem cells (HSC) [1,2]. The first hematopoietic cells emerge within the blood islands of the yolk sac (YS), an extra-embryonic site. Most of the cells belong to the primitive erythroid lineage, but a few myeloid cells are also generated [3,4]. In a second wave, hematopoietic progenitors emerge from the mesoderm of the paraaortic-splanchnopleura (Sp), an intra-embryonic site, which later gives rise to the aorta, gonads and mesonephros, and has been named AGM region [1]. Data show that almost all long-term definitive progenitors derive from the AGM region, as those originated in the YS fail to properly reconstitute the adult bone marrow of a lethally irradiated animal [5]. However, when cultured under the right combination of cytokines, cells derived from the mesoderm of the YS, in which blood islands originate, can be instructed to become long-term hematopoietic progenitors [6]. In these conditions, even higher numbers of progenitors could be found in the YS compared to the AGM region. This shows that, in vivo, the YS niche does not hold proper conditions to promote full commitment of the hematopoietic progenitors.

After the vascular system is established in the embryo, hematopoietic progenitors migrate and colonize the fetal liver. Fetal liver is the main hematopoietic organ during embryo's development [7,8] and its hematopoiesis requires exogenous colonization. So far, no data indicate that new progenitors emerge in the fetal liver. All hematopoietic cells are derived from the YS and AGM region [8–10].

At this stage, fetal liver microenvironment is responsible for two very important tasks: full commitment of mesoderm derived progenitors to long-term HSC, and their increase in numbers (higher numbers). Although AGM derived progenitors are able to fully reconstitute a lethally irradiated animal in an experimental model, well-defined HSC can only be observed in the embryo a few days after fetal liver colonization. It has been shown that, to become adult long-term HSC, progenitors derived from the AGM region must go through the fetal liver microenvironment for proper instructions [11]. A few progenitors from the yolk sac become long-term HSC, but those originated from the AGM region are far more predominant. Long-term HSC in the fetal liver are highly proliferative and self-renewable. In mouse embryo, in five days, there could be an increase in 30 times the original number of HSC [7,9]. Different from what would be expected, not a huge number of progenitors colonize the fetal liver. Only a few are necessary. Commitment to HSC and their expansion require two distinct niches in the fetal liver microenvironment, at the same time or at different maturation stages of the liver. This requires further investigation.

The fetal liver remains hematopoietic until birth, or even a short period after. By the time the organ starts to acquire its metabolic properties, HSC are then progressively transferred to their final destination: the bone marrow [9]. In contrast to the fetal liver hematopoietic activity, bone marrow main assignment is blood production – not HSC expansion (only). All types of blood cells are produced in the bone marrow, except for the T lymphocytes, produced in the thymus. Despite its high dynamics, the bone marrow microenvironment is organized, in order to guarantee a finely tuned hierarchical differentiation cascade. Hematopoietic system organization in the marrow cavity follows an also organized distribution of the stromal cells. Different stromal cell types – osteoblasts, reticular cells, perivascular cells, endothelial cells, macrophages – interact with different groups of hematopoietic cells, creating distinct niches in bone marrow microenvironment to harbor. This is the way the differentiation cascade is controlled as hematopoietic cells at different stages of differentiation demand distinct combinations of factors for their proliferation and differentiation [12–15].

Based on cells behavior, at least three niches can be identified in the marrow microenvironment: one responsible for HSC maintenance (self-renewal) throughout life; a second to induce intermediate progenitors expansion; and a third to guarantee hematopoietic cells full commitment and differentiation to the lineages.

In humans, during childhood, almost all bones in our body hold a "hematopoietically" active bone marrow (red bone marrow). After reaching maturity, active bone marrow is restricted to the sternum, ribs, vertebrae, ilium, and femurs' heads. The rest the bones are filled with "inactive" bone marrow, which is called yellow bone marrow, due to the high number of fat storing cells.

1.1.2. Inside the bone marrow

In spite of its high dynamic, the hematopoietic system, in the bone marrow cavity, is widely hierarchical and hematopoietic cells are not randomly distributed. As mentioned before, specific niches control HSC self-renewal and their engagement to a differentiation cascade.

The concept that different niches would compose the bone marrow microenvironment was envisioned already in the 70's [16]. Based on a stereological study, it was proposed that bone marrow microenvironment could be subdivided into, at least, four niches: endosteal, subendosteal, central, and perisinusoidal [13,16]. Histological and functional assays showed that HSC and primitive progenitors preferentially colonize the endosteal and subendosteal regions – close to the bone surface. Intermediate progenitors and differentiated cells are distributed in the central and perisinusoidal niches, respectively [13,14,16–21]. Due to their close range, endosteal and subendosteal regions are usually identified as one niche, named "endosteal niche". However, these two regions harbor very distinct stromal cells [15,22,23] and must then be considered as two different niches, as they play distinct roles on HSC behavior.

Based on the expression of different surface markers [24,25] one can isolate the long-term HSC separately from other progenitors. Under physiological conditions, 20%-30% of the HSC are in a quiescent stage. Studies have shown that slow-cycling HSC are found in association with endosteal osteoblasts [19,21,26]. On the other hand, most of the fast-cycling HSC are found in close association with perivascular cells of the blood vessels distributed in the subendosteal zone [27,28]. This has been described *in vivo* in long-term BrdU retaining assays and myeloablation models. In experimental in vivo myelosuppressive models, HSC colonizing the vascular niches in the subendosteal region are mostly ablated. Almost all HSC in contact with endosteal osteoblasts are preserved [20,28]. By the time this chapter has been written, the existence of the two separate yet complimentary niches is still questioned by a few authors based on technical issues arguments.

The role of endosteal osteoblasts on the HSC maintenance and self-renewal was first proposed in vitro by Taichman and Emerson [14,29,30] and later evidenced in vivo by others [31–33]. In transgenic animals, increased numbers of osteoblasts results in an increased number of long-term HSC, without affecting any other hematopoietic subpopulation in the bone marrow [31, 32,34]. Furthermore, when osteoblasts are removed from the marrow cavity, HSC numbers reduces drastically [33]. This is evidence that osteoblasts play a crucial role in HSC maintenance and behavior.

On the same study mentioned before, Lambertsen and Weiss [13] showed that most of the perivascular niches harboring HSC are distributed in the subendosteal zone. In the perivascular niche [19,20,27], HSC reside on the abluminal side of bone marrow sinusoids, and are supported by the endothelial and perivascular reticular cells. HSC residing in the perivascular niche are in close association with reticular cells, which express high levels of CXCL12, a chemokine required for HSC maintenance and lodging [17,27]. Most of the cells creating the proliferative niche express CXCL12. In situ observation demonstrated that most of hematopoietic stem cells are concentrated in the trabecular zone of the marrow cavity, which also harbors high numbers of niche osteblasts, sinusoids, and CXCL12-positive reticular cells.

Nonetheless, HSC maintenance by both endosteal and perivascular niches are, at least in part, mediated by Jagged-Notch and angiopoietin-1-Tie2 interactions [20,27,31,32].

So, all in one thought, in the bone marrow, there are two distinct niches to harbor HSC, referred as to "proliferative niche" and "quiescent niche", which are composed by perivascular cells and endosteal osteoblasts, respectively [15,22,31,32,35,36]. The real conversation between these niches, and how other elements, such as the immune system, would contribute to the niche formation, organization and dynamics are still to be understood.

The technique to isolate and culture separately endosteal osteoblasts and subendosteal reticular/perivascular cells from the marrow cavity of murine long bones was established [15] and global gene analyses data suggest that both endosteal and subendosteal stroma contribute to the formation of both niches in the marrow.

2. T cells as messengers from the periphery to the hematopoietic bone marrow

2.1. An overview of the immune system

The immune system is composed of hematopoietic cells, which we can be characterized according to the way they recognize and respond to antigens.

The innate immune system, phylogenetically, arises before the adaptive immune system and is so called because its ability to respond to antigens is ready and immediate. Characteristically, the innate immune cells recognize antigen through Pathogen Recognition Receptors (PRR), which are evolutionary conserved and can be common to different cell types. PRRs recognize defined molecular patterns from a pathogen [37] or something that is 'dangerous' to the body [38]. These molecular patterns, which are named Pathogen or Danger Associated Molecular Patterns (PAMPs or DAMPs), are poorly present or even absent in healthy mammals and are rich in or characteristics of bacteria, fungi, virus and so on. The cellular composition of the innate immune system is represented by phagocytes (granulocytes, monocytes/macrophages and dendritic cells) which deals with antigen, ultimately, eliminating it by phagocytosis or secretion of the internal granules content, and some lymphocytes as Natural Killer (NK) cells, $\gamma\delta$ T cells and B1 cells. In common, all these cellular types promptly respond to antigen and will do so in the same time frame and efficiency regardless their previous experience with the same antigen.

The adaptive immune system is so called because its components do not mount an immediate response to antigen. They need to be stimulated in order to mature their effectors functions and these take 3-5 days to happen, and will only be clinically effective after 7 days. Although it takes a while for the adaptive immune response to occur, it does so only once-on the first encounter with the antigen. On the following and subsequent encounters with the same antigen, the response will be fast occurring in less than 24h, revealing the existing memory response. Characteristically, the antigen recognition is done by antigen recognition receptors, which are diverse at the population level and clonal and unique at individual cell level. These clonal receptors are not conserved and are generated by gene rearrangements during ontogeny of T-and B-lymphocytes, the cellular components of the adaptive immune system.

Innate immune cells and cells from the adaptive immune system mostly differentiate within the adult bone marrow, except for T lymphocytes that differentiate inside the thymus but also arise from hematopoietic progenitors. Although we can didactically separate the immune system into two categories, an effective immune response depends on both innate and adaptive cells. For T cells to be activated, they depend to see antigen complexed to the Major Histocompatibility Molecules (MHC) presented to them by antigen presenting cells (APC), having the dendritic cells (DC) as the most important cell type to initiate the response, or to prime the adaptive immune response. Also, the cytokines secreted by DC at the moment of antigen presentation, will define the fate of the T cell, meaning the cytokines these T cells will present or their specialization on Th1, Th2, Th17, etc for CD4+T cells or Tc1, Tc2, etc for CD8+cells. Also, the antibody class produced by B cells – IgG, IgA, IgE, etc – will be defined by cytokines produced by a given CD4T cell which will 'help' the given B cell at the moment of its activation. So, indirectly, it depends upon the APC which will modulate the CD4 fate. On the other hand, although the cells from the innate immune system can play their role independently, T helper cells can efficiently modulate it and, for example, optimize the microbicidal function of macrophages or even down modulate it. Also, through their role on dictating the immunoglobulin isotype to B cells, T cells will indirectly act on opsonization which will ultimately be effective through the innate system by optimizing phagocytosis and activation of phagoctyes and granulocytes, all actors of the so called innate response.

The important point to have in mind here, before getting into the T cells inside the bone marrow, is that the effective immune response depends on T cells and an important part of the immune effector mechanisms rely on the collaboration between adaptive and innate responses, with the innate response being in many cases, the main players at the effector phase.

2.2. Bone marrow : A hospitable environment for T cells

After maturation inside the thymus, T cells gain the peripheral blood circulation and enter the secondary lymphoid organs (SLO) - lymph nodes and spleen - where they can be activated. Classically, these two SLO are considered the sites of naive T cell activation given their architecture, which allows concentration of antigen, DCs and naive T cells in the same neighborhood. This architecture is extremely important given the low frequency of antigen specific T cells making it difficult to meet with antigen, by chance, anywhere in the body.

Primed T cells will generate effectors cells, which will deal with the incoming antigen in the short-term response and will be vanished after antigen clearance. Primed T cells will also generate memory cells, which will be kept, even in the absence of antigen. Memory cells can be found in the SLO and in tissues as different memory cell subpopulations. Those in the tissues, are the effectors memory cells, which respond rather quickly after antigen exposure and those in the SLO are considered the central memory cells, responsible for keeping the memory pool and they take a little longer than effector memory cells to respond to antigen [39].

However, the above mentioned circulation pattern and activation sites of mature T cells had been challenged and revitalized by studies on BM T cells in the last decade.

T cells account for only 3-8% of total BM cells, what seems a small number, but in fact it is estimated to be close to or even higher than the number of T cells in the spleen when all hematopoietic bones are considered [40]. Moreover, the CD4 to CD8 ratio is 1:2 instead of the 2:1 ratio found in peripheral blood, indicating a local microenvironmental regulation of these cell subsets. Interestingly, these cells do not seem to be BM resident cells nor depend on antigen presence for its location in the bone marrow. Naive as well as memory T cells carry CXCR4, a receptor for CXCL12 (SDF1) a critical chemokine produced by stromal cells in the bone marrow, which play a significant role on HSC migration into the BM and its specific niche. Although both, memory and naive cells can respond to CXCL12 in migration assays, memory or activated cells respond more efficiently [41]. Besides, parabiont studies had shown that activated/memory CD4 and CD8 T cells recirculate and distribute equally through SLO and BM between the two animals [42,43] indicating that T cells recirculate through the BM.

Antigen recognition in the bone marrow could be one important requirement to keep them there not only as spectators but as active cells influencing the microenvironment. Of note is the fact BM CD8 T cells are extremelly active, with a proliferation rate in vivo higher than the ones in spleen and lymph nodes [44,45] Similarly, BM CD4 T cells produce high amounts of cytokine in the absence of intentional stimulation [46,47]. However, in the case of CD8 cells, when taken out from the BM, their behavior in vitro is similar to the one from splenic cells, indicating that this is not an intrinsic characteristic of BM cells, but is a modulation imposed by the BM microenvironment [40]. The presence of antigen is actually possible, as bone marrow DC were shown to present blood born antigens to naive CD4 and CD8 T cells [48]. Moreover, not only DCs, but other myeloid cells can also present antigen to naive BM T cells, what is not observed in spleen where T cell primming depends mostly on DCs [49]. Another curious fact about the bone marrow environment and T cells is that antigen specific cells are found in several diseases but do not always relate to the presence of antigen, neither in the bone marrow nor in the periphery [40]. Memory CD8 T cells are maintained by IL-7 and IL-15 which are produced in copious amounts by stromal cells in the BM. On the other hand, memory CD4 cells do not need recognition of MHC with the cognate peptide, but depend on the presence of MHC and IL-7 to be maintained.

So, it seems that the BM environment have all the requirements to attract and eventually keep T cells active: BM DC and other myeloid cells can present antigen and prime T cells, the stroma produces IL-7 and IL-15 necessary for memory CD8 maintenance, and hematopoietic cells in the marrow express MHC molecules fulfilling the requirements to maintain memory CD4 cells.

But what are these cells doing in the bone marrow, since the majority of infections are not BM specific, and in pathological conditions such as cancer, antigen specific cells are found there in the absence of the pathogenic antigen? (although the antigen peptide might be presented by BM DCs as mentioned above)

2.3. T cell help for hematopoiesis

BM is especialized in generating hematopoietic cells and T cells localize in the hematopoietic niche, on the perivascular regions where pericytes are present being one of the stem cell niches [50]. These data show that T cells are in close physical contact with the hematopoietic environment. In infectious situations, it appears that T cell amplification of hematopoiesis is required to clear pathogens [51–53]. These can be achieved by local secretion of cytokines by T cells (at the expense of antigen recognition *in situ*) including GM-CSF, IL-3, IL-4, IL-5, IL-6, IL-13, IL17 and oncostatin M, which all contribute to amplify granulocyte generation inside the bone marrow. However, the role of T cells in "normal" hematopoiesis has not been extensively considered as normal hematopoiesis is considered an innate immune phenomena independent of antigen recognition.

The relationship between hematopoiesis and T cells was first suggested almost 40 years ago when it was shown that 1-day thymectomized mice were anemic, showed arrested erythroid maturation and reduction in the number of spleen colony-forming units in the bone marrow and spleen [54,55]. In addition, intravenous injection of live thymocytes accelerated hematopoietic reconstitution in sublethally irradiated mice [56]. In the 90's it was suggested that singeneic T cells could stimulate the growth of hematopoietic progenitors [57]. Much more recently, it was clearly shown that T cell deficient mice (nude and SCID mice) have a severe reduction in the number of granulocytes in peripheral blood, despite the high frequency of granulo-monocytic progenitors in the bone marrow. By injecting CD4 T cells into these animals, the peripheral cytopenia was corrected and the number of progenitors accumulated in the bone marrow diminished to levels similar to the ones found in normal euthymic animals [46]. Moreover, studies with TCR transgenic mice in the RAGKO background, i.e., in the absence of endogenous gene rearrangements to guarantee that the only T cell specificity in the animals was the one from the transgenic receptor, evidenced the same altered hematopoiesis present in the T cell deficient mice: accumulation of immature myeloid-monocytic progenitors in the BM and granulopenia in the peripheral blood. Also, their BM T cells did not show the characteristic activated phenotype found in regular animals. Strikingly after injection of the cognate antigen, BM T cells became activated and the abnormal hematopoietic phenotype was corrected.

All these results show that T cells, as antigen recognition entities present in the periphery, act as messengers to the hematopoietic bone marrow. They traffic to the bone mainly after activation, recognize antigen within the bone cavity and help hematopoiesis so that it acquire the so-called 'normal' configuration. Normal hematopoiesis is not a phenomenon independent of the adaptive immune response, but is a response to an immunological insult, instructed by T cells (Figure 1).

At the end, this makes a lot of sense since the optimal effector immune mechanisms relies on innate immune cells acting on its best with components of the adaptive responses. And if memory response need to be fast and precise, T cells need to be rapidly activated and find their way to the bone marrow to instruct hematopoiesis to produce more of effectors. In the absence of enough phagocytes, and these in the absence of T cell help and immunoglobulins, the response will not be as efficient as necessary to counteract an invasive pathogen.

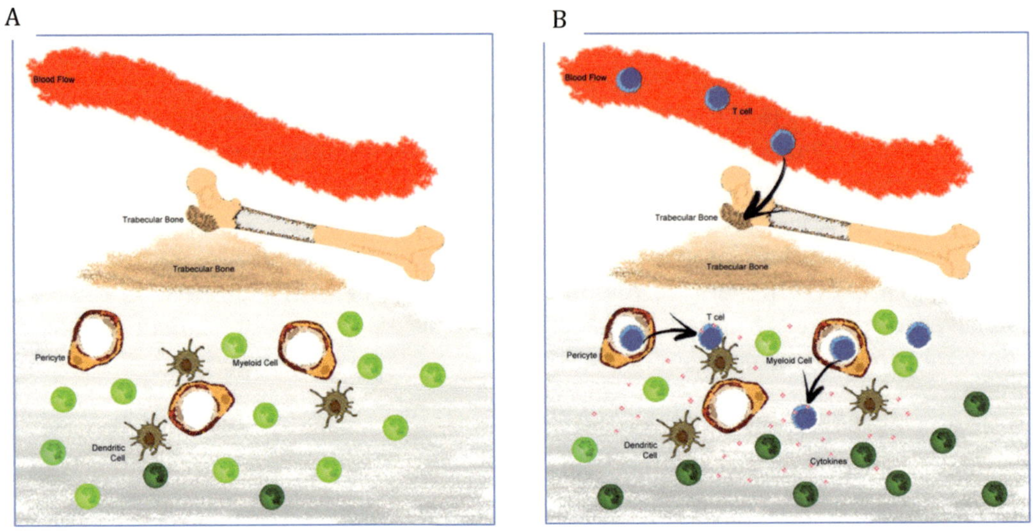

Figure 1. T cells help myeloid cell differentiation-In A, in the absence of activated T cells, myeloid progenitor cells (light green) accumulate in the bone marrow. When T cells are activated and migrate to the bone cavity, as shown in B, they most probably interact with APC such as DCs, secrete cytokines which will help the terminal differentiation of myeloid cells (dark green cells) giving rise to the normal cells counts in the peripheral blood.

3. Roommates in the bone cavity: Tumor cells, HSC and T cells

3.1. The crosstalk between T cells and bone: An overview of osteoimmunology

First of all, bone marrow is in close contact with bone tissue formed by the organized deposits of type I collagen and hydroxyapatite, a calcium phosphate salt, in which bone cells are distributed. Rather than being an inert matrix, bone undergoes a continuous turnover: osteoblast activity resulting in bone deposition is counteracted by osteoclast mediated bone resorption. Osteoblasts are cells of mesenchymal origin, whereas osteoclasts are of hematopoietic origin — multinucleated giant cells, derived from monocytes/macrophages progenitors expressing CD11b–c, CD14 and receptor activator of nuclear factor, (RANK). Curiously, several factors regulating bone homeostasis are also molecular players of the immune response. For example, the TNF family member RANK ligand (RANKL) (also called TRANCE, OPGL, ODF), a potent regulator of osteoclast activation and differentiation, is expressed not only by osteoblasts, but also by monocytes, neutrophils, dendritic cells, B cells and activated CD4 and CD8 T cells [58–61]. RANKL mediates its biological effects by binding to RANK, expressed by osteoclast progenitors, mature osteoclasts, DCs and neutrophils [60,62]. RANKL can also bind to the soluble protein osteoprotegerin (OPG), which acts as an inhibitory decoy receptor and can be produced by osteoblasts, DCs and B cells [58–61]. By binding to RANK, RANKL strongly stimulates bone resorption, contributes to lymph node organogenesis, prolongs DC survival and augments DC adjuvant properties [63].

Once both immune and skeletal systems share many regulatory elements, including some which are key in bone remodelling, it seems reasonable to think that these two systems interact with each other. Indeed, the new and complex interdisciplinary field of osteoimmunology implies the concept that bone, and its cavity, crosstalk with the immune system. Osteoimmunology investigates the interactions between these two systems, since bone marrow stromal cells express surface molecules essential for hematopoiesis— from which all cells of the mammalian immune system derive — and stimulate immune cells, which produce various regulatory cytokines that influence the bone fate [58,64].

It is clear that immune cells producing pro-inflammatory cytokines contribute to bone damage by potentiating the effects of RANK/RANKL/OPG pathway. The cytokines TNF-α, interleukin (IL)-1, IL-3, IL-6, IL-7, IL-11, IL-15 IL-17A/F, and prostaglandin-E2 (as well hormones and related peptides as parathyroid hormone, parathyroid hormone-related protein, and glucocorticoids, besides 1,25(OH)2 vitamin D3) potentiate bone loss either by increasing osteoclast generation and activation or by inducing RANKL expression by the osteoblasts [60]. On the other hand, IL-4, IL-5, IL-10, IL-12, IL-13, IL-18 and interferon (IFN)-α, IFN-β and IFN-γ are inhibitors of osteoclastogenesis by blocking RANKL signalling, either directly or indirectly [58,59]. Interestingly, IL-1 is a stimulator of TRAF6 expression on the osteoclast, thereby potentiating RANK/RANKL signaling cascade, whereas IFN-γ is known to downregulate TRAF6 by proteossomal degradation aborting osteoclast formation [58,65]. In contrast with this effect, IFN-γ has also been implicated in osteoclast formation and bone resorption, underlining the controversial role of IFN-γ in osteoclastogenesis [59,61,63,66].

The crosstalk mechanisms between T cells and osteoclasts have been extensively documented. The key role of Th 17 CD4 T cells, — an osteoclastogenic subclass of T cells expressing membrane and soluble RANKL, IL-17A/F and TNFa — on exacerbated and uncontrolled function of osteoclasts, has been investigated in models of inflammatory diseases, such as autoimmune rheumathoid arthritis [58,59], periodontitis [67], multiple myeloma [61] and breast tumor skeletal metastasis [68]. Nevertheless, the literature of this field has also been showing that these exacerbated osteoclast pattern, might also be controlled by other T cells subsets as regulatory CD4 T (Treg) cells. T reg cells produce anti-osteoclastogenic cytokines such as IL-4, IL-10 and TGF-β and express CTLA-4 inhibiting bone destruction [69,70]. Besides that, recently, it was showed that osteoclasts can present antigenic peptides to CD8 T cells, apart from CD4 T cells, resulting in FoxP3 expression. In this way, CD8 FoxP3+cells function as CD8 Treg cells, able to cause an inappropriate activation of the immune response through reciprocal interactions between CD137/CD137L and RANK/RANKL pathways. CD137, expressed on T cells, is a co-stimulatory molecule induced by TCR activation and its ligand, CD137L, is expressed on DCs and osteoclasts precursors. Once T cell CD137 binds to CD137L on osteoclasts precursors, multinucleation of osteoclasts is suppressed.. However, CD137/CD137L will signal simultaneously with RANKL/RANK on the Tcell/Osteoclast pair and this might lead to increased apoptosis by T cells [71]. Therefore, in pathological conditions, the effects of T cells on osteoclastogenesis will depend on the balance between positive and negative factors that they express.

In addition to inflammatory pathological conditions, increasing evidence supports the notion that T cells are also involved in post-menopausal osteoporosis [66]. Experiments in mice showed that, in the absence of estrogens, higher numbers of TNF-α producing T cells were found in the bone marrow, stimulating directly osteoclasts activity and augmenting their response to RANKL. By comparing peripheral blood mononuclear cells from pre and post-menopausal women, it was observed that estrogen deficiency was associated with an increased production of TNF-α. The action of TNF-α is not limited to the induction of local inflammation, but is both directly and indirectly involved in the activation of osteoclasts. Although further work is necessary to clarify the complex changes leading to post-menopausal osteoporosis in women, a pro-osteoclastogenetic contribution of T cells has to be taken into account.

Finally, it was documented that T cells have a protective role on bone turnover under physiological conditions [66,72]. Hints that this modulation may occur came from in vitro studies showing that osteoclastogenesis was inhibited by CD8 T cells. Moreover, after CD3 and CD28 activation, mouse lymph node CD8 T cells showed a delayed kinetics of RANKL expression, as compared with corresponding CD4 T cells. Culture of bone marrow cells from CD4 and CD8 T cell depleted mice showed enhanced osteoclastogenesis in response to 1,25(OH)2 vitamin D3 stimulation, suggesting that T cells had a suppressive effect in this system [6]. Moreover, the protective role of T cells on bone metabolism was also documented by in vivo studies, showing that both B cell-and T cell-deficient mice have decreased bone mineral density. A detailed analysis demonstrated that osteoporosis was prevented by osteoprotegerin produced by bone marrow resident B cells stimulated by T cells through CD40L/CD40 interactions. In contrast, IL-17A does not play any relevant role in physiological bone homeostasis, as IL17A-deficient mice show normal bone mineral density and skeletal development [6]. Taken together, these findings support the notion that bone marrow derived CD4 and CD8 T cells play a protective role in physiological bone homeostasis, using pathways different from those associated with inflammatory bone diseases.

3.2. Inflammatory bone diseases: Rheumatoid arthritis as a model of T cell involvement in bone diseases

Rheumatoid arthritis (RA) is the prototype of chronic inflammatory joint diseases and is characterized by persistent inflammation and progressive bone erosions, leading to functional disability and high morbidity. In this disease it is clear that the pro-inflammatory cytokines IL-17A, TNF-α, IL-1 and IL-6 are involved in the perpetuation of the inflammatory condition. The RANK/RANKL/OPG pathway is also strongly involved in RA pathology and it was observed that the RANKL/OPG ratio is increased, leading to bone erosions [58,59]. This effect is mainly dependent on osteoclastic activity and is expressed by two main mechanisms: i) destruction of the organic matrix (type I collagen) by osteoclast cathepsin K, and ii) dissolution of the mineralized component (hydroxyapatite crystals) by the acidic microenvironment generated by the osteoclastic proton pump.

In RA, T cell derived RANKL was initially proposed to be the main contributor to exacerbated osteoclastogenesis, but Th17 RANKL+ subset T cells from RA joints also produce IFN-γ, an anti-osteoclastogenic cytokine, which counterbalance the action of RANKL [58–61]. Thus, in such model, it is not clear whether Th17 RANKL+ subset T cells exert a direct effect on osteoclastogenesis. It is more likely that Th17 RANKL+ subset T cells contribute indirectly through IL-17A activity over sinovial fibroblasts which produce RANKL and directly stimulates osteoclastogenesis. Another T cell shown to be potentially envolved is the CD4 Treg subset. In fact, an increase of Treg CD4 T cell number improves clinical signs of arthritis and suppressed local and systemic bone destruction. Synovial tissues of patients with RA also produce many factors regulating bone resorption, such as TNF-α, IL-1 and IL-6, which amplify osteoclast differentiation, activation and consequent bone destruction. Inhibitors that target TNF-α, IL-1 and IL-17A pro-inflammatory and osteoclastogenic cytokines have been approved for the treatment of RA.

More recently, investigators also demonstrated that RANKL plus macrophage colony-stimulating factor can induce transdifferentiation of immature dendritic cells into the osteoclastogenic lineage and that this process is significantly enhanced by RA synovial fluid [73]. Dendritic cells are antigen presenting cells, but they could function as osteoclasts precursors in inflammatory conditions. We can conclude that since dendritic cells modulate T cell activity through the RANK/RANKL pathway and other cytokines associated with osteoclastogenesis, as mentioned earlier, it can function as an osteoimmune interface, contributing to bone loss in inflammatory diseases.

Although T cells clearly contribute to RA pathology, they do so in the outer face of the bone, outside the bone marrow cavity. By the same token, periodontal disease, which had been shown to be dependent of Th17 RANKL+ T cells activated by bacteria present in the oral cavity also lead to extra medullary lesions [67]. Similar mechanisms might act in pathological situations arising within the bone marrow cavity such as post-menopausal osteoporosis and cancers, as myeloma and solid tumor metastasis. In either case it is clear the potential for the adaptive immune system to interact with the bone remodelling system.

3.3. Cancer: Multiple myeloma as an example of bone marrow derived tumors

Multiple myeloma (MM) is a clonal B-cell malignancy characterized by an accumulation of mature plasma cells in the bone marrow, leading to bone destruction and failure of normal hematopoiesis. However the cancer induced osteolytic disease in this case, may count on T cell activity inside the bone marrow, in addition to the presence of the tumor itself [61]. In multiple myeloma patients with lytic bone disease, it was observed an increase in bone marrow Th17 T cells expressing high levels of RANKL that can directly stimulate osteoclasts [49,74]; moreover, increased production of T cell derived IL-3, occurring in this disease, can inhibit osteoblast generation and facilitate hematopoiesis.

This is one the few, if not the only, malignant bone marrow disease associated with bone loss where T cell activity has been studied, and actually shown to be concordant with osteolytic activity. So, instead of having the osteolytic disease induced by cancer cells only, it is proposed the participation of Th17 cells in the pathogenesis of lytic lesions in bone marrow malignancies. In fact, as reported for other human malignacies [75] and in accordance to the phenotype of BM T cells, the number of memory/activated T cells in MM patients is increased, their activity is enhanced and they proliferate much more efficiently than blood derived T cells. These data suggest that BM T cells in pathological, non-infectious conditions such as cancer, can also migrate to the bone marrow and in addition to its effects over hematopoiesis, as discussed above, they can also influence bone remodelling.

3.4. Breast tumor skeletal metastasis: the case of osteolytic bone disease in the absence of tumor cells

Bone metastases, present in 70% of patients with metastatic breast cancer, lead to skeletal disease, fractures and intense pain, which are all believed to be mediated by tumor cells. Engraftment of tumor cells is supposed to be preceded by changes in the target tissue to create a permissive microenvironment, the pre-metastatic niche, for the establishment of the metastatic foci. In bone metastatic niche, metastatic cells stimulate bone consumption resulting in the release of growth factors that feed the tumor, establishing a vicious cycle between the bone remodelling system and the tumor itself [76]. Yet, how the pre-metastatic niches arise in the bone tissue remains unclear.

As already mentioned before, CD4 and CD8 T cells have been shown to unbalance the bone remodeling process in inflammatory osteolytic diseases, however, little is known about their role in cancer induced bone disease, a process that differs from inflammatory diseases as in the former it happens in the bone cavity and not on the periosteal surface.

It had been shown, in an experimental model, that tumor specific T cells have a pro-osteoclastogenic phenotype, i.e., Th17 producers of IL-17F and RANKL among others, when tumors are highly metastatic. On the other hand, the T cell phenotype was not pro-osteoclastogenic, and even rich in anti-osteoclastogenic cytokines as IFN-γ and IL-10, if the tumor was localized to the breast and incapable of sending metastasis to any distant organ, including the BM. This suggest that T cells activity is modulated by the tumor since sibling cell lines, with different metastatic characteristics and sharing the same cognate T cell antigen, trigger different T cell phenotypes.

The pro-osteoclastogenic T cell phenotype observed with metastatic tumor was evident inside the BM, and preceded bone metastatic colonization. Also, osteolytic lesions were already present very early on disease evolution, and again, before metastatic colonization. By transfering BM T cells from animals bearing the highly aggressive tumors, before metastasis started, to athymic mouse which never saw tumors, led to an intense osteolytic disease, similar in kinetics and intensity to the one observed in the tumor bearing donor animals. These indicate that T cells can induce osteolytic disease which precedes metastatic colonization. In vivo inhibition of RANKL production by Th17 CD4 T cells, but not of IL-17F, completely protects mice from osteolytic disease and, surprisingly, completely abolishes the development of bone metastases, suggesting that CD4 T cells prepare the metastatic niche for further establishment of tumor cells. In conclusion, these results unveil an unexpected role for RANKL derived from T cells in setting the pre-metastatic niche and promoting tumor spread, an extra role for T cells only recently explored (Figure 2).

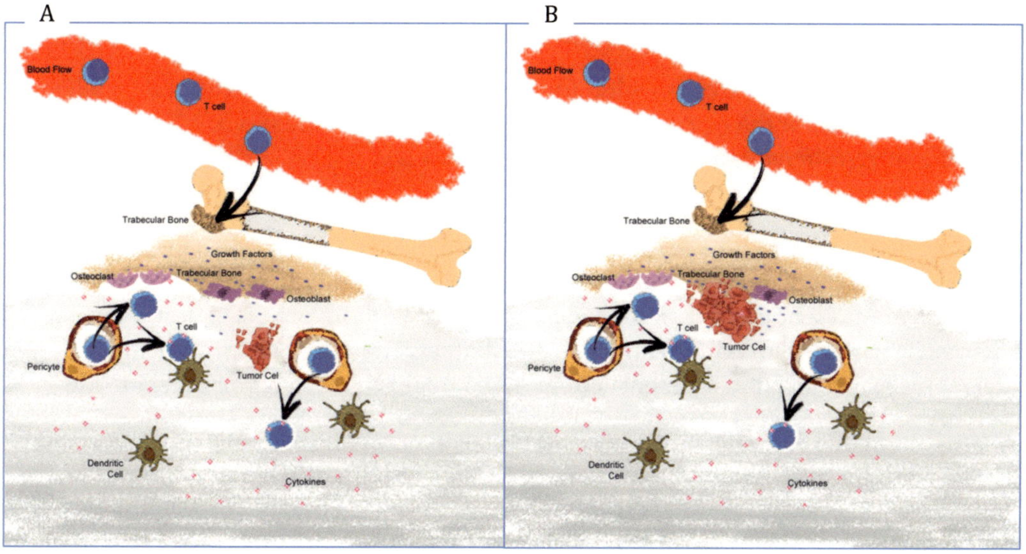

Figure 2. T cells induce osteolytic disease which is critical for bone metastatsis establishment. In A ; Tumor primed T cells, inside the BM, produce pro-osteolcastogenic cytokines (mainly IL-17F and RANKL) which promotes osteoclastogenesis and activate osteoclasts. Osteoclast activity generates a pro-metastatic environment, with the release of growth factors from the bone matrix. In B, with the hospitable environment prepared, tumor cells can establish themselves in the marrow, and after that, they will progress by themselves, directly regulating bone remodeling.

In summary, for tumor cells to first establish in the bone marrow niche, growth factors need to be available favoring its hostage, in other words, a pre-metastatic niche needs to be prepared. On that sense, again, T lymphocytes work as messengers from the periphery as they migrate to the BM after they get primed by tumor antigens. In the BM, they signal and alter bone homeostasis as to prepare the pre-metastatic niche instructed by the tumor modulation of T cell phenotype in the periphery [68].

4. Conclusions and perspectives

From all of the above it can be depicted that once the hematopoietic niche depends upon the bone lining osteoblasts, it is reasonable to expect an interplay between bone and hematopoietic regulation. On the other hand, it is known that T lymphocytes also communicate with the hematopoietic and bone tissues adding more complexity to the whole balance of these systems.

T cells in the bone marrow are compatible with memory cells and found in their activated state. In the absence of T cells, hematopoiesis is altered and a maturation arrest is observed in the bone marrow, were high numbers of immature myelo-monocytic progenitors are found accompanied by peripheral cytopenia. When T cells are replenished, the bone marrow arrested progenitors progress, differentiate and migrate to the periphery, correcting the myelogram and the peripheral cytopenia. This is conceptually important since what we use to understand as "normal" hematopoiesis, which should be an antigen independent activity, is already the result of the adaptive immune response, which, in fact, need the innate immune cells to operate!

The localization of active T cells within the marrow cavity coincides with the "proliferative niche" of adult HSC, or the perivascular niche. In fact, the evidences favor a T cell function on the proliferative/differentiative phase of myelopoiesis and not on stem cell maintenance, as stated above.

Hematopoietic stem cells are supported and regulated by stromal cells covering the inner surface of bones, or the endosteum. The endosteum also supports bone remodeling and osteoblasts are present there. The subendosteal region harbors pre-osteoblast, reticular cells and mesenchymal stem cells, with higher hematopoietic supporting role, characterizing different niches involved in different activities that might be cross-regulated somehow.

Activated T cells are able to interact with hematopoietic system, apparently on the proliferative niche, and this is in close contact with the bone remodeling system. When activated, BM T cells can increase osteoclastogenesis and this will favor, in case of a bone metastatic disease, the establishment of bone colonization by the malignant cell. Curiously, metastatic cells in the BM use the same niche as hematopoietic progenitor cells [77] and might well be regulated by antigen specific activated T cells, which are able to prepare the pre-metastatic niche and interfere with the hematopoietic niche. At one side, increase in osteoclasts numbers might be the result of increased proliferation of myeloid progenitors, but it also result from direct T cell signaling to stimulate osteoclast differentiation (figure 3).

Whether or not alterations on bone remodeling are directly linked to alterations in hematopoiesis and vice-versa, and the dependency on the T cell adaptive immune response is still a theme of debate.

Studies in germ free mice might elucidate the subject, since the absence of intentional stimulation is not synonymous of no stimulation making it difficult to address the immune regulation of blood and bone.

It is now believed that the commensal microbiota plays an important role on basically every system related to immune response. Very recently, it was shown that Germ Free mice are cytopenic not only in the peripheral blood, but also in the bone marrow. The cytopenia is reversed after intestinal colonization with commensal bacteria [78] and this provide an optimal tune to the immune system to fight infection. On the other hand, Germ Free mice were recently shown to be osteopetrotic and this is reversed by colonization with commensal microbiota [79]. Moreover, in this report, the number and activation state of immune cells was analyzed and in the BM, the number of T cells was diminished in GF mice and the frequency of osteoclast precursors was also deficient. These are in accordance with the view that in the steady state, recognition of antigen by T cells tune the bone remodelling system and this might be related to hematopoietic activity.

Altogether, we provide evidence that, inside the BM there are at least two co-existing systems-bone and hematopoietic-which can be regulated by T cells as they bring messages from the periphery to the BM, resulting in hematopoietic/cancer niche and bone remodeling regulation (figure 3). Whether or not, regulation of one system interferes with the other one, and most important, to which extend, is still a matter of debate.

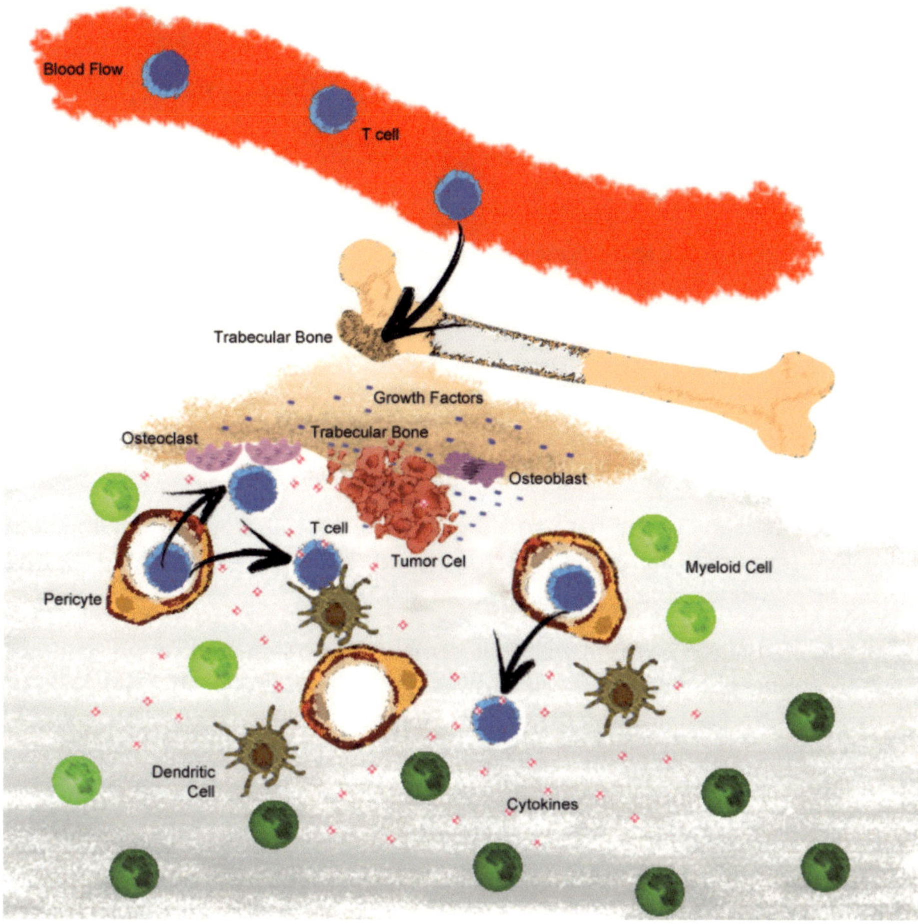

Figure 3. T cells modulate the bone and hematopoietic system after activation:. Antigen primed T cells, inside the BM, produce cytokines which can act in both, hematopoiesis and bone remodeling. Depending on the stimuli, each of the two systems can receive positive or negative signals. In the case exemplified here, positive signals for osteolytic disease and myeloid cell expansion and maturation is shown. On this case, T cell help metastasis establishment and boost hematopoiesis. (dark green : mature myeloid cells, light green: immature myeloid cells, Blue: T cells, brown: dendritic cells, pink: osteoclas, red: tumor cells, pruple: osteoblast.

Acknowledgements

We are deeply indebted with Romulo G. Galvani, who kindly did all the schematics presented in this chapter

Author details

Adriana Bonomo[1,2*], Ana Carolina Monteiro[2] and Alex Balduíno[3,4]

*Address all correspondence to: acbonomo@gmail.com

1 Department of Immunology, Federal Universisty of Rio de Janeiro, Rio de Janeiro, RJ, Brazil

2 Laboratory of Thymus Research, Oswaldo Cruz Institute, Oswaldo Cruz Foundation, Rio de Janeiro, RJ, Brazil

3 Technology and Research Center, Veiga de Almeida University, Rio de Janeiro, RJ, Brazil

4 Excellion Biomedical Services Lab, Rio de Janeiro, RJ, Brazil

References

[1] Godin I, Cumano A. The hare and the tortoise: an embryonic haematopoietic race. Nat Rev Immunol [Internet]. 2002 Aug [cited 2011 Apr 18];2(8):593–604. Available from: http://www.ncbi.nlm.nih.gov/pubmed/12154378

[2] Ikuta K, Weissman IL. Evidence that hematopoietic stem cells express mouse c-kit but do not depend on steel factor for their generation. Proc Natl Acad Sci U S A [Internet]. 1992 Feb [cited 2011 Apr 18];89(4):1502–6. Available from: http://www.pubmedcentral.nih.gov/articlerender.fcgi?artid=48479&tool=pmcentrez&rendertype=abstract

[3] Moore MA, Metcalf D. Ontogeny of the haemopoietic system: yolk sac origin of in vivo and in vitro colony forming cells in the developing mouse embryo. Br J Haematol [Internet]. 1970 Mar [cited 2014 Jun 23];18(3):279–96. Available from: http://www.ncbi.nlm.nih.gov/pubmed/5491581

[4] Dzierzak E, Medvinsky A. Mouse embryonic hematopoiesis. Trends Genet [Internet]. 1995 Sep [cited 2014 Jun 23];11(9):359–66. Available from: http://www.ncbi.nlm.nih.gov/pubmed/7482788

[5] Cumano A, Ferraz JC, Klaine M, Di Santo JP, Godin I. Intraembryonic, but not yolk sac hematopoietic precursors, isolated before circulation, provide long-term multilineage reconstitution. Immunity [Internet]. 2001 Sep [cited 2014 Jun 23];15(3):477–85. Available from: http://www.ncbi.nlm.nih.gov/pubmed/11567637

[6] Inlay MA, Serwold T, Mosley A, Fathman JW, Dimov IK, Seita J, et al. Identification of Multipotent Progenitors that Emerge Prior to Hematopoietic Stem Cells in Embryonic Development. Stem cell reports [Internet]. 2014 Apr 8 [cited 2014 Jun 2];2(4):457–

72. Available from: http://www.pubmedcentral.nih.gov/articlerender.fcgi?artid=3986503&tool=pmcentrez&rendertype=abstract

[7] Ema H, Nakauchi H. Expansion of hematopoietic stem cells in the developing liver of a mouse embryo. Blood [Internet]. 2000 Apr 1 [cited 2014 Jul 4];95(7):2284–8. Available from: http://www.ncbi.nlm.nih.gov/pubmed/10733497

[8] Morrison SJ, Hemmati HD, Wandycz AM, Weissman IL. The purification and characterization of fetal liver hematopoietic stem cells. Proc Natl Acad Sci U S A [Internet]. 1995 Oct 24 [cited 2014 Jun 23];92(22):10302–6. Available from: http://www.pubmedcentral.nih.gov/articlerender.fcgi?artid=40784&tool=pmcentrez&rendertype=abstract

[9] Christensen JL, Wright DE, Wagers AJ, Weissman IL. Circulation and chemotaxis of fetal hematopoietic stem cells. PLoS Biol [Internet]. 2004 Mar [cited 2011 Apr 18]; 2(3):E75. Available from: http://www.pubmedcentral.nih.gov/articlerender.fcgi?artid=368169&tool=pmcentrez&rendertype=abstract

[10] Medvinsky A, Dzierzak E. Definitive hematopoiesis is autonomously initiated by the AGM region. Cell [Internet]. 1996 Sep [cited 2011 Mar 26];86(6):897–906. Available from: http://www.ncbi.nlm.nih.gov/pubmed/8808625

[11] Kieusseian A, Brunet de la Grange P, Burlen-Defranoux O, Godin I, Cumano A. Immature hematopoietic stem cells undergo maturation in the fetal liver. Development [Internet]. 2012 Oct [cited 2014 Jun 12];139(19):3521–30. Available from: http://www.ncbi.nlm.nih.gov/pubmed/22899849

[12] Tokoyoda K, Egawa T, Sugiyama T, Choi B-I, Nagasawa T. Cellular niches controlling B lymphocyte behavior within bone marrow during development. Immunity [Internet]. 2004 Jun [cited 2014 Jun 24];20(6):707–18. Available from: http://www.ncbi.nlm.nih.gov/pubmed/15189736

[13] Lambertsen RH, Weiss L. A model of intramedullary hematopoietic microenvironments based on stereologic study of the distribution of endocloned marrow colonies. Blood [Internet]. 1984 Feb [cited 2011 Nov 13];63(2):287–97. Available from: http://www.ncbi.nlm.nih.gov/pubmed/6692036

[14] Taichman RS, Emerson SG. Human osteoblasts support hematopoiesis through the production of granulocyte colony-stimulating factor. J Exp Med [Internet]. 1994 May [cited 2011 Nov 13];179(5):1677–82. Available from: http://www.pubmedcentral.nih.gov/articlerender.fcgi?artid=2191506&tool=pmcentrez&rendertype=abstract

[15] Balduino A, Hurtado SP, Frazão P, Takiya CM, Alves LM, Nasciutti L-E, et al. Bone marrow subendosteal microenvironment harbours functionally distinct haemosupportive stromal cell populations. Cell Tissue Res [Internet]. 2005 Feb [cited 2011 Aug 5];319(2):255–66. Available from: http://www.ncbi.nlm.nih.gov/pubmed/15578225

[16] Lord BI, Testa NG, Hendry JH. The relative spatial distributions of CFUs and CFUc in the normal mouse femur. Blood [Internet]. 1975 Jul [cited 2011 Nov 13];46(1):65–72. Available from: http://www.ncbi.nlm.nih.gov/pubmed/1131427

[17] Jung Y, Wang J, Schneider A, Sun Y-X, Koh-Paige AJ, Osman NI, et al. Regulation of SDF-1 (CXCL12) production by osteoblasts; a possible mechanism for stem cell homing. Bone [Internet]. 2006 Apr [cited 2011 Nov 13];38(4):497–508. Available from: http://www.ncbi.nlm.nih.gov/pubmed/16337237

[18] Nilsson SK, Johnston HM, Coverdale JA. Spatial localization of transplanted hemopoietic stem cells: inferences for the localization of stem cell niches. Blood [Internet]. 2001 Apr [cited 2011 Nov 13];97(8):2293–9. Available from: http://www.ncbi.nlm.nih.gov/pubmed/11290590

[19] Kiel MJ, Iwashita T, Yilmaz OH, Morrison SJ. Spatial differences in hematopoiesis but not in stem cells indicate a lack of regional patterning in definitive hematopoietic stem cells. Dev Biol [Internet]. 2005 Jul [cited 2011 Sep 15];283(1):29–39. Available from: http://www.ncbi.nlm.nih.gov/pubmed/15913595

[20] Arai F, Hirao A, Ohmura M, Sato H, Matsuoka S, Takubo K, et al. Tie2/angiopoietin-1 signaling regulates hematopoietic stem cell quiescence in the bone marrow niche. [Internet]. Cell. 2004 [cited 2011 Jun 27]. p. 149–61. Available from: http://www.ncbi.nlm.nih.gov/pubmed/15260986

[21] Lo Celso C, Fleming HE, Wu JW, Zhao CX, Miake-Lye S, Fujisaki J, et al. Live-animal tracking of individual haematopoietic stem/progenitor cells in their niche. Nature [Internet]. 2009 Jan [cited 2011 Jul 21];457(7225):92–6. Available from: http://www.pubmedcentral.nih.gov/articlerender.fcgi?artid=2820276&tool=pmcentrez&rendertype=abstract

[22] Taichman RS, Wang Z, Shiozawa Y, Jung Y, Song J, Balduino A, et al. Prospective identification and skeletal localization of cells capable of multilineage differentiation in vivo. Stem Cells Dev [Internet]. 2010 Oct [cited 2011 Jul 5];19(10):1557–70. Available from: http://www.pubmedcentral.nih.gov/articlerender.fcgi?artid=3089003&tool=pmcentrez&rendertype=abstract

[23] Balduino A, Mello-Coelho V, Wang Z, Taichman RS, Krebsbach PH, Weeraratna AT, et al. Molecular signature and in vivo behavior of bone marrow endosteal and subendosteal stromal cell populations and their relevance to hematopoiesis. Exp Cell Res [Internet]. 2012 Nov 15 [cited 2014 Jun 24];318(19):2427–37. Available from: http://www.pubmedcentral.nih.gov/articlerender.fcgi?artid=3445703&tool=pmcentrez&rendertype=abstract

[24] Logan AC, Weissman IL, Shizuru JA. The road to purified hematopoietic stem cell transplants is paved with antibodies. Curr Opin Immunol [Internet]. 2012 Oct [cited 2014 Jul 5];24(5):640–8. Available from: http://www.ncbi.nlm.nih.gov/pubmed/22939368

[25] Oguro H, Ding L, Morrison SJ. SLAM family markers resolve functionally distinct subpopulations of hematopoietic stem cells and multipotent progenitors. Cell Stem Cell [Internet]. 2013 Jul 3 [cited 2014 May 31];13(1):102–16. Available from: http://www.pubmedcentral.nih.gov/articlerender.fcgi?artid=3736853&tool=pmcentrez&rendertype=abstract

[26] Nombela-Arrieta C, Pivarnik G, Winkel B, Canty KJ, Harley B, Mahoney JE, et al. Quantitative imaging of haematopoietic stem and progenitor cell localization and hypoxic status in the bone marrow microenvironment. Nat Cell Biol [Internet]. 2013 May [cited 2014 May 23];15(5):533–43. Available from: http://www.ncbi.nlm.nih.gov/pubmed/23624405

[27] Sugiyama T, Kohara H, Noda M, Nagasawa T. Maintenance of the hematopoietic stem cell pool by CXCL12-CXCR4 chemokine signaling in bone marrow stromal cell niches. Immunity [Internet]. 2006 Dec [cited 2011 Jul 5];25(6):977–88. Available from: http://www.ncbi.nlm.nih.gov/pubmed/17174120

[28] Kiel MJ, Radice GL, Morrison SJ. Lack of evidence that hematopoietic stem cells depend on N-cadherin-mediated adhesion to osteoblasts for their maintenance. Cell Stem Cell [Internet]. 2007 Aug [cited 2011 Jun 11];1(2):204–17. Available from: http://www.ncbi.nlm.nih.gov/pubmed/18371351

[29] Taichman RS, Reilly MJ, Emerson SG. Human osteoblasts support human hematopoietic progenitor cells in vitro bone marrow cultures. Blood [Internet]. 1996 Jan [cited 2011 Nov 13];87(2):518–24. Available from: http://www.ncbi.nlm.nih.gov/pubmed/8555473

[30] Taichman RS, Emerson SG. The role of osteoblasts in the hematopoietic microenvironment. Stem Cells [Internet]. 1998 Jan [cited 2011 Nov 13];16(1):7–15. Available from: http://www.ncbi.nlm.nih.gov/pubmed/9474743

[31] Calvi LM, Adams GB, Weibrecht KW, Weber JM, Olson DP, Knight MC, et al. Osteoblastic cells regulate the haematopoietic stem cell niche. Nature [Internet]. 2003 Oct [cited 2011 Aug 8];425(6960):841–6. Available from: http://www.ncbi.nlm.nih.gov/pubmed/14574413

[32] Zhang J, Niu C, Ye L, Huang H, He X, Tong W-G, et al. Identification of the haematopoietic stem cell niche and control of the niche size. Nature [Internet]. 2003 Oct [cited 2011 Aug 8];425(6960):836–41. Available from: http://www.ncbi.nlm.nih.gov/pubmed/14574412

[33] Visnjic D, Kalajzic Z, Rowe DW, Katavic V, Lorenzo J, Aguila HL. Hematopoiesis is severely altered in mice with an induced osteoblast deficiency. Blood [Internet]. 2004 May [cited 2011 Aug 30];103(9):3258–64. Available from: http://www.ncbi.nlm.nih.gov/pubmed/14726388

[34] Kuznetsov SA, Riminucci M, Ziran N, Tsutsui TW, Corsi A, Calvi L, et al. The interplay of osteogenesis and hematopoiesis: expression of a constitutively active PTH/

PTHrP receptor in osteogenic cells perturbs the establishment of hematopoiesis in bone and of skeletal stem cells in the bone marrow. J Cell Biol [Internet]. 2004 Dec 20 [cited 2012 Mar 7];167(6):1113–22. Available from: http://www.pubmedcentral.nih.gov/articlerender.fcgi?artid=2172616&tool=pmcentrez&rendertype=abstract

[35] Wilson A, Trumpp A. Bone-marrow haematopoietic-stem-cell niches. Nat Rev Immunol [Internet]. 2006 Feb [cited 2011 Jun 16];6(2):93–106. Available from: http://www.ncbi.nlm.nih.gov/pubmed/16491134

[36] Parmar K, Mauch P, Vergilio J-A, Sackstein R, Down JD. Distribution of hematopoietic stem cells in the bone marrow according to regional hypoxia. Proc Natl Acad Sci U S A [Internet]. 2007 Mar [cited 2011 Sep 12];104(13):5431–6. Available from: http://www.pubmedcentral.nih.gov/articlerender.fcgi?artid=1838452&tool=pmcentrez&rendertype=abstract

[37] Medzhitov R, Janeway CA. How does the immune system distinguish self from nonself? Semin Immunol [Internet]. 2000 Jun [cited 2014 Jul 5];12(3):185–8; discussion 257–344. Available from: http://www.ncbi.nlm.nih.gov/pubmed/10910738

[38] Matzinger P. Tolerance, danger, and the extended family. Annu Rev Immunol [Internet]. 1994 Jan [cited 2014 May 28];12:991–1045. Available from: http://www.ncbi.nlm.nih.gov/pubmed/8011301

[39] Sallusto F, Geginat J, Lanzavecchia A. Central memory and effector memory T cell subsets: function, generation, and maintenance. Annu Rev Immunol [Internet]. 2004 Jan [cited 2014 May 29];22:745–63. Available from: http://www.ncbi.nlm.nih.gov/pubmed/15032595

[40] Di Rosa F. T-lymphocyte interaction with stromal, bone and hematopoietic cells in the bone marrow. Immunol Cell Biol [Internet]. 2009 Jan [cited 2014 Jun 4];87(1):20–9. Available from: http://www.ncbi.nlm.nih.gov/pubmed/19030018

[41] Cinamon G, Shinder V, Alon R. Shear forces promote lymphocyte migration across vascular endothelium bearing apical chemokines. Nat Immunol [Internet]. 2001 Jun [cited 2014 May 24];2(6):515–22. Available from: http://www.ncbi.nlm.nih.gov/pubmed/11376338

[42] Klonowski KD, Williams KJ, Marzo AL, Blair DA, Lingenheld EG, Lefrançois L. Dynamics of blood-borne CD8 memory T cell migration in vivo. Immunity [Internet]. 2004 May [cited 2014 Jul 5];20(5):551–62. Available from: http://www.ncbi.nlm.nih.gov/pubmed/15142524

[43] Nemoto Y, Kanai T, Makita S, Okamoto R, Totsuka T, Takeda K, et al. Bone marrow retaining colitogenic CD4+T cells may be a pathogenic reservoir for chronic colitis. Gastroenterology [Internet]. 2007 Jan [cited 2014 Jul 5];132(1):176–89. Available from: http://www.ncbi.nlm.nih.gov/pubmed/17241870

[44] Parretta E, Cassese G, Barba P, Santoni A, Guardiola J, Di Rosa F. CD8 cell division maintaining cytotoxic memory occurs predominantly in the bone marrow. J Immu-

nol [Internet]. 2005 Jun 15 [cited 2014 Jul 5];174(12):7654–64. Available from: http://www.ncbi.nlm.nih.gov/pubmed/15944266

[45] Becker TC, Coley SM, Wherry EJ, Ahmed R. Bone marrow is a preferred site for homeostatic proliferation of memory CD8 T cells. J Immunol [Internet]. 2005 Feb 1 [cited 2014 Jul 5];174(3):1269–73. Available from: http://www.ncbi.nlm.nih.gov/pubmed/15661882

[46] Monteiro JP, Benjamin A, Costa ES, Barcinski MA, Bonomo A. Normal hematopoiesis is maintained by activated bone marrow CD4+T cells. Blood [Internet]. 2005 Feb 15 [cited 2013 Dec 15];105(4):1484–91. Available from: http://www.ncbi.nlm.nih.gov/pubmed/15514013

[47] Tokoyoda K, Zehentmeier S, Hegazy AN, Albrecht I, Grün JR, Löhning M, et al. Professional memory CD4+T lymphocytes preferentially reside and rest in the bone marrow. Immunity [Internet]. 2009 May [cited 2014 May 24];30(5):721–30. Available from: http://www.ncbi.nlm.nih.gov/pubmed/19427242

[48] Feuerer M, Beckhove P, Garbi N, Mahnke Y, Limmer A, Hommel M, et al. Bone marrow as a priming site for T-cell responses to blood-borne antigen. Nat Med [Internet]. 2003 Sep [cited 2014 Jun 3];9(9):1151–7. Available from: http://www.ncbi.nlm.nih.gov/pubmed/12910264

[49] Milo I, Sapoznikov A, Kalchenko V, Tal O, Krauthgamer R, van Rooijen N, et al. Dynamic imaging reveals promiscuous crosspresentation of blood-borne antigens to naive CD8+T cells in the bone marrow. Blood [Internet]. 2013 Jul 11 [cited 2014 Jun 3];122(2):193–208. Available from: http://www.ncbi.nlm.nih.gov/pubmed/23637125

[50] Sipkins DA, Wei X, Wu JW, Runnels JM, Côté D, Means TK, et al. In vivo imaging of specialized bone marrow endothelial microdomains for tumour engraftment. Nature [Internet]. 2005 Jun 16 [cited 2014 May 24];435(7044):969–73. Available from: http://www.pubmedcentral.nih.gov/articlerender.fcgi?artid=2570168&tool=pmcentrez&rendertype=abstract

[51] Ieki R, Furuta T, Asano S, Mori S, Kudoh S, Kimura H, et al. Effect of recombinant human granulocyte colony-stimulating factor on Pneumocystis carinii infection in nude mice. Jpn J Exp Med [Internet]. 1989 Apr [cited 2014 Jul 5];59(2):51–8. Available from: http://www.ncbi.nlm.nih.gov/pubmed/2787431

[52] Hamood M, Chatelain C, Fondu P, Symann M. In vivo elaboration of CSF in acute inflammation: proportionality to the intensity of the inflammatory stimulus and requirement of T lymphocytes. Eur J Haematol [Internet]. 1990 Nov [cited 2014 Jul 5];45(5):244–9. Available from: http://www.ncbi.nlm.nih.gov/pubmed/2261949

[53] Magee DM, Williams DM, Wing EJ, Bleicker CA, Schachter J. Production of colony-stimulating factors during pneumonia caused by Chlamydia trachomatis. Infect Immun [Internet]. 1991 Jul [cited 2014 Jul 5];59(7):2370–5. Available from: http://

www.pubmedcentral.nih.gov/articlerender.fcgi?artid=258020&tool=pmcentrez&rendertype=abstract

[54] Trainin N, Resnitzky P. Influence of neonatal thymectomy on cloning capacity of bone marrow cells in mice. Nature [Internet]. 1969 Mar 22 [cited 2014 Jul 5];221(5186): 1154–5. Available from: http://www.ncbi.nlm.nih.gov/pubmed/4975282

[55] Resnitzky P, Zipori D, Trainin N. Effect of neonatal thymectomy on hemopoietic tissue in mice. Blood [Internet]. 1971 Jun [cited 2014 Jul 5];37(6):634–46. Available from: http://www.ncbi.nlm.nih.gov/pubmed/5162309

[56] Lord BI, Schofield R. The influence of thymus cells in hemopoiesis: stimulation of hemopoietic stem cells in a syngeneic, in vivo, situation. Blood [Internet]. 1973 Sep [cited 2014 Jul 5];42(3):395–404. Available from: http://www.ncbi.nlm.nih.gov/pubmed/4725601

[57] Bonomo AC, el-Cheikh MC, Borojevic R, Cavalcante LA, DosReis GA. Comparative analysis of splenic cell proliferation induced by interleukin 3 and by syngeneic accessory cells (syngeneic mixed leukocyte reaction): evidence that autoreactive T-cell functioning instructs hematopoietic phenomena. Cell Immunol [Internet]. 1990 Jan [cited 2014 Jul 5];125(1):210–24. Available from: http://www.ncbi.nlm.nih.gov/pubmed/1967158

[58] Takayanagi H. Osteoimmunology: shared mechanisms and crosstalk between the immune and bone systems. Nat Rev Immunol [Internet]. 2007 Apr [cited 2014 Jun 3]; 7(4):292–304. Available from: http://www.ncbi.nlm.nih.gov/pubmed/17380158

[59] Takayanagi H. Osteoimmunology and the effects of the immune system on bone. Nat Rev Rheumatol [Internet]. 2009 Dec [cited 2014 Jun 4];5(12):667–76. Available from: http://www.ncbi.nlm.nih.gov/pubmed/19884898

[60] Takayanagi H. New developments in osteoimmunology. Nat Rev Rheumatol [Internet]. 2012 Nov [cited 2014 Jul 6];8(11):684–9. Available from: http://www.ncbi.nlm.nih.gov/pubmed/23070645

[61] Danks L, Takayanagi H. Immunology and bone. J Biochem [Internet]. 2013 Jul [cited 2014 Jul 6];154(1):29–39. Available from: http://www.ncbi.nlm.nih.gov/pubmed/23750028

[62] Chakravarti A, Raquil M-A, Tessier P, Poubelle PE. Surface RANKL of Toll-like receptor 4-stimulated human neutrophils activates osteoclastic bone resorption. Blood [Internet]. 2009 Aug 20 [cited 2014 May 25];114(8):1633–44. Available from: http://www.ncbi.nlm.nih.gov/pubmed/19546479

[63] D'Amico L, Roato I. Cross-talk between T cells and osteoclasts in bone resorption. Bonekey Rep [Internet]. 2012 Jan [cited 2014 Jul 6];1:82. Available from: http://www.pubmedcentral.nih.gov/articlerender.fcgi?artid=3727728&tool=pmcentrez&rendertype=abstract

[64] Arron JR, Choi Y. Bone versus immune system. Nature [Internet]. 2000 Nov 30 [cited 2014 Jul 6];408(6812):535–6. Available from: http://www.ncbi.nlm.nih.gov/pubmed/11117729

[65] Takayanagi H, Ogasawara K, Hida S, Chiba T, Murata S, Sato K, et al. T-cell-mediated regulation of osteoclastogenesis by signalling cross-talk between RANKL and IFN-gamma. Nature [Internet]. 2000 Nov 30 [cited 2014 Jul 6];408(6812):600–5. Available from: http://www.ncbi.nlm.nih.gov/pubmed/11117749

[66] Pacifici R. The immune system and bone. Arch Biochem Biophys [Internet]. 2010 Nov 1 [cited 2014 Jul 6];503(1):41–53. Available from: http://www.pubmedcentral.nih.gov/articlerender.fcgi?artid=2943054&tool=pmcentrez&rendertype=abstract

[67] Graves DT, Oates T, Garlet GP. Review of osteoimmunology and the host response in endodontic and periodontal lesions. J Oral Microbiol [Internet]. 2011 Jan [cited 2014 Jul 6];3. Available from: http://www.pubmedcentral.nih.gov/articlerender.fcgi?artid=3087239&tool=pmcentrez&rendertype=abstract

[68] Monteiro AC, Leal AC, Gonçalves-Silva T, Mercadante ACT, Kestelman F, Chaves SB, et al. T cells induce pre-metastatic osteolytic disease and help bone metastases establishment in a mouse model of metastatic breast cancer. PLoS One [Internet]. 2013 Jan [cited 2013 Dec 15];8(7):e68171. Available from: http://www.pubmedcentral.nih.gov/articlerender.fcgi?artid=3730734&tool=pmcentrez&rendertype=abstract

[69] Kiesel JR, Buchwald ZS, Aurora R. Cross-presentation by osteoclasts induces FoxP3 in CD8+T cells. J Immunol [Internet]. 2009 May 1 [cited 2014 Jul 6];182(9):5477–87. Available from: http://www.ncbi.nlm.nih.gov/pubmed/19380796

[70] Wang M, Tian T, Yu S, He N, Ma D. Th17 and Treg cells in bone related diseases. Clin Dev Immunol [Internet]. 2013 Jan [cited 2014 Jul 6];2013:203705. Available from: http://www.pubmedcentral.nih.gov/articlerender.fcgi?artid=3800633&tool=pmcentrez&rendertype=abstract

[71] Senthilkumar R, Lee H-W. CD137L-and RANKL-mediated reverse signals inhibit osteoclastogenesis and T lymphocyte proliferation. Immunobiology [Internet]. 2009 Jan [cited 2014 Jul 6];214(2):153–61. Available from: http://www.ncbi.nlm.nih.gov/pubmed/19167994

[72] Weitzmann MN, Pacifici R. T cells: unexpected players in the bone loss induced by estrogen deficiency and in basal bone homeostasis. Ann N Y Acad Sci [Internet]. 2007 Nov [cited 2014 Jun 4];1116:360–75. Available from: http://www.ncbi.nlm.nih.gov/pubmed/18083938

[73] Rivollier A, Mazzorana M, Tebib J, Piperno M, Aitsiselmi T, Rabourdin-Combe C, et al. Immature dendritic cell transdifferentiation into osteoclasts: a novel pathway sustained by the rheumatoid arthritis microenvironment. Blood [Internet]. 2004 Dec 15 [cited 2014 Jul 6];104(13):4029–37. Available from: http://www.ncbi.nlm.nih.gov/pubmed/15308576

[74] Braga WMT, Atanackovic D, Colleoni GWB. The role of regulatory T cells and TH17 cells in multiple myeloma. Clin Dev Immunol [Internet]. 2012 Jan [cited 2014 Jun 27]; 2012:293479. Available from: http://www.pubmedcentral.nih.gov/articlerender.fcgi?artid=3318676&tool=pmcentrez&rendertype=abstract

[75] Choi C, Witzens M, Bucur M, Feuerer M, Sommerfeldt N, Trojan A, et al. Enrichment of functional CD8 memory T cells specific for MUC1 in bone marrow of patients with multiple myeloma. Blood [Internet]. 2005 Mar 1 [cited 2014 Jun 10];105(5):2132–4. Available from: http://www.ncbi.nlm.nih.gov/pubmed/15561890

[76] Mundy GR. Metastasis to bone: causes, consequences and therapeutic opportunities. Nat Rev Cancer [Internet]. 2002 Aug [cited 2013 Dec 15];2(8):584–93. Available from: http://www.ncbi.nlm.nih.gov/pubmed/12154351

[77] Shiozawa Y, Pedersen EA, Havens AM, Jung Y, Mishra A, Joseph J, et al. Human prostate cancer metastases target the hematopoietic stem cell niche to establish footholds in mouse bone marrow. J Clin Invest [Internet]. 2011 Apr [cited 2013 Dec 15]; 121(4):1298–312. Available from: http://www.pubmedcentral.nih.gov/articlerender.fcgi?artid=3069764&tool=pmcentrez&rendertype=abstract

[78] Lee YK, Mazmanian SK. Microbial learning lessons: SFB educate the immune system. Immunity [Internet]. 2014 Apr 17 [cited 2014 May 28];40(4):457–9. Available from: http://www.ncbi.nlm.nih.gov/pubmed/24745329

[79] Sjögren K, Engdahl C, Henning P, Lerner UH, Tremaroli V, Lagerquist MK, et al. The gut microbiota regulates bone mass in mice. J Bone Miner Res [Internet]. 2012 Jun [cited 2014 May 23];27(6):1357–67. Available from: http://www.pubmedcentral.nih.gov/articlerender.fcgi?artid=3415623&tool=pmcentrez&rendertype=abstract

Section 2

Stem Cell Niches and Reproductive Systems

Chapter 5

Male stem Cell Niche and Spermatogenesis in the *Drosophila* testis — A Tale of Germline-Soma Communication

Fani Papagiannouli

Additional information is available at the end of the chapter

http://dx.doi.org/10.5772/58756

1. Introduction

A fundamental question in biology is how communication and exchange of short-range signals shape the microenvironment for setting up functional tissues. In all adult tissues and organs harboring stem cells, tissue homeostasis and repair relies on the proper communication of stem cells and their differentiating daughter cells with the local tissue microenvironment that homes them [1, 2]. Stem cell research has made outstanding contributions on the factors that maintain stem cells or drive them to generate differentiated daughter cells. The use of stem cells in the development of cell-based medicine and in repairing malformed, damaged or aging tissues demands a better understanding of stem cells at a molecular level and of how they behave in their physiological context.

The basic principles controlling stem cell self-renewal versus differentiation are strikingly conserved during evolution and their regulatory logic is often very similar among homologous stem cell niches. Since the signaling pathways and their regulatory circuits are highly complex in the mammalian system with significant molecular redundancy, they are often difficult to study. Therefore, using a simpler model system such as the *Drosophila* testis allows us to elucidate the underlying cellular and molecular mechanisms of stem cell maintenance and differentiation in a straightforward way.

The *Drosophila* testis provides an excellent system to study *in vivo* how two closely apposed cell types communicate and coordinate their reciprocal interaction. Recent advances in spermatogenesis have shown that testis morphogenesis is achieved through the physical contact and diffusible signals exchanged between the germline and the somatic cell populations [3]. Moreover, the *Drosophila* testis provides a powerful system to study germline-soma communication as it is possible to identify the different cell populations with specific markers, study them within the context of their wild type surrounding and trace them after genetic manipulations [2, 4]. Although several signaling molecules, cytoskeletal and other factors have been so far identified, many aspects of the coordination of these events remain unsolved. Using well-established genetic tools, cell-type specific markers and imaging techniques we can manipulate cell function in a spatio-temporal specific way within the germline-soma micro-environment and decode how signal transmission and polarity are established, maintained and coordinated on the mechanistic level. Therefore, elucidating the mechanisms and factors that regulate these processes is crucial for understanding cell communication and coordination *per se,* which is a prerequisite for the therapeutic applications in other stem cell systems and in various tissue contexts.

The proposed chapter gives an overview of the *Drosophila* male stem cell niche and its importance as a model system for understanding stem cell function. The chapter starts with an introduction to the system, focusing on the importance of soma-germline communication, mutual coordination and progressive co-differentiation. As next, follows the role of the stem cell niche and signaling pathways in balancing stem cell maintenance and differentiation. The specification and positioning of the stem cell niche is discussed, in view of recent data in the field, which put the way we understand stem cell niche establishment and maintenance into a new perspective. Finally, the role of septate junctions and cortical polarity components in the somatic lineage is presented, together with open questions and challenges of the current research in the field.

2. The *Drosophila* testis

Organogenesis of the *Drosophila* testis, a structure first made by the coalesce of germ cells and somatic gonadal cells in late embryogenesis, proceeds continuously throughout embryonic and larval stages, to reach maturation in adult stages. The embryonic gonad results from the coalescence of the germ cells that completed migration and the somatic gonadal precursors (SGPs). SPGs are mesodermal cells specified in bilateral clusters within the *eve* domain of abdominal parasegments [5] 10 to 13 [6-9]. The development of male and female gonads already differs at the time of gonad coalescence. In the male gonads three SGP populations are identifiable by their different gene expression: the posterior-SGPs, the posterior male-specific SGPs which die by apoptosis in females [6] and the anterior-SGPs which will give rise to the hub, the core of the testicular niche which will recruit and organize the anterior-most germ cells to become germline stem cells (GSCs) [10]. Therefore, it becomes evident that the different SGP populations joining the male gonad orchestrate testis morphogenesis since the germ cells represent a uniform population at that time. The SGPs are specified initially through the function of Zinc-finger homeodomain protein 1 (Zfh-1) within the cluster of the lateral mesoderm (PS2-14) which work together the homeobox protein Tinman to promote germ cell migration to the lateral mesoderm. Subsequently, Zfh-1 restriction in PS10-13 correlates with the specification of these cells as SPGs.

The first signs of testis organogenesis are detected in 1st instar larvae (L1) and a testis with a mature stem cell niche and all premeiotic stages is detected at 3rd instar larvae (L3). The *Drosophila* testis contains two types of stem cells: the germline stem cells (GSCs) and the somatic cyst stem cells (CySCs). Each GSC is surrounded by two somatic cyst stem cells (CySCs) and both types of stem cells are maintained through their association to the hub cells, a cluster of non-dividing cells forming the niche organizer. Upon asymmetric cell division, each GSC produces a new GSC attached to the hub and a distally located gonialblast (Gb), whereas each CySC pair divides to generate two CySCs remaining associated with the hub and two distally located post-mitotic daughter somatic cyst cells (SCCs) [1, 11]. Upon asymmetric stem cell division, each GSC produces a new GSC attached to the hub and a distally located gonialblast, whereas each CySC pair divides to generate two CySCs and two somatic cyst cells (SCCs) [1, 12]. GSCs divide asymmetrically with the mitotic spindle orientated perpendicular to the hub [13, 14]. After division the GSC remains in contact with the hub and inherits the mother centriole whereas the gonialblast, inherits the daughter centriole and initiates differentiation [15]. However, upon starvation-or genetically-induced GSC loss, the GSC population can be renewed both by symmetric renewal and de-differentiation of transient amplifying spermatogonia, which repopulate the niche and reestablish contact to the hub [16].The gonialblast divides mitotically four more times to give rise to 16 interconnected spermatogonial cells, forming a cyst surrounded by the two SCCs (Fig.1). As germ cells enter their differentiation program of four transient amplifying divisions followed by pre-meiotic gene expression and meiotic divisions, the SCCs grow enormously in size, elongate and wrap the germ cells creating cysts [17] outside "sealed" by extracellular matrix (ECM) [18]. After the growth phase, the spermatocytes undergo meiosis and differentiate into elongated spermatids.

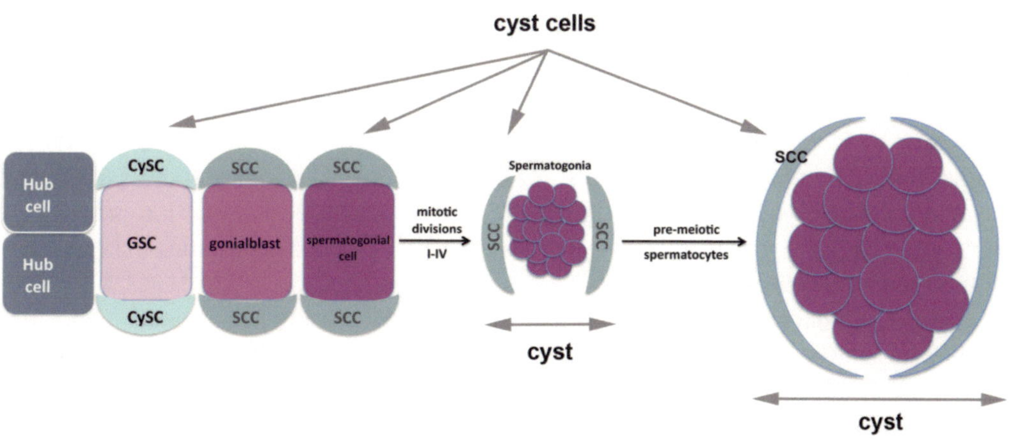

Figure 1. Diagram depicting early spermatogenesis in *Drosophila*. GSC: germline stem cell, CySC: somatic cyst stem cell, SCC: somatic cyst cell. For simplicity reasons CySCs and SCCs are collectively called cyst cells. Testicular cysts comprise of a pair of cyst cells flanking the germline (GSCs, spermatogonia or spermatocytes).

Testis organogenesis is completed during pupal stages. For the formation of a mature testis and a functional reproductive tract, the *Drosophila* testis contacts the seminal vesicle growing out of the genital disc during metamorphosis. The outer sheath of the male reproductive tract develops from two populations of cells: the pigment cells of the testis and the precursors of smooth muscle cells from the genital disc [19]. First, the muscle progenitor cells of the genital disc contact the basal surface of the pigment cells of the testis. Then, migration of muscle and pigment cells proceeds in opposite directions until the gonad and the seminal vesicle have each acquired an inner layer of muscle tissue and an outer layer of pigment cells [19]. It is the addition of the acto-myosin sheath, which gives to the adult testis its characteristic coiled-shape. The pigment cells are responsible for the yellow color of the testis sheath and seminal vesicle [17]. *wnt2*, expressed in the SGPs, is required for the correct development of pigment cells [19], and in *wnt2* mutant embryos pigment cells are not specified and *Sox100B* is not expressed in pigment cell precursors [20, 21].

2.1. Cyst cells: The safeguards of the Germline

Critical for testis differentiation and morphogenesis is the cyst microenvironment created by the cyst cells (CySCs and SCCs) that enclose the germline cells, accompany them throughout their differentiation steps up to sperm individualization and maintain cyst integrity and architecture [22, 23]. Although it is well established that soma-germline physical contact is critical for the cell communication and for promoting their mutual development and differentiation [3], it remains so far elusive how these tightly packed cysts coordinate adhesion and cell shape changes with signaling and membrane addition on a mechanistic level.

The thin and squamous cyst cells lack the columnar epithelial structure of e.g. the ovarian follicular epithelium, which caught the attention of scientists analyzing apico-basal polarity many years ago. For this reason, several questions concerning cyst cell architecture, apical-basal polarity and sub-cellular localization of cytoskeletal proteins such as Dlg, Integrin and Talin remained unclear. Preliminary data show that cyst cells are polarized with an inner-apical surface phasing the germline (Fig. 2E; arrowheads) [22] and an outer-basal surface surrounded by ECM [18]. Critical cytoskeletal and polarity components localize at cyst cells, such as Rho1, Bazooka (Baz), Fasciclin II (FasII), Integrin-linked kinase (ILK), βPS-Integrin (encoded by the *myospheroid* gene) (Fig. 2B-F′), as well as the septate junction proteins Dlg, Scrib and Lgl (Fig.3 A-D). Moreover, cyst cells are able to extend projections in between the germline spermatogonia (small insets of Fig.3 A-C) and spermatocytes (Fig.2 C-C′, E-F′; yellow arrowheads), similar to what was previously observed in the embryonic gonads [24]. On the morphological level, the orientation of the SCCs flanking the germ cells changes in comparison to their mother CySCs via a not yet uncovered mechanism. The two CySCs flanking the same GSC are arranged parallel to the testis anterior-posterior axis (A/P) and attach to the hub whereas their post-mitotic daughter SCCs change their orientation perpendicular to the A/P testis axis (Fig.1). During terminal differentiation, the two cyst cells of the same cyst acquire different identities followed by morphological changes [25]: the forward SCC becomes the "head cyst cell" (HCC) onto which all 64 spermatid heads are anchored shortly after meiosis, and the posterior one becomes the much larger "tail cyst" (TCC) that surrounds the spermatid tails of 1.8 mm length [26]. This results in creating polarized cysts across the testis A/P axis and towards the direction (A→P) of differentiation. The HCC finally is engulfed by cells of the terminal epithelium to allow coiling of the spermatid bundles towards the testis base [27].

So far the main evidence for cyst cell (CySCs and SCCs) function came from the analysis of individual signal transduction pathways that establish a cross talk between the soma and the germline. In this chapter recent findings critically affecting germline-soma communication and coordination will be highlighted, with emphasis on the role of cytoskeletal and scaffolding components such as integrins and adaptor proteins, ECM and the septate junction components. Interestingly, the *Drosophila* testis cyst cells show striking similarities with the Sertoli cells, the supportive cells of the mammalian germline, in terms of cytoskeletal and scaffolding components [2]. Moreover, the genes presented in this study show high degree of conservation to their vertebrate homologues [18, 23]. Accordingly, although we use *Drosophila* spermatogenesis as a model for its powerful genetic tools, accessible imaging and the wealth of underlying prior knowledge on which to built on, the regulatory mechanisms discovered in the *Drosophila* testis provide paradigms for regulatory strategies in spermatogenesis and allow us to discern the complexity of niche and testis homeostasis in other organisms and stem cell systems in other tissues, which will eventually advance the basic knowledge required for stem cell applications.

2.2. Niche Homeostasis: Signaling regulation of stemness vs. differentiation

Tissue specific stem cells are the lifetime source of many types of differentiated cells. They reside in microenvironments, the stem cells niches that have an important role in stem cell behavior [28]. Gamete development requires a coordinated soma-germ line interaction that keeps the balance between germline stem cell renewal and differentiation. The balance between stem cell identity and differentiation at the *Drosophila* testicular niche results from signals exchanged among the hub, GSCs and CySCs. The Janus-kinase transducer and activator of transcription (JAK-STAT) pathway was the first signaling pathway found to regulate GSC and CySC maintenance in the *Drosophila* testis [29, 30]. The hub cells secrete the ligand Unpaired (Upd), which activates the JAK-STAT pathway in adjacent GSCs and CySCs [29-31]. In the absence of JAK-STAT signaling the GSCs differentiate and are unable to self-renew, whereas ectopic expression of *upd* in the germline greatly expands the population of GSCs and CySCs in adult as well as in the larval testis [29, 30]. In GSCs, STAT is required so that E-cadherin (E-cad) maintains the connection of the GSC to the hub and ectopic E-cad partially rescues the maintenance of STAT-depleted GSCs [32]. Another STAT target in GSCs is *chickadde*, the homologue of the *Drosophila* profilin. Chic is required cell autonomously to maintain GSCs by facilitating GSC-hub contact possibly via E-cad whereas Chic in the SCCs is affecting germ cell enclosure and restricting trans-amplifying (TA) spermatogonial divisions [33]. When GSCs divide, their daughter cells displaced from the hub are thought to receive lower levels of hub-derived signals and therefore differentiate. In CySCs, STAT is critical for maintaining their stem cell character and the activation of targets essential for their identity such as *zfh-1* and *chinmo* [32, 34]. *zfh-1* is expressed predominantly in CySCs and their immediate SCCs, and ectopic expression in late SCCs outside the niche leads in accumulation of GSC-and CySCs-like cells which fill in the whole testis. Similarly, *chinmo* is expressed in comparable levels in CySCs and early SCCs, is required for CySCs and not GSC renewal, and ectopic expression causes accumulation of GSCs-and CySCs-like cells. Furthermore, *zfh-1* and *chinmo* are not expressed in GSCs meaning that STAT can activate distinct downstream cascades in the GSC vs. CySCs. *ken* and *barbie* (*ken*) is another gene necessary and sufficient to promote CySC identity, yet in a STAT independent manner and with similar ectopic phenotypes like *zfh-1* and *chinmo* [35]. At the same time, Suppressor of cytokine signaling 36E (Socs36E) suppresses Jak-Stat signaling in the CySCs preventing them from outcompeting the GSCs and thereby maintains the proper balance of GSCs and CySCs, in a manner that depends on the adhesion protein integrin [36].

Interestingly, very recent findings revealed that the Hedgehog (Hh) ligand secreted from the hub cells activates the Hh signaling in CySCs (and not in the GSCs) with critical function in CySC maintenance [37-40]. Hh overexpression leads in increased number of CySCs, identified as Zfh-1 positive cyst cells outside the niche, which can still proliferate in contrast to the normal post-mitotic SCCs. Furthermore, rescue of STAT depleted testis by Hh signaling activation in the CySCs can rescue the CySCs but GSC and germline maintenance is still impaired, as these Zfh-1 positive CySCs are not able to induce the GSC over-proliferation phenotype observed in SCCs ectopic Zfh-1 activation [38]. This suggests that [1] *zfh-1* expression relies on inputs from both Hh and JAK-STAT signaling pathways and that [2] apart from Zfh-1 other STAT regulated factors are necessary for allowing the CySC-to-GSC communication, which promotes GSC maintenance.

Notably, BMP seems to be the primary pathway leading to GSC self-renewal in the *Drosophila* testis [41-44]. BMP ligands and the BMP modulator *magu*, are expressed in the hub and CySCs that serve as the GSC niche and their loss results in reduced GSC numbers and *bam* derepression, whereas the hub and CySCs remain unaffected [42-44]. This could also suggest that expansion of GSC population by the JAK-STAT signaling could be due to its activation in the CySCs that consequently leads to enhanced expression of BMP ligands from CySCs [32] that finally drive GSC expansion. The BMP pathway is also negatively regulated in the course of testis morphogenesis along embryonic-larval-adult stages via Smurf (SMAD ubiquitination regulatory factor) [45]. High BMP levels are required at the initial steps of niche establishment when the hub cells attract the nearby germ cells to become GSCs in late embryogenesis up to early 3rd instar larval stages. Apparently, BMP signaling is spatially and temporally downregulated in stem cells and early germline cells in late 3rd instar larval and pupal testes through Smurf proteolytic activity. The described BMP downregulation seems to be critical for the normal decrease in stem cell number during pupal development, for restricting TA spermatogonia proliferation and control of the testis size. This dynamic regulation indicates the requirement for fine trimming the BMP signaling intensity during subsequent developmental stages and might even suggest a difference between establishment vs. maintenance of certain cell populations across different stages. Yet, another recent story revealed that GSC characteristics can be maintained over time even after ablating the CySC and SCCs [46]. Without CySCs and SCCs, early germ cells away from the hub failed to initiate differentiation and maintained their GSC-like characteristics. Therefore, it becomes evident that the interactions between different stem cell populations and how one stem cell population influences the other can be indeed very complex. Finally, antagonistic functions between the *Drosophila* β-catenin Armadillo (Arm) and the microRNAs-(miR-) 310-313 suggest that modulation of the Wingless signaling activity is important to buffer germ cell and somatic differentiation in the *Drosophila* testis [47].

Critical for germ cell differentiation is the expression of *bag of marbles (bam)* and *benign gonial cell neoplasm (bgcn)* in dividing spermatogonial cells in order to regulate their proliferation [48]. *bam* transcription is negatively regulated by the cooperation of the Glass bottom boat (Gbb) and Decapentaplegic (Dpp) signaling pathways emanating from the hub and CySCs to maintain the GSC identity [42]. Bam is required cell autonomously in TA spermatogonia to stop proliferation and enter the spermatocyte differentiation program [49]. The switch from TA proliferation to differentiation is mediated by translational control: Mei-P26 facilitates the accumulation of Bam in TA cells whereas Bam and Bcgn bind *mei-P26* 3′ untranslated region and repress translation of mei-P26 in late TA cells. Thus, germ cells progress through subsequent regulatory states that is: from a Mei-P26 on/Bam off to a Bam on/Mei-P26 off state.

Another signaling pathway restricting GSC proliferation is mediated by Epidermal Growth Factor Receptor (EGFR), whose inactivation in SCCs leads to an expansion of male GSCs [50]. In *Drosophila* testis, the major ligand of the EGFR pathway, Spitz (Spi) is secreted from the germline cells to stimulate the EGFR on cyst cells (CySCs and SCCs) [25]. Removal of either *spi* or *stet* from the germline cells, or removal of the EGFR from the cyst cells resulted in increased division frequencies of GSCs but did not affect the division frequencies of CySCs, suggesting that EGF signaling downregulates GSC divisions. Likewise, Raf, an EGFR downstream component, is required in SCCs to limit GSC expansion [51-53]. In testes mutated for the *rhomboid* homologue *stet*, the germ cells fail to associate with SCCs. Furthermore, germ cells recruit CySCs via the ligand Spitz, which binds to EGFR, and acts through the nucleotide exchange factor Vav to regulate the activity of Rac1, a downstream component of the EGFR pathway. Taken together, EGF signaling from the germline cells produces differential Rac-and Rho-activities across the cyst cells that leads to a directional growth of the cyst cells around the germline cells [25]. Finally, Zero population growth (Zpg), the *Drosophila* gap junction Innexin 4, is localized to the spermatogonia surface, primarily on the sides adjacent to SCCs [54] and is required for the survival and differentiation of early germ cells in both sexes [55, 56].

3. The male stem cell niche: Specification and positioning

The somatic cells of the hub form the organizing center, a cluster of non-dividing cells, at the anterior part of the embryonic male gonad originating, as already discussed, from SGPs [10]. However, not only the hub but also the cyst cells are specified from the SGPs and the common origin between hub and CySCs has been shown by lineage tracing experiments [57]. This is further supported by the fact that both cell types can be traced using the same cell markers such as Zfh-1 and Traffic Jam (TJ) [25]. Hub cell fate vs. cyst cell fate is specified prior to gonad coalesce in a subset of somatic gonadal precursor cells (SGPs) upon Notch signaling activation [57]. In a next step, the *abdominal A (abd-A)* and *Abdominal B (Abd-B) Hox* genes promote the distinct identities of the SGP clusters: anterior SGP identity (PS10-11) is specified by *Abd-A* and repressed by *Abd-B*, a combination of *Abd-A* and *Abd-B* specifies the posterior SPGs (PS12) and *Abd-B* alone specifies the male-specific [58] SPGs (PS13) [9, 10, 20, 59]. Thus, *Abd-A* and *Abd-B* pattern the A/P axis of the formed gonad. Moreover, *Abd-B* can control the correct hub positioning by upregulating the tyrosine-kinase *sevenless (sev)* in the ms-SGPs. Sev is activated by the Boss ligand emanating from the primordial germ cells to represses ectopic hub differentiation [60] whereas the Epidermal growth factor receptor (EGFR) signaling represses hub formation in the rest of the SGPs [61]. Specification of CySCs vs. hub cell fate relies as well on the antagonistic function of *lines (lin)* and *brother of odd with entrails limited (bowl)*. Bowl is a zinc finger transcription factor required in the hub cells and its antagonist Lin is a cytoplasmic protein with catalytic activity whereas Drumstick (Drm) competes with Lin for binding to Bowl [25, 62]. This regulatory network was supported by analysis of mutant phenotypes: *bowl* mutant gonads had fewer hub cells, *lines* mutant gonads had increased number of hub cells, whereas *lines* depleted CySCs acquired some hub-like properties and markers [57]. Once specified, the hub cells are able to recruit the anterior-most germ cells to become the germline stem cells (GSCs) [63], giving rise to the male stem cell niche [64].

We have discussed how the posteriorly expressed Hox genes *AbdA* and *AbdB* promote the distinct identities of the SGP clusters in the embryonic male gonad and how the diffusible signals and physical contact of germ and somatic cells keep the balance between stem cell renewal and differentiation in the larval and adult testis. However, it is interesting to understand how the male stem cell niche is maintained from its initial specification up to the adult stages and how this morphogenetic process is coordinated. In order to ensure normal niche function in the *Drosophila* testis, the hub cells not only need to be properly specified but also need to be correctly placed. Integrin-mediated adhesion is important for maintaining the correct position of the embryonic hub cells during gonad morphogenesis. In the absence of integrin-mediated adhesion, the hub cells still form a cluster, but instead of remaining at the anterior part of the gonad they migrate to the middle part of the developing gonad [65]. Disruption of integrin-mediated adhesion in adult testis by knocking down *talin/rhea,* an integrin-binding and essential focal adhesion protein of the Integrin-cytoskeleton link [66, 67], results in GSC loss and gradual hub disappearance, a phenotype, which becomes more severe as adult males age [67]. As in *talin*-depleted adult testis the hub is progressively lost, the signals that normally emanate from the hub to instruct stem cell renewal are absent, driving the balance between stem cell maintenance and differentiation towards more differentiation and progressive stem cell loss [65]. A similar hub displacement phenotype is observed by depleting adult testis of Lasp [68], an actin-binding protein. From the vertebrate system we know that Lasp interacts genetically with Integrin [69] and in blood platelets Lasp requires Integrin for its proper localization to the cytoskeleton [70]. Moreover, expression levels of Integrin and Talin are critical for occupation of the niche as CySCs with enhanced integrin-mediated adhesion are able to compete and displace their neighboring GSCs [36].

http://dx.doi.org/10.5772/58756

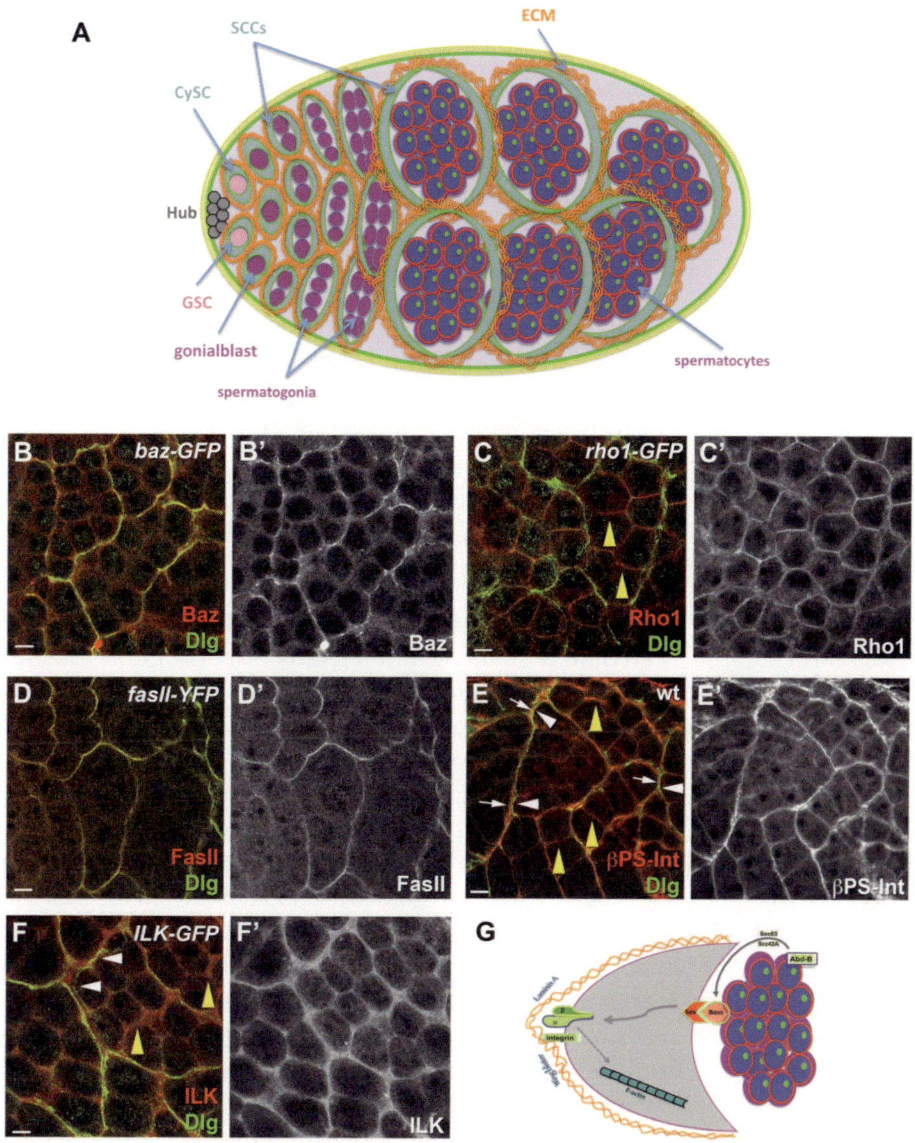

Figure 2. Somatic cyst cells are thin, elongated cells with apical and basal surfaces surrounded by ECM. (A) Schematic diagram of early *Drosophila* spermatogenesis. Somatic cyst cells (SCC) are thin, squamous cells and wrap the germline creating cysts surrounded by ECM (orange). (B-F') Components of cyst cells (red) co-stained with Dlg (green). Here, only the spermatocyte region is shown. Baz (B, B') and FasII (D, D') co-localize with Dlg. Rho1 and Dlg decorate the SCCs but do not co-localize (C, C'). In SCCs Dlg and Integrin are not co-localizing, with Dlg being apical (inner side; white arrowheads) and PS-Integrin more basal (facing outside; white arrows) (E, E'). ILK decorates the SCC cytoplasm and Dlg decorates SCCs facing the germline (white arrowheads) (F, F'). Yellow arrowheads in (C), (E) and (F) show SCC cellular projections growing in between the germ cells. Testes are oriented anterior left. Scale Bar: 10 mm. (G) Schematic diagram depicting a close up of a spermatocyte cyst with key players involved in niche positioning (for simplicity only one somatic cyst cell is shown). Within the spermatocytes, red line indicates the spermatocyte nuclear membrane, green dots illustrate Abd-B in the nucleolus and blue represents the nucleoplasm.

3.1. Some function, different mechanisms: How the Boss/Sev-AbdB cross-talk regulates niche positioning and integrity

As already mentioned, the Boss/Sev signaling pathway plays an important role in hub positioning in the *Drosophila* embryonic male gonads by preventing ectopic niche differentiation in the posterior gonadal somatic cells. *Abd-B*, upstream of this cascade, activates *sev* in the posterior SGPs [60] and consistent with the fact that weak *Abd-B* mutant alleles result in hub expansion and integrity defects in embryonic gonads [10]. A very recent study revealed a new role for the posterior *Hox* gene *Abd-B* in the larval and adult testis. Analysis of the role of the Hox protein Abd-B in the *Drosophila* testis revealed that Abd-B present in the germline spermatocytes acts upstream of the Boss/Sev pathway to regulate hub positioning and integrity, which finally leads to loss of Integrin and Actin localization in the neighboring cyst cells [18]. Analysis of the genetic interactions of *Abd-B* with *integrin* and focal adhesion proteins, revealed that male stem cell niche positioning is regulated by a number of factors, which link Integrin to the extracellular matrix (ECM) and actin filaments. Interestingly, the incorrect placement of the niche in *Abd-B* depleted testes, results in cell non-autonomous centrosome mispositioning and reduced GSC divisions, leading to a dramatic reduction of the pre-meiotic stages of the adult testis, a hallmark of aging in testis [14, 71].

Taken together these studies show that the same players, AbdB, Boss, Sev and Integrin, are used in larval stages to preserve hub positioning and integrity after the initial establishment at embryonic stages but using a slightly variable mechanism: (a) In embryonic gonads, *Abd-B* from the male-specific SGPs regulates *sev* expression in the same cells, whereas Boss signals from the germ cells signals to the Sev expressing cells to ensure that the niche develops in the anterior region of the gonad [60]. Integrin is also required in the somatic cells of the embryonic gonads for anterior positioning of the hub [65]. (b) In larval testes, Abd-B regulates the same process from the germline spermatocytes and via the Boss/Sev pathway controls integrin localization in the neighboring SCCs. This expression switch of Abd-B from the somatic to the germline lineage not only highlights that the mechanism of Abd-B dependent hub positioning is different between embryonic and larval stages but also raises the interesting questions of why and how Abd-B changes its expression and thus the mechanism of hub positioning. During adult stages when testis morphogenesis is completed with the addition of the acto-myosin sheath originating from the genital disc [19], hub positioning and integrity is regulated by Sev, Boss [60] and Integrin [68] whereas Abd-B regulates hub positioning from a different cell type in comparison to embryonic and larval stages which is this time the cells of the acto-myosin sheath, originating from the genital disc. It seems that the occurrence of new cell types and cell interactions in the course of testis organogenesis made it necessary to adapt the whole stem cell system to the new cellular conditions by reusing the same main players of niche positioning in an alternative manner.

3.1.1. Boss mediates, in a Dynamin-and Src-dependent way, germline-soma signaling in larval testis

Drosophila Boss is an atypical G-protein coupled receptor membrane protein that was first identified as a ligand of the Sevenless (Sev) tyrosine kinase involved in eye differentiation. Previous studies in the eye showed that upon binding of the transmembrane protein Boss to its receptor Sev, Boss becomes internalized in the *sev*-expressing cell (Cagan et al., 1992; Kramer, 1993; Kramer et al., 1991] whereas in the fat body, in response to stimulation by glucose, Boss becomes enclosed in internalized vesicles (Kohyama-Koganeya et al., 2008]. In the *Drosophila* testis, Boss is found in the germline spermatocytes, primarily in vesicles (Fig. 3G), whereas Sev localizes in the cyst cells enclosing them. Abd-B performs its function by affecting Boss internalization in the germline, as Boss is lost from internalized vesicles in *Abd-B* depleted testes [18]. Expression of activated Sev in cyst cells of Abd-B depleted testes could fully rescue the phenotype, meaning the Boss exerts its function via Sev activation. Similarly, a partial rescue of hub positioning and integrin localization was observed by expressing the *shibire* (*shi*) gene [72, 73], which is critical for the endocytic uptake of receptors from the plasma membrane [74, 75] in spermatocytes of *Abd-B* depleted testes. This further suggested that Boss functions in a dynamin-dependent way for its endocytic recycling.

In order to elucidate how the Hox transcription factor Abd-B affects Boss localization, genes directly regulated by Abd-B in the *Drosophila* testis were identified by mapping Abd-B binding sites *in vivo* using the DNA adenine methyltransferase identification (DamID) technology [76-79]. This analysis resulted in the identification of 1804 Abd-B binding regions in larval testes, which are associated with 2771 genes. To determine over-representation of GO terms, GO terms were grouped using their annotated Biological Process and subsequently the over-representation of GO term groups among the identified genes was analyzed [18]. Since Abd-B controls signaling between the germline and somatic lineage by regulating genes required for Boss receptor recycling or trafficking, further analysis focused on genes involved in trafficking processes. Two genes, one encoding the non-receptor tyrosine kinase Src oncogene at 42A (Src42A) and another one encoding the putative signal recognition binding protein Sec63, were identified as potential mediators of Boss function in the larval testis. In support of a direct regulatory interaction between *src42A* and Abd-B in the larval testis, *src42A* mRNA levels [80] were found to be significantly downregulated in spermatocytes of $AbdB^{RNAi}$*::T100* animals (with *T100-GAL4* driving expression of UAS-$AbdB^{RNAi}$ in germline spermatocytes), and likewise the activity of the protein tyrosine kinase Src42A was dramatically reduced [18]. Importantly, functional analysis revealed that *src42A* depleted testes mimic the loss of *Abd-B* function: in contrast to wild-type testes, Boss protein was not detected in vesicles, the hub was mispositioned and βPS-integrin was not properly localized in somatic cyst cells of *src42A* depleted testes. Same results were obtained for *sec63*.

4. Dlg, Scrib & Lgl: New functions in the *Drosophila* testis

The *discs large (dlg), scribble (scrib) and lethal* [2] *giant larvae (lgl)* genes were initially identified in *Drosophila* as tumor suppressor genes (TSGs) whose mutations lead to neoplastic transformation, such as imaginal disc overgrowth and brain tumors [81-84]. Mutant flies die after an extended larval life as "giant" larvae without pupariation. In these tumors the overproliferating cells lose their typical epithelial apico-basal polarity, fail to organize an epithelial monolayer and terminally differentiate [84-86]. Therefore, all three TSGs are additionally classified as "cell polarity genes" [83, 84, 87-89]. Since their initial discovery, *dlg, scrib* and *lgl* have been recognized as having important roles also in other forms of polarity as well as in regulation of the actin cytoskeleton, cell signaling and vesicular trafficking [86, 90].

Dlg belongs to the MAGUK (membrane-associated guanylate kinases) protein family, a class of scaffolding proteins that recruit signaling molecules into localized multimolecular complexes [83, 91]. Dlg localizes at the cytoplasmic side of septate junctions between adjacent epithelial cells (the equivalent of vertebrate tight junctions), as well as in neuromuscular junctions (NMJs). It contains three PDZ domains involved in protein-protein interactions with membrane or cytoskeletal proteins, an SH3 domain and a GUK domain. Scrib is also a septate junctional protein of the LAP protein family, containing four PDZ domains and leucine-rich repeats (LRRs) [85, 87, 91, 92]. Lgl is a cytosolic protein containing two WD40 motifs, involved in protein-protein interactions [87]. Lgl can bind to non-muscle myosin II and to the cytoskeleton matrix, along the baso-lateral portion of the plasma membrane of epithelial cells to affect cell polarization [93]. All three proteins, often referred to as the Dlg-polarity module, are highly conserved in sequence among different species and growing evidence suggests that they are functionally conserved to a large degree since the vertebrate homologues can rescue the polarity defects and tumorous overgrowth of the respective *Drosophila* mutants [94-96].

4.1. Dlg, Scrib & Lgl: Multitasking proteins in common pathways in various tissues

Research over several years, defined *dlg, scrib* and *lgl* as key players in numerous tissues contents and malignancies at different time points throughout development, and revealed their multitasking role in: polarity and septate junction establishment; nervous system and brain development; organ development; cancer initiation, progression and metastasis; and mechanism of cooperation with various signaling pathways (Ras, Salvador-Warts-Hippo, Dpp, JNK, Wg, EGFR etc) [22, 97-104]. Some of their common modes of action across different tissues and organisms are analyzed below.

4.1.1. Polarity establishment in various cellular contexts

The Dlg polarity module works in cooperation with the Crumbs-(Crb, Pals1 & Patj) and the Par-(Bazooka/Par3, Par6, αPKC) polarity complexes to control polarity in several tissues. In epithelial cells, polarity is established in a finely balanced process involving cooperative and antagonistic interactions among the apical Par-and Crumbs-complexes and the basolateral Dlg-complex, which restrict the activity of each complex to its specific membrane domain [85, 86]. In neuroblast asymmetric cell division Dlg, Scrib and Lgl cooperate with the Par and Inscutable-Pins complexes whereas microtubules induce Pins & Gαi cortical polarity through Dlg and Khc-73 interactions [86, 105, 106]. In the *Drosophila* ectoderm, phosphorylation of αPKC is required for Lgl to establish the lateral domain and to prevent apical Lgl recruitment. Lgl homologues genetically interact with Par components to regulate apicobasal polarity in *Xenopus* and MDCK epithelial cells, and in partitioning cell fate determinants in *C.elegans* [85, 90, 91, 107]. Finally, the Dlg polarity module has critical functions also in *Drosophila* dorsal closure formation, in patterning anterior and posterior follicle cells, in wound healing processes, in planar cell polarity, in formation of synapses and in NMJs together with other polarity, scaffolding and receptor complexes [86, 102, 108].

4.1.2. Vesicle and membrane trafficking

Several pieces of evidence suggest that Dlg, Scrib and Lgl are involved in vesicle and membrane trafficking [86, 102]: i) Dlg and Strabismus (VanGogh) form a complex that allows membrane deposition during cellularization in *Drosophila* embryos [109] ii) Dlg regulates membrane proliferation of the subsynaptic reticulum (SSR) in NMJs by binding the t-SNARE protein Gtaxin [110, 111] iii) Dlg and Lgl genetically interact with Exo84 which is required for membrane addition [112] iv) the yeast Lgl homologues Sro7p and Sro77p interact directly with Exo84p and Sec9p traggicking components [113], v) mammalian Lgl binds Syntaxin-4 (t-SNARE) to direct protein trafficking [114], and vi) mammalian Scrib regulates exocytosis by binding to the β–Pix-GIT1 complex [115].

4.1.3. Gene regulation and signaling output

Recent studies associate Dlg, Scrib and Lgl with transcriptional response and signaling output since they can regulate the shuttling of critical components between junctional complexes and the nucleus. Such a shuttling mechanism has been described for the Dlg and Scrib vertebrate homologues [116, 117]. In *Drosophila* salivary glands, Lgl together with non-muscle myosin regulate in the cytoplasm access to chromatin modifiers, remodeling and transcription factors necessary for salivary gland degeneration [118]. In wild type salivary glands, chromatin remodeling factors are localized in the nucleus to bind chromatin whereas in the absence of Lgl they accumulate in the cytoplasm and the cortical nuclear zone but cannot bind to chromatin to regulate secondary gene expression [118].

Taken together, Dlg, Scrib and Lgl emerge as dynamic cytoskeletal components which affect polarity, cell structure and behavior by directing the trafficking of proteins to proper plasma membrane surfaces of the cell, and by organizing and stabilizing supramolecular adhesion and signaling complexes through their action as scaffolding adaptor molecules [83-86, 89-91, 109, 111].

4.2. Dlg, Scrib & Lgl in testis somatic cells promote cyst cell function & testis homeostasis

Septate junctions are primary candidates for cyst integrity and coordination, as apart from acting as sealing junctions in epithelia and neurons by mediating cell-cell adhesion, they act as scaffolding networks together with multiple pathways to promote organ morphogenesis [120]. Although the function of Dlg, Scrib and Lgl as TSGs has been intensively studied, their role in testis development has been largely overlooked, as mutations in their coding genes do not result in testis tumors. Moreover, the fact that testes lack an easy to study columnar epithelium, which facilitates analysis of apicobasal polarity genes, didn't favor the analysis of these genes in this stem cell system for many years. The last years a number of studies addressed the role of *scrib*, *dlg* and *lgl* scaffolding proteins in the *Drosophila* male gonad, testis architecture and homeostasis [22-24, 119, 121]. Prompted by the observation that the septate junction protein Scrib [122] is expressed in the newly formed embryonic *Drosophila* gonads [88], Scrib dynamics in the embryonic gonads was investigated [24]. During gonad formation Scrib forms a polygonal network around the germ cells and is present primarily in the somatic gonadal cells, the so-called gonadal mesoderm, that surrounds them. Scrib synthesis in the gonadal mesoderm is cell autonomous, since analysis of agametic gonads and pseudo-gonads made of aggregated germ cells revealed that Scrib in the germ cells requires a direct contact to the gonadal mesoderm [24].

As Dlg, Scrib and Lgl act cooperatively in several tissue contexts [23, 84], their function during male gonad and testis development was analyzed in a comparable way [22, 119]. This work revealed that cell autonomous *scrib* and *dlg* expression in the gonadal mesoderm affects critically the internal structure of the gonads by establishing the intimate contacts of the germ cells to the gonadal mesoderm [24, 119]. At later stages, *dlg*, *scrib* and *lgl* expression in the hub, CySCs and SCCs (Fig.3 A-C) is indispensable for testis development and homeostasis, as depletion of these genes results in extremely small testes with reduced number of germline stem cells and impaired differentiation (Fig.3 E-H). Moreover, Dlg localization in CySCs establishes a tight connection between GSCs and CySCs, and thereby preserves the niche architecture. In late SCCs *dlg* expression is critical for their survival, growth, expansion and for maintaining the integrity of the cysts [22]. This is supported by the observation that the Eya-positive SCCs present in the wild-type testes (Fig.3I; arrowheads) are lost in *dlg* testes (Fig.3J) and die due to apoptosis [22]. Similar to *dlg*, *lgl* testes also lose Eya-positive SCCs (Fig.3L), whereas in *scrib* testes late SCCs are still present (Fig.3K; arrowheads) but the size of these Eya-positive nuclei and of overall testis size is significantly reduced [119]. In contrast to the overgrowth phenotypes observed in imaginal discs and brain hemispheres, the extensive defects in *dlg*, *scrib* and *lgl* mutant testes underline the importance of the somatic lineage in the establishment of a tight soma-germline adhesion and cyst integrity, which is a prerequisite for a functional male stem cell niche and proper testis differentiation [2, 23, 119].

Another striking finding was the formation of wavy and ruffled plasma membrane upon *dlg* over-expression in somatic cyst cells capping the spermatocyte cysts. Up to now, there is no mechanism describing how cyst cells in *Drosophila* testis grow enormously, elongate and ensheath the germ cells of spermatogonial and spermatocyte cysts or how spermatid differentiation and individualization is guided by the polarized head and tail SCC. From other systems we know that Dlg regulates membrane proliferation in a subset of NMJs in a dose-dependent fashion [123] and is an important player in the process of polarized membrane insertion during cellularization [109, 124-126].

Another way to interpret this result would be to consider that Dlg regulates the intensity of germ cell encapsulation through the Egfr pathway, which is the major signaling pathway active at the microenvironment of the spermatogonial cysts [50, 51]. Membrane ruffling, detected in somatic cells upon *dlg* over-expression, is highly reminiscent of the formation of lammellipodia-like structures, formed upon up-regulation of Rac1 in SCCs [53]. Rac1 is a downstream component of the Egfr pathway and acts antagonistically to Rho in order to regulate germ cell encapsulation; moreover, Rho activation perturbates TJ function in various experimental systems [129]. It has already been shown that Dlg regulates membrane proliferation in a subset of NMJs in a dose-dependent fashion [123] and is an important player in the process of polarized membrane insertion during cellularization [109,

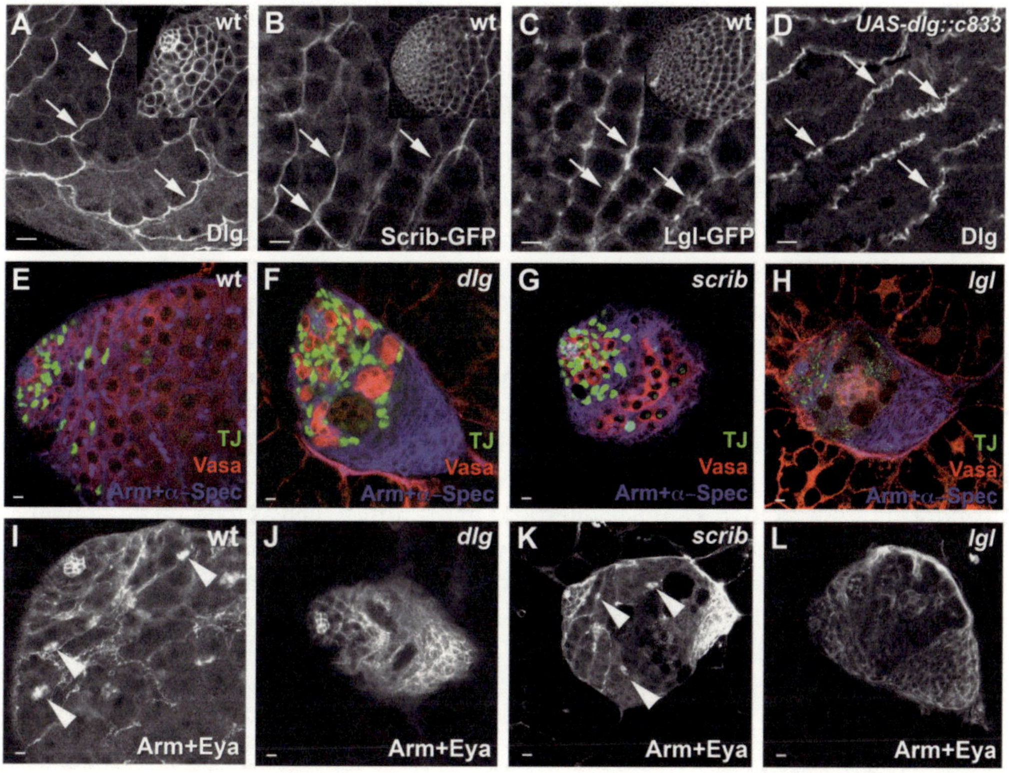

Figure 3. Dlg, Scrib and Lgl in the somatic lineage have critical functions in niche architecture, testis differentiation and homeostasis. (A-C) Dlg, Scrib and Lgl localize in somatic hub, somatic stem and cyst cells in *Drosophila* testis. (D) Dlg overexpression leads to ruffled membranes of somatic cyst cells, showing that Dlg promotes somatic cyst cell growth and membrane addition. (E-H) *dlg, scrib* and *lgl* mutant testes are extremely small, with reduced number of germline stem cells and impaired differentiation with only few spermatogonial cysts. Traffic Jam (TJ) marks the somatic stem cell and cyst cell nuclei, Vasa the germline, Armadillo (Arm) the hub and somatic stem and cyst cells, a-Spectrin the fusome growing through the interconnected spermatogonia and spermatocytes. (I-L) In *dlg* and *lgl* testes late somatic cyst cells are lost as no Eyes Absent (Eya)-positive cyst cells are observed and the tight connection between the cyst cells and the germline is lost. In *scrib* testes Eya-positive somatic cyst cells are present, however testes are small and underdeveloped. Arrows point at the somatic cyst cell membrane. Arrowheads point at Eya-positive late somatic cyst cells. Testis hub is oriented towards the left. Scale Bar: 10mm

124-126]. The fact that membrane proliferation is also involved in mechanisms such as tissue spreading and cell surface extensions, including membrane ruffles [127, 128] and combined with our results on SCCs membrane ruffling upon Dlg overexpression it can be suggested that polarized membrane insertion, mediated by Dlg, might conduct SCCs growth, expansion and spreading over the germ cells of testicular cysts.

5. Conclusions and future perspectives

Cell polarity and signaling are fundamental biological processes that impact stem cell function, cancer, cell migration, tissue morphogenesis and response to pathogenic infections. Growing scientific evidence suggests that these processes are intimately linked. Moreover, shuttling of signaling complexes into specific intracellular regions happens via their recruitment in subcellular domains guided by polarity scaffolds. The microenvironment of the male testis cysts, built by the cyst cell-germline intimate connection, provides an ideal model system to investigate how soma-germline adhesion and cell morphological changes are coordinated with cell communication and exchange of short-range signals.

So far the main evidence for cyst cell (CySCs and SCCs) function came from the analysis of individual signal transduction pathways that establish a cross-talk between the soma and the germline. Now we know that cyst cells are crucially important for soma-germline cyst integrity, overall rigidity and for setting up a functional cyst microenvironment. To this end, it is important (a) to investigate the requirement of the somatic lineage, the cyst cells, as safeguard of germline function, and (b) to characterize the local soma-germline communication within the cysts with focus on how polarity scaffolds and signaling platforms promote this. Resolving the basic features of cyst's microenvironment and soma-germline coordination will allow the study of more complex questions in the future such as long-range signaling at the level of cyst-cyst communication. Moreover, the use of a combination of genetic, genomic and high-resolution microscopy techniques to approach these questions will enable us to adapt tools, already successfully established in other tissues and model systems (such as FRAP, FRET and organ cultures) to the *Drosophila* testis.

Acknowledgements

The author wishes to thank the *Drosophila* community for providing generously fly stocks and antibodies and apologizes to all whose work has not been sited due to space limitations.

Author details

Fani Papagiannouli*

Centre for Organismal Studies (COS), University of Heidelberg, Germany

References

[1] Fuller MT, Spradling AC. Male and female Drosophila germline stem cells: two versions of immortality. Science. 2007;316(5823):402-4.

[2] Papagiannouli F, Lohmann I. Shaping the niche: Lessons from the Drosophila testis and other model systems. Biotechnol J. 2012.

[3] Walker MR, Patel KK, Stappenbeck TS. The stem cell niche. J Pathol. 2009;217(2): 169-80.

[4] Papagiannouli F, Mechler BM. Modeling tumorigenesis in Drosophila: current advances and future perspectives. In: Cheng Y, editor. "Future Aspects of Tumor Suppressor Genes": InTech publications; 2013. p. 97-128.

[5] Lang D, Lu MM, Huang L, Engleka KA, Zhang M, Chu EY, et al. Pax3 functions at a nodal point in melanocyte stem cell differentiation. Nature. 2005;433(7028):884-7.

[6] DeFalco TJ, Verney G, Jenkins AB, McCaffery JM, Russell S, Van Doren M. Sex-specific apoptosis regulates sexual dimorphism in the Drosophila embryonic gonad. Dev Cell. 2003;5(2):205-16.

[7] Riechmann V, Rehorn KP, Reuter R, Leptin M. The genetic control of the distinction between fat body and gonadal mesoderm in Drosophila. Development. 1998;125(4): 713-23.

[8] Moore LA, Broihier HT, Van Doren M, Lehmann R. Gonadal mesoderm and fat body initially follow a common developmental path in Drosophila. Development. 1998;125(5):837-44.

[9] Boyle M, DiNardo S. Specification, migration and assembly of the somatic cells of the Drosophila gonad. Development. 1995;121(6):1815-25.

[10] Le Bras S, Van Doren M. Development of the male germline stem cell niche in Drosophila. Dev Biol. 2006;294(1):92-103.

[11] Yamashita YM, Fuller MT, Jones DL. Signaling in stem cell niches: lessons from the Drosophila germline. J Cell Sci. 2005;118(Pt 4):665-72.

[12] Wong MD, Jin Z, Xie T. Molecular mechanisms of germline stem cell regulation. Annu Rev Genet. 2005;39:173-95.

[13] Yamashita YM, Jones DL, Fuller MT. Orientation of asymmetric stem cell division by the APC tumor suppressor and centrosome. Science. 2003;301(5639):1547-50.

[14] Cheng J, Turkel N, Hemati N, Fuller MT, Hunt AJ, Yamashita YM. Centrosome misorientation reduces stem cell division during ageing. Nature. 2008;456(7222): 599-604.

[15] Yamashita YM, Mahowald AP, Perlin JR, Fuller MT. Asymmetric inheritance of mother versus daughter centrosome in stem cell division. Science. 2007;315(5811): 518-21.

[16] Sheng XR, Matunis E. Live imaging of the Drosophila spermatogonial stem cell niche reveals novel mechanisms regulating germline stem cell output. Development. 2011;138(16):3367-76.

[17] Fuller MT. Spermatogenesis. Arias MBAM, editor. Cold Spring Harbor, New York: Cold Spring Harbor Laboratory Press; 1993. 71-147 p.

[18] Papagiannouli F, Schardt L, Grajcarek J, Ha N, Lohmann I. The hox gene abd-B controls stem cell niche function in the Drosophila testis. Dev Cell. 2014;28(2):189-202.

[19] Kozopas KM, Samos CH, Nusse R. DWnt-2, a Drosophila Wnt gene required for the development of the male reproductive tract, specifies a sexually dimorphic cell fate. Genes Dev. 1998;12(8):1155-65.

[20] DeFalco T, Camara N, Le Bras S, Van Doren M. Nonautonomous sex determination controls sexually dimorphic development of the Drosophila gonad. Dev Cell. 2008;14(2):275-86.

[21] Nanda S, DeFalco TJ, Loh SH, Phochanukul N, Camara N, Van Doren M, et al. Sox100B, a Drosophila group E Sox-domain gene, is required for somatic testis differentiation. Sex Dev. 2009;3(1):26-37.

[22] Papagiannouli F, Mechler BM. discs large regulates somatic cyst cell survival and expansion in Drosophila testis. Cell Res. 2009;19(10):1139-49.

[23] Papagiannouli F, Mechler BM. discs large in the Drosophila testis: An old player on a new task. Fly (Austin). 2010;4(4).

[24] Marhold J, Papagiannouli F, Li M, Patel A, Mechler BM. Requirements for scribble expression in newly formed gonads of Drosophila embryos. Gene Expr Patterns. 2003;3(2):143-6.

[25] Zoller R, Schulz C. The Drosophila cyst stem cell lineage: Partners behind the scenes? Spermatogenesis. 2012;2(3):145-57.

[26] Lindsley DL, Tokuyasu KT. Spermatogenesis. In: Ashburner M, Wright TRF, editors. Genetics and Biology of Drosophila. 2D. 2nd ed. New York: Academic Press; 1980. p. 225-94.

[27] Tokuyasu KT, Peacock WJ, Hardy RW. Dynamics of spermiogenesis in Drosophila melanogaster. II. Coiling process. Z Zellforsch Mikrosk Anat. 1972;127(4):492-525.

[28] Schofield R. The relationship between the spleen colony-forming cell and the haemopoietic stem cell. Blood Cells. 1978;4(1-2):7-25.

[29] Tulina N, Matunis E. Control of stem cell self-renewal in Drosophila spermatogenesis by JAK-STAT signaling. Science. 2001;294(5551):2546-9.

[30] Kiger AA, Jones DL, Schulz C, Rogers MB, Fuller MT. Stem cell self-renewal specified by JAK-STAT activation in response to a support cell cue. Science. 2001;294(5551):2542-5.

[31] Zeng X, Singh SR, Hou D, Hou SX. Tumor suppressors Sav/Scrib and oncogene Ras regulate stem-cell transformation in adult Drosophila malpighian tubules. J Cell Physiol. 2010;224(3):766-74.

[32] Leatherman JL, Dinardo S. Germline self-renewal requires cyst stem cells and stat regulates niche adhesion in Drosophila testes. Nat Cell Biol. 2010;12(8):806-11.

[33] Shields AR, Spence AC, Yamashita YM, Davies EL, Fuller MT. The actin-binding protein profilin is required for germline stem cell maintenance and germ cell enclosure by somatic cyst cells. Development. 2014;141(1):73-82.

[34] Flaherty MS, Salis P, Evans CJ, Ekas LA, Marouf A, Zavadil J, et al. chinmo is a functional effector of the JAK/STAT pathway that regulates eye development, tumor formation, and stem cell self-renewal in Drosophila. Dev Cell. 2010;18(4):556-68.

[35] Issigonis M, Matunis E. The Drosophila BCL6 homolog Ken and Barbie promotes somatic stem cell self-renewal in the testis niche. Dev Biol. 2012;368(2):181-92.

[36] Issigonis M, Tulina N, de Cuevas M, Brawley C, Sandler L, Matunis E. JAK-STAT signal inhibition regulates competition in the Drosophila testis stem cell niche. Science. 2009;326(5949):153-6.

[37] Michel M, Kupinski AP, Raabe I, Bokel C. Hh signalling is essential for somatic stem cell maintenance in the Drosophila testis niche. Development. 2012;139(15):2663-9.

[38] Amoyel M, Sanny J, Burel M, Bach EA. Hedgehog is required for CySC self-renewal but does not contribute to the GSC niche in the Drosophila testis. Development. 2013;140(1):56-65.

[39] Zhang Z, Lv X, Jiang J, Zhang L, Zhao Y. Dual roles of Hh signaling in the regulation of somatic stem cell self-renewal and germline stem cell maintenance in Drosophila testis. Cell Res. 2013;23(4):573-6.

[40] Zhang Z, Pan C, Zhao Y. Hedgehog in the Drosophila testis niche: what does it do there? Protein & cell. 2013.

[41] Shivdasani AA, Ingham PW. Regulation of stem cell maintenance and transit amplifying cell proliferation by tgf-beta signaling in Drosophila spermatogenesis. Curr Biol. 2003;13(23):2065-72.

[42] Kawase E, Wong MD, Ding BC, Xie T. Gbb/Bmp signaling is essential for maintaining germline stem cells and for repressing bam transcription in the Drosophila testis. Development. 2004;131(6):1365-75.

[43] Schulz C, Kiger AA, Tazuke SI, Yamashita YM, Pantalena-Filho LC, Jones DL, et al. A misexpression screen reveals effects of bag-of-marbles and TGF beta class signaling on the Drosophila male germ-line stem cell lineage. Genetics. 2004;167(2):707-23.

[44] Zheng Q, Wang Y, Vargas E, DiNardo S. magu is required for germline stem cell self-renewal through BMP signaling in the Drosophila testis. Dev Biol. 2011;357(1):202-10.

[45] Chang YJ, Pi H, Hsieh CC, Fuller MT. Smurf-mediated differential proteolysis generates dynamic BMP signaling in germline stem cells during Drosophila testis development. Dev Biol. 2013;383(1):106-20.

[46] Lim JG, Fuller MT. Somatic cell lineage is required for differentiation and not maintenance of germline stem cells in Drosophila testes. Proc Natl Acad Sci U S A. 2012;109(45):18477-81.

[47] Pancratov R, Peng F, Smibert P, Yang S, Jr., Olson ER, Guha-Gilford C, et al. The miR-310/13 cluster antagonizes beta-catenin function in the regulation of germ and somatic cell differentiation in the Drosophila testis. Development. 2013;140(14): 2904-16.

[48] Gonczy P, Matunis E, DiNardo S. bag-of-marbles and benign gonial cell neoplasm act in the germline to restrict proliferation during Drosophila spermatogenesis. Development. 1997;124(21):4361-71.

[49] Insco ML, Bailey AS, Kim J, Olivares GH, Wapinski OL, Tam CH, et al. A self-limiting switch based on translational control regulates the transition from proliferation to differentiation in an adult stem cell lineage. Cell Stem Cell. 2012;11(5):689-700.

[50] Kiger AA, White-Cooper H, Fuller MT. Somatic support cells restrict germline stem cell self-renewal and promote differentiation. Nature. 2000;407(6805):750-4.

[51] Tran J, Brenner TJ, DiNardo S. Somatic control over the germline stem cell lineage during Drosophila spermatogenesis. Nature. 2000;407(6805):754-7.

[52] Schulz C, Wood CG, Jones DL, Tazuke SI, Fuller MT. Signaling from germ cells mediated by the rhomboid homolog stet organizes encapsulation by somatic support cells. Development. 2002;129(19):4523-34.

[53] Sarkar A, Parikh N, Hearn SA, Fuller MT, Tazuke SI, Schulz C. Antagonistic roles of Rac and Rho in organizing the germ cell microenvironment. Curr Biol. 2007;17(14): 1253-8.

[54] Tazuke SI, Schulz C, Gilboa L, Fogarty M, Mahowald AP, Guichet A, et al. A germline-specific gap junction protein required for survival of differentiating early germ cells. Development. 2002;129(10):2529-39.

[55] Gilboa L, Forbes A, Tazuke SI, Fuller MT, Lehmann R. Germ line stem cell differentiation in Drosophila requires gap junctions and proceeds via an intermediate state. Development. 2003;130(26):6625-34.

[56] Bohrmann J, Zimmermann J. Gap junctions in the ovary of Drosophila melanogaster: localization of innexins 1, 2, 3 and 4 and evidence for intercellular communication via innexin-2 containing channels. BMC Dev Biol. 2008;8:111.

[57] Dinardo S, Okegbe T, Wingert L, Freilich S, Terry N. lines and bowl affect the specification of cyst stem cells and niche cells in the Drosophila testis. Development. 2011;138(9):1687-96.

[58] Redline RW, Williams AJ, Patterson P, Collins T. Human HOX4E: a gene strongly expressed in the adult male and female urogenital tracts. Genomics. 1992;13(2):425-30.

[59] DeFalco T, Le Bras S, Van Doren M. Abdominal-B is essential for proper sexually dimorphic development of the Drosophila gonad. Mech Dev. 2004;121(11):1323-33.

[60] Kitadate Y, Shigenobu S, Arita K, Kobayashi S. Boss/Sev signaling from germline to soma restricts germline-stem-cell-niche formation in the anterior region of Drosophila male gonads. Dev Cell. 2007;13(1):151-9.

[61] Kitadate Y, Kobayashi S. Notch and Egfr signaling act antagonistically to regulate germ-line stem cell niche formation in Drosophila male embryonic gonads. Proc Natl Acad Sci U S A. 2010;107(32):14241-6.

[62] Hatini V, Green RB, Lengyel JA, Bray SJ, Dinardo S. The Drumstick/Lines/Bowl regulatory pathway links antagonistic Hedgehog and Wingless signaling inputs to epidermal cell differentiation. Genes Dev. 2005;19(6):709-18.

[63] Sheng XR, Posenau T, Gumulak-Smith JJ, Matunis E, Van Doren M, Wawersik M. Jak-STAT regulation of male germline stem cell establishment during Drosophila embryogenesis. Dev Biol. 2009;334(2):335-44.

[64] Casper A, Van Doren M. The control of sexual identity in the Drosophila germline. Development. 2006;133(15):2783-91.

[65] Tanentzapf G, Devenport D, Godt D, Brown NH. Integrin-dependent anchoring of a stem-cell niche. Nat Cell Biol. 2007;9(12):1413-8.

[66] Cram EJ, Schwarzbauer JE. The talin wags the dog: new insights into integrin activation. Trends Cell Biol. 2004;14(2):55-7.

[67] Tanentzapf G, Martin-Bermudo MD, Hicks MS, Brown NH. Multiple factors contribute to integrin-talin interactions in vivo. J Cell Sci. 2006;119(Pt 8):1632-44.

[68] Lee S, Zhou L, Kim J, Kalbfleisch S, Schock F. Lasp anchors the Drosophila male stem cell niche and mediates spermatid individualization. Mech Dev. 2008;125(9-10): 768-76.

[69] Suyama R, Jenny A, Curado S, Pellis-van Berkel W, Ephrussi A. The actin-binding protein Lasp promotes Oskar accumulation at the posterior pole of the Drosophila embryo. Development. 2009;136(1):95-105.

[70] Traenka J, Hauck CR, Lewandrowski U, Sickmann A, Gambaryan S, Thalheimer P, et al. Integrin-dependent translocation of LASP-1 to the cytoskeleton of activated platelets correlates with LASP-1 phosphorylation at tyrosine 171 by Src-kinase. Thromb Haemost. 2009;102(3):520-8.

[71] Boyle M, Wong C, Rocha M, Jones DL. Decline in self-renewal factors contributes to aging of the stem cell niche in the Drosophila testis. Cell Stem Cell. 2007;1(4):470-8.

[72] Kitamoto T. Conditional modification of behavior in Drosophila by targeted expression of a temperature-sensitive shibire allele in defined neurons. J Neurobiol. 2001;47(2):81-92.

[73] Rogat AD, Miller KG. A role for myosin VI in actin dynamics at sites of membrane remodeling during Drosophila spermatogenesis. J Cell Sci. 2002;115(Pt 24):4855-65.

[74] Urrutia R, Henley JR, Cook T, McNiven MA. The dynamins: redundant or distinct functions for an expanding family of related GTPases? Proc Natl Acad Sci U S A. 1997;94(2):377-84.

[75] McPherson PS, Kay BK, Hussain NK. Signaling on the endocytic pathway. Traffic. 2001;2(6):375-84.

[76] van Steensel B, Henikoff S. Identification of in vivo DNA targets of chromatin proteins using tethered dam methyltransferase. Nat Biotechnol. 2000;18(4):424-8.

[77] van Steensel B, Delrow J, Henikoff S. Chromatin profiling using targeted DNA adenine methyltransferase. Nat Genet. 2001;27(3):304-8.

[78] Vogel MJ, Peric-Hupkes D, van Steensel B. Detection of in vivo protein-DNA interactions using DamID in mammalian cells. Nat Protoc. 2007;2(6):1467-78.

[79] Choksi SP, Southall TD, Bossing T, Edoff K, de Wit E, Fischer BE, et al. Prospero acts as a binary switch between self-renewal and differentiation in Drosophila neural stem cells. Dev Cell. 2006;11(6):775-89.

[80] Toledano H, D'Alterio C, Loza-Coll M, Jones DL. Dual fluorescence detection of protein and RNA in Drosophila tissues. Nat Protoc. 2012;7(10):1808-17.

[81] Gateff E SH. Developmental capacities of benign and malignant neoplasms of Drosophila. Wilhelm Roux' Archiv. 1974;176(23-65).

[82] Mechler BM, McGinnis W, Gehring WJ. Molecular cloning of lethal(2)giant larvae, a recessive oncogene of Drosophila melanogaster. EMBO J. 1985;4(6):1551-7.

[83] Woods DF, Hough C, Peel D, Callaini G, Bryant PJ. Dlg protein is required for junction structure, cell polarity, and proliferation control in Drosophila epithelia. J Cell Biol. 1996;134(6):1469-82.

[84] Bilder D, Li M, Perrimon N. Cooperative regulation of cell polarity and growth by Drosophila tumor suppressors. Science. 2000;289(5476):113-6.

[85] Humbert P, Russell S, Richardson H. Dlg, Scribble and Lgl in cell polarity, cell proliferation and cancer. Bioessays. 2003;25(6):542-53.

[86] Elsum I, Yates L, Humbert PO, Richardson HE. The Scribble-Dlg-Lgl polarity module in development and cancer: from flies to man. Essays in biochemistry. 2012;53:141-68.

[87] Wodarz A. Tumor suppressors: linking cell polarity and growth control. Curr Biol. 2000;10(17):R624-6.

[88] Li M, Marhold J, Gatos A, Torok I, Mechler BM. Differential expression of two scribble isoforms during Drosophila embryogenesis. Mech Dev. 2001;108(1-2):185-90.

[89] Goode S, Perrimon N. Inhibition of patterned cell shape change and cell invasion by Discs large during Drosophila oogenesis. Genes Dev. 1997;11(19):2532-44.

[90] Harris BZ, Lim WA. Mechanism and role of PDZ domains in signaling complex assembly. J Cell Sci. 2001;114(Pt 18):3219-31.

[91] Bilder D. PDZ proteins and polarity: functions from the fly. Trends Genet. 2001;17(9): 511-9.

[92] Mathew D, Gramates LS, Packard M, Thomas U, Bilder D, Perrimon N, et al. Recruitment of scribble to the synaptic scaffolding complex requires GUK-holder, a novel DLG binding protein. Curr Biol. 2002;12(7):531-9.

[93] Strand D, Jakobs R, Merdes G, Neumann B, Kalmes A, Heid HW, et al. The Drosophila lethal(2)giant larvae tumor suppressor protein forms homo-oligomers and is associated with nonmuscle myosin II heavy chain. J Cell Biol. 1994;127(5):1361-73.

[94] Thomas U, Phannavong B, Muller B, Garner CC, Gundelfinger ED. Functional expression of rat synapse-associated proteins SAP97 and SAP102 in Drosophila dlg-1 mutants: effects on tumor suppression and synaptic bouton structure. Mech Dev. 1997;62(2):161-74.

[95] Grifoni D, Garoia F, Schimanski CC, Schmitz G, Laurenti E, Galle PR, et al. The human protein Hugl-1 substitutes for Drosophila lethal giant larvae tumour suppressor function in vivo. Oncogene. 2004;23(53):8688-94.

[96] Dow LE, Brumby AM, Muratore R, Coombe ML, Sedelies KA, Trapani JA, et al. hScrib is a functional homologue of the Drosophila tumour suppressor Scribble. Oncogene. 2003;22(58):9225-30.

[97] Moreau MM, Piguel N, Papouin T, Koehl M, Durand CM, Rubio ME, et al. The planar polarity protein Scribble1 is essential for neuronal plasticity and brain function. J Neurosci. 2010;30(29):9738-52.

[98] Brumby AM, Goulding KR, Schlosser T, Loi S, Galea R, Khoo P, et al. Identification of Novel Ras-Cooperating Oncogenes in Drosophila melanogaster: A RhoGEF/Rho-Family/JNK Pathway Is a Central Driver of Tumorigenesis. Genetics. 2011;188(1): 105-25.

[99] Brumby AM, Richardson HE. scribble mutants cooperate with oncogenic Ras or Notch to cause neoplastic overgrowth in Drosophila. EMBO J. 2003;22(21):5769-79.

[100] Brumby AM, Richardson HE. Using Drosophila melanogaster to map human cancer pathways. Nat Rev Cancer. 2005;5(8):626-39.

[101] Cordero JB, Macagno JP, Stefanatos RK, Strathdee KE, Cagan RL, Vidal M. Oncogenic Ras diverts a host TNF tumor suppressor activity into tumor promoter. Dev Cell. 2010;18(6):999-1011.

[102] Humbert PO, Grzeschik NA, Brumby AM, Galea R, Elsum I, Richardson HE. Control of tumourigenesis by the Scribble/Dlg/Lgl polarity module. Oncogene. 2008;27(55): 6888-907.

[103] Pagliarini RA, Xu T. A genetic screen in Drosophila for metastatic behavior. Science. 2003;302(5648):1227-31.

[104] Grzeschik NA, Parsons LM, Allott ML, Harvey KF, Richardson HE. Lgl, aPKC, and Crumbs regulate the Salvador/Warts/Hippo pathway through two distinct mechanisms. Curr Biol. 2010;20(7):573-81.

[105] Wirtz-Peitz F, Knoblich JA. Lethal giant larvae take on a life of their own. Trends Cell Biol. 2006;16(5):234-41.

[106] Albertson R, Doe CQ. Dlg, Scrib and Lgl regulate neuroblast cell size and mitotic spindle asymmetry. Nat Cell Biol. 2003;5(2):166-70.

[107] Humbert PO, Dow LE, Russell SM. The Scribble and Par complexes in polarity and migration: friends or foes? Trends Cell Biol. 2006;16(12):622-30.

[108] Thomas U, Ebitsch S, Gorczyca M, Koh YH, Hough CD, Woods D, et al. Synaptic targeting and localization of discs-large is a stepwise process controlled by different domains of the protein. Curr Biol. 2000;10(18):1108-17.

[109] Lee OK, Frese KK, James JS, Chadda D, Chen ZH, Javier RT, et al. Discs-Large and Strabismus are functionally linked to plasma membrane formation. Nat Cell Biol. 2003;5(11):987-93.

[110] Gorczyca D, Ashley J, Speese S, Gherbesi N, Thomas U, Gundelfinger E, et al. Postsynaptic membrane addition depends on the Discs-Large-interacting t-SNARE Gtaxin. J Neurosci. 2007;27(5):1033-44.

[111] Chen K, Featherstone DE. Discs-large (DLG) is clustered by presynaptic innervation and regulates postsynaptic glutamate receptor subunit composition in Drosophila. BMC Biol. 2005;3:1.

[112] Blankenship JT, Fuller MT, Zallen JA. The Drosophila homolog of the Exo84 exocyst subunit promotes apical epithelial identity. J Cell Sci. 2007;120(Pt 17):3099-110.

[113] Zhang X, Wang P, Gangar A, Zhang J, Brennwald P, TerBush D, et al. Lethal giant larvae proteins interact with the exocyst complex and are involved in polarized exocytosis. J Cell Biol. 2005;170(2):273-83.

[114] Musch A, Cohen D, Yeaman C, Nelson WJ, Rodriguez-Boulan E, Brennwald PJ. Mammalian homolog of Drosophila tumor suppressor lethal (2) giant larvae interacts

with basolateral exocytic machinery in Madin-Darby canine kidney cells. Mol Biol Cell. 2002;13(1):158-68.

[115] Audebert S, Navarro C, Nourry C, Chasserot-Golaz S, Lecine P, Bellaiche Y, et al. Mammalian Scribble forms a tight complex with the betaPIX exchange factor. Curr Biol. 2004;14(11):987-95.

[116] Petit MM, Meulemans SM, Alen P, Ayoubi TA, Jansen E, Van de Ven WJ. The tumor suppressor Scrib interacts with the zyxin-related protein LPP, which shuttles between cell adhesion sites and the nucleus. BMC Cell Biol. 2005;6(1):1.

[117] Carr HS, Cai C, Keinanen K, Frost JA. Interaction of the RhoA exchange factor Net1 with discs large homolog 1 protects it from proteasome-mediated degradation and potentiates Net1 activity. J Biol Chem. 2009;284(36):24269-80.

[118] Farkas R, Kucharova-Mahmood S, Mentelova L, Juda P, Raska I, Mechler B. Cytoskeletal proteins regulate chromatic access of BR-C transcription factor and Rpd3-Sin3A histone deacetylase complex in Drosophila salivary glands. Nucleus. 2011:(in print).

[119] Papagiannouli F. The internal structure of embryonic gonads and testis development in Drosophila melanogaster requires scrib, lgl and dlg activity in the soma. Int J Dev Biol. 2013;57(1):25-34.

[120] Yi P, Johnson AN, Han Z, Wu J, Olson EN. Heterotrimeric G proteins regulate a noncanonical function of septate junction proteins to maintain cardiac integrity in Drosophila. Dev Cell. 2008;15(5):704-13.

[121] Papagiannouli F, Mechler BM. Refining the role of Lgl, Dlg and Scrib in tumor suppression and beyond: Learning from the old time classics. In: Cheng Y, editor. "Tumor Suppressor Genes": InTech publications; 2012. p. 181-220.

[122] Bilder D, Perrimon N. Localization of apical epithelial determinants by the basolateral PDZ protein Scribble. Nature. 2000;403(6770):676-80.

[123] Budnik V, Koh YH, Guan B, Hartmann B, Hough C, Woods D, et al. Regulation of synapse structure and function by the Drosophila tumor suppressor gene dlg. Neuron. 1996;17(4):627-40.

[124] Lecuit T, Wieschaus E. Polarized insertion of new membrane from a cytoplasmic reservoir during cleavage of the Drosophila embryo. J Cell Biol. 2000;150(4):849-60.

[125] Dudu V, Pantazis P, Gonzalez-Gaitan M. Membrane traffic during embryonic development: epithelial formation, cell fate decisions and differentiation. Curr Opin Cell Biol. 2004;16(4):407-14.

[126] Strickland LI, Burgess DR. Pathways for membrane trafficking during cytokinesis. Trends Cell Biol. 2004;14(3):115-8.

[127] Lecuit T, Pilot F. Developmental control of cell morphogenesis: a focus on membrane growth. Nat Cell Biol. 2003;5(2):103-8.

[128] Albertson R, Riggs B, Sullivan W. Membrane traffic: a driving force in cytokinesis. Trends Cell Biol. 2005;15(2):92-101.

[129] Fischer A, Stuckas H, Gluth M, Russell TD, Rudolph MC, Beeman NE, et al. Impaired tight junction sealing and precocious involution in mammary glands of PKN1 transgenic mice. J Cell Sci. 2007;120(Pt 13):2272-83.

Chapter 6

Adult Stem Cell Niches — Stem Cells in the Female Reproductive System

Mirjana Kessler, Rike Zietlow and Thomas F. Meyer

Additional information is available at the end of the chapter

http://dx.doi.org/10.5772/58842

1. Introduction

The female genital tract is a complex and physiologically dynamic system of organs which undergo continuous and profound changes during the reproductive years. Each of its components – ovaries, fallopian tubes and uterus - has unique and indispensable roles in reproduction. Each month, under stimulation from the pituitary gland, the ovaries produce and release a mature oocyte, which moves into the neighboring fallopian tube. Conception takes place in the lumen of the tube and the mucosal epithelium lining plays a critical part in the transport of the gametes and the successful transfer of the zygote to the uterus, where it implants 6-12 days after fertilization. Considering the importance of successful reproduction for species survival, there is strong evolutionary pressure to make the process robust and to respond quickly to any cellular damage with effective repair mechanisms. Also, the inner layer of the uterus, the endometrium, is subjected to monthly shedding and regeneration in order to sustain a suitable environment for implantation of a potential embryo. Similar to other tissues like intestine and hair, which continue to undergo rapid cellular turnover throughout adult life, there is an increasing number of studies describing the existence of adult stem cells in the genital tract that ensure tissue renewal throughout life.

Historically, adult stem cells have been described in vivo as rare, slow-cycling cells that maintain self-renewal by asymmetric division and can differentiate into different progenies. The traditional method of identification has been BrdU labeling, and designated stem cells have been described as label retaining cells (LRC), although there has been some doubt concerning the accuracy of the underlying premise that stem cells efficiently incorporate BrdU [1]. Since BrdU is a mutagen and toxic, this prevents successful recovery and in vitro analysis of labeled cells. However, both old and new experimental approaches recently confirmed the existence of adult stem cells in the female reproductive system using the mouse model system with BrdU pulse labeling [2,3], as well as with transgenic animals with fluorescently labeled histone 2B [4], although it is not yet clear if the two methods do in fact label the same cells.

Deregulation of the adult stem cell niche, which is the main custodian of homeostasis in healthy tissues, is considered a potentially significant step in the etiology of cancer as well as other proliferative disorders, such as endometriosis. Therefore, basic research into the biology of adult stem cells is a new and promising field in the search for novel therapeutic strategies for these diseases. This chapter will provide an overview of the current understanding of adult stem cell niches in the female genital tract, and how new evidence regarding its molecular regulation changes our perspective of analyzing and treating some of its most common pathologies: endometriosis as well as endometrial and ovarian cancer. Based on the available evidence, it is safe to conclude that the female reproductive tract harbors adult stem cells. However, in contrast to an already very comprehensive and detailed insight into the structure and regulation of the adult stem cell niche in the gastro-intestinal tract, hair follicle or hematopoetic tissue, details of the niche organization in the genital tract mucosa remain at best sketchy. Experimental data on adult stem cells from the genital tract almost exclusively originate from in vitro studies of primary culture isolates and so-called "functional assays" describing clonality assays, sphere formation and differentiation capacity for the small population of presumptive stem cell candidates. Still, many questions remain unanswered regarding the molecular mechanisms of epithelial renewal in a system which during the average reproductive period undergoes more than 400 cycles of phenotypical changes in response to shifts in hormonal stimulation. Of particular importance is the relationship between adult stem cells in the healthy tissue and so-called cancer stem cells and we will address the most important developments in this area of research.

We will also briefly review contentious recent evidence for a putative reserve of germline stem cells (GSCs) in the ovary, which would represent a further population of adult stem cells, akin to spermatogonia in men. One of the pillars of reproductive medicine and infertility treatments is the dogma that women are born with all potential oocytes in place, arrested in the first meiotic prophase. Thus, the available "ovarian reserve" is a limiting factor in infertility treatments, and egg donation remains the only option for patients who show diminished parameters for remaining primordial follicles. Following on from the postulation that somatic, mitotically active cells in Drosophila melanogaster can act as GSCs [5] several groups have provided evidence for the existence of GSCs in mouse [6,7] as well as human ovaries [8]. However, they are yet to find final acceptance in the scientific community. If they do indeed exist, they would without doubt revolutionize the field of reproductive medicine.

Before we review the current "state of the art" of adult stem cell research in the female genital tract, it is useful to summarize its basic anatomy and histopathology in order to understand the environment in which stem cells function.

http://dx.doi.org/10.5772/58842

2. Anatomy and histopathology of cervix, uterine endometrium and fallopian tube and ovary

Distinct portions of the mucosa along the tract are morphologically and functionally specialized to facilitate the successful completion of the reproductive process: from the oocyte maturation (ovary) through the transport of gametes (fallopian tube) to the implantation of the embryo in the endometrium and establishment of a viable pregnancy (uterus). The ovarian surface epithelium (OSE) is built of simple, flat, cuboidal cells, without mature adhesion junctions or prominent polarity (Fig. 1A).

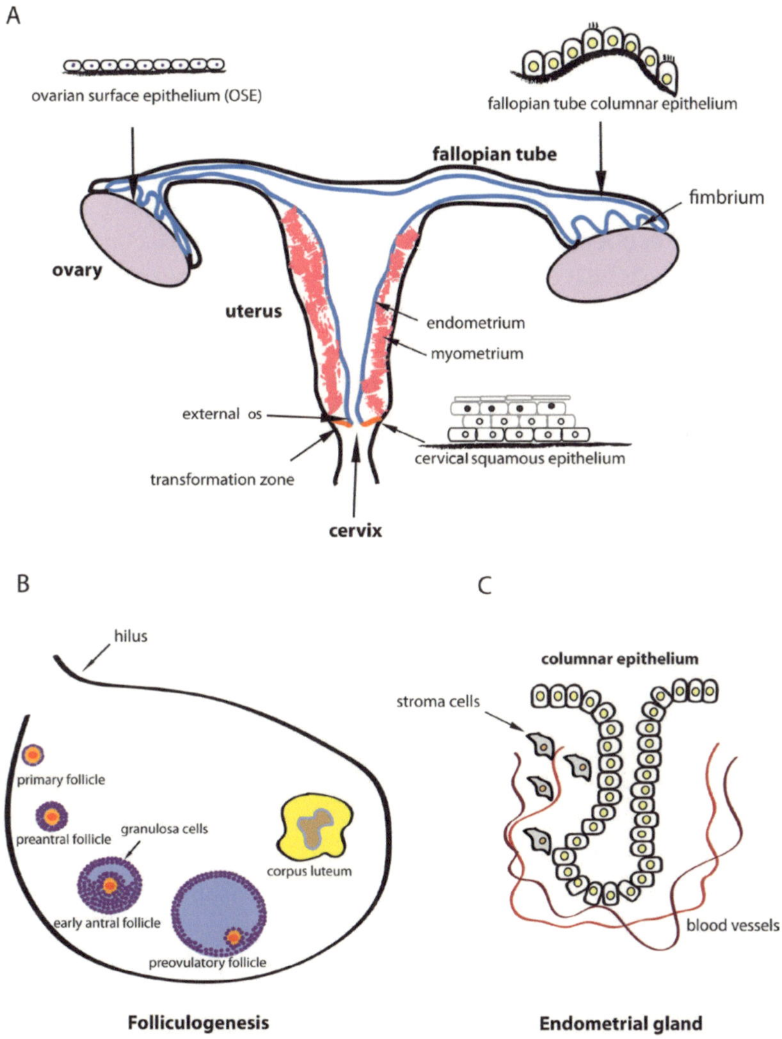

Figure 1. Overview of the histology of the female genital tract

The stroma of the ovary is filled with luteinized stromal cells, decidual cells, neuroendocrine cells, fat and muscle cells and a population of endometrial-like stromal cells. In addition, the ovary contains the pool of primordial follicles, which consist of immature primary oocytes, arrested at birth in the prophase of meiosis 1, surrounded by a densely packed shell of somatic granulosa cells, which are of key importance for successful growth and maturation of the ovum. The developing oocyte and neighboring granulosa cells represent a perfect example of the "niche" where cell-cell interaction and signaling from surrounding tissue compartments determine cell fate and differentiation (Fig. 1B). The process of ovulation, or specifically follicule rupture, creates a pro-inflammatory environment high in reactive oxygen species [9, 10], that requires extensive repair mechanisms and tissue remodeling to avoid permanent damage to the ovary. Indeed OSE cells, though simple in their cellular phenotype, are uniquely adapted to not only respond to stress conditions, but also to actively participate in the tissue breakdown and remodeling that enables rupture of the follicle through the ovarian surface [11]. They express proteolytic enzymes such are metalloprotease 2 and 9 [12], and can undergo epithelial mesenchymal transition (EMT) [13], which may facilitate efficient repair of the injured ovarian surface during the post-ovulatory phase of the menstrual cycle. The ovaries also produce hormones that control the cellular changes associated with the menstrual cycle. The granulosa cells of growing follicles secrete exponentially increasing amounts of estradiol and progesterone. After oocyte release, the remaining follicular cells undergo transformation into the corpus luteum, which immediately starts producing progesterone under the control of pituitary LH pulses. These two main phases of the menstrual cycle – estradiol-driven follicular phase and progesterone-driven luteal phase – determine cyclical homeostasis of all mucosal surfaces in the lower genital tract including fallopian tube, uterus and cervix.

The fallopian tube, or salpinx, is ~ 10 cm long and anatomically divided into three segments: the isthmus, which connects to the uterus, the ampulla, which constitutes the middle part, and the infundibulum, proximal to the ovary. The infundibulum terminates in relatively large opening, the ostium, which has many fine projections, or fimbriae, that capture the oocyte upon follicule rupture and guide it towards the ampulla, where fertilization takes place. The histology of the tube, starkly different from the ovarian surface, is characterized by the presence of an epithelial monolayer of highly differentiated columnar cells with two distinct cell types: secretory – producing tubular fluid – and ciliated – enabling transport along the lumen (Fig. 1A). Numerous mucosal folds in the distal and ampullar regions provide a suitable environment for the early stages of blastocyst development. Contractility of the tube is ensured by the muscular layer surrounding the epithelium. Although, unlike uterine endometrium, fallopian epithelium does not undergo extensive monthly shedding, it responds to the follicular phase hormonal environment by proliferation [14]. Changes in homeostasis are supported by global gene expression data from fallopian tube mucosal samples, which show marked differences between follicular and luteal phase [15]. The tubal epithelium is also exposed to the pro-inflammatory environment associated with ovulation, leading for instance to an increase in double stranded DNA breaks marked by phospho-γH2A.X in a mouse model in vivo [16]. Exposing human fallopian tube isolates to follicular fluid ex vivo increases expression of inflammation-related genes and DNA repair components [17]. In particular, there is a noticeable accumulation of p53 it the nuclei of tubal cells, which is thought to be a key step in the development of premalignant lesions. As well as the potentially genotoxic effects of follicular fluid, their connection to the uterus also renders the fallopian tubes vulnerable to ascending infections by sexually transmitted pathogens. Chlamydia trachomatis and Neisseria gonorrhea are major causes of the inflammatory disease salpingitis, which is marked by scarring and tissue injury and dramatically increases the risk for tubal occlusion and infertility. In addition, there is increasing evidence that at least some STDs may also have a pro-malignant effect on host cells [18-20].

The uterus is a muscular organ with great capacity for growth and physiological transformation, which is necessary to support pregnancy. It consists of three layers: the outer serosal layer, or perimetrium, the myometrium, which contributes the most to the volume and mass of the organ, and the inner mucosal layer, or endometrium, which is vital for the initiation of pregnancy. The structure of the endometrial layer has traditionally been divided into the stratum basalis and the stratum functionalis, which is subjected to monthly cyclical renewal, differentiation and shedding. The stratum functionalis changes greatly during each menstrual cycle and is further divided into stratum compactum and stratum spongiosum. The stratum functionalis consists of glandular epithelium (Fig. 1C) residing on the supportive connective tissue and blood vessels which supply nutrients. It is at its thinnest at the beginning of the follicular phase, and proliferates strongly under stimulation by estradiol. The estradiol peak coincides with the maximum follicle diameter prior to ovulation and with maximal thickness of the proliferative endometrium. Following ovulation, rising progesterone levels, trigger differentiation of the stratum functionalis into "secretory" endometrium. At the cellular level, endometrial glands, which resemble narrow straight tubes during the follicular phase, begin to swell as progesterone stimulates the columnar epithelial cells to produce and secrete glycogen granules. It is assumed that this glycogen-rich environment serves as an energy depot for the implanting blastocyst. In addition, stromal cells undergo a profound change, converting from a fibroblast-like phenotype into rounded cells that produce prolactin and insulin growth factor binding protein [21,22]. This complex transformation of the functional layer of the endometrium, called decidualization, is a prerequisite for successful attachment and invasion of the trophoblast and thereby initiation of the pregnancy. Progesterone secretion by the corpus luteum is limited to around two weeks, after which it disintegrates if no chorionic gonadotropin from an implanted embryo is present in the circulation to rescue its function. In the absence of progesterone, the endometrial stratum functionalis, epithelium and supportive connective tissue is shed and expelled by menstrual bleeding.

The uterus is separated from the vaginal canal by the narrow muscular cervix. Due to its physiological elasticity, the cervix tightens under the influence of progesterone, and essentially seals the uterus from the outside environment during the second part of the cycle. This protective barrier is enhanced in the case of pregnancy by the formation of a mucus plug that fills the endocervical canal. The entrance to the cervix is called external orifice of the uterus, or external os. The endocervix is lined by simple columnar epithelium, with a similar structure to the endometrial monolayer, while the outer part is lined by stratified squamous epithelium. The segment in between – the squamo-columnar junction (SCJ) – is a dynamic zone, which does not have a fixed location but migrates under the influence of major hormonal changes, such as puberty, pregnancy and menopause. Prior to puberty, squamous epithelium covers the outer segment of the cervix and the lower part of the canal. During puberty and first pregnancy, the columnar epithelium of the endocervix expands and covers the outer rim of the external os. Through contact with the low pH of the vagina, the columnar epithelium undergoes metaplasia over time and converts back towards a squamous phenotype (squamous metaplasia). It is widely accepted that over 90% of malignancies of the cervix originate from cellular changes which are initiated in this region of intense tissue remodeling-thus the SCJ is frequently labelled "transformation zone" (TZ).

3. Putative female GSCs

Before devoting our attention to the somatic adult stem cells of the female genital tract, we will review the rather contentions field of presumptive GSCs in the adult ovary. Since the 1950s, the accepted dogma in reproductive biology has been that in mammals primordial germ cell (PGC)-derived oogonia cease proliferation shortly after birth and differentiate into primary oocytes, which arrest in prophase of meiosis I until fertilization triggers the completion of meiosis [23]. As a consequence, the pool of available oocytes is finite – and exhaustion of the pool of resulting follicles is believed to be responsible for menopause. In the last decade, however, this assumption has been challenged by a series of papers reporting the existence of GSCs in the ovaries of mice and humans. First hints came from observations by Tilly's group [6], who tried to assess the dynamics of germ cell loss during adult life. Their estimation that up to a third of immature follicles in mice are degenerating at any given time led them to postulate that this unexpectedly high rate is incompatible with the slow rate of decline observed in the follicle reserve. In trying to resolve this contradiction, they identified cells on the ovarian surface of young mice that express the meiotic entry marker SCP3, as well as BrdU-incorporating cells that simultaneously express the germ cell marker Ddx4, suggesting that these cells may be proliferating GSCs responsible for replenishing the follicle pool.

Further evidence for ovarian GSCs that are capable of proliferation followed by differentiation into oocytes came from the observation that immunomagnetically isolated $Ddx4^{+}$ cells from both mouse and human ovaries can be expanded in vitro and give rise to oocytes following transplantation into donor ovarian tissue [7,8]. In the mouse, GFP-labelled putative Ddx4+GSCs were able to give rise to offspring following transplantation, with transmission through the germline to subsequent generations [7]. Using transplantation of premeiotic female PGCs Zhang et al [24] further provided proof of principle that the adult ovary is able to support oogenesis. Tilly's group subsequently showed that following extensive doxyrubin-induced loss, follicle numbers recover within 36 h. Reasoning that the small numbers of presumed GSCs they had previously identified in the adult ovary would not be sufficient to support this rapid recovery, they transplanted bone marrow from mice expressing GFP under the Oct 4 promoter into germ cell-deficient $Atm^{-/-}$mice and identified GFP-labelled primordial follicles [25]. They concluded that a subpopulation of circulating bone marrow stem cells that express PGC marker genes are responsible for the observed oocyte replenishment.

Putative GSCs have been identified by co-labelling with proliferative and germ cell markers and described either as small cells within the ovarian cortex [6,26-28] or clusters of cells which also include somatic cells [29]. Other studies supported the discovery of ovarian GSCs [26,30] but the considerable differences in marker expression, in vitro phenotype and differentiation potential observed between these studies make the case somewhat controversial. In addition, other groups have failed to repeat these findings and challenged their validity [24,31,32]. Kerr et al described the expulsion of oocytes through the OSE into the peritoneum during the post-natal phase of oocyte reduction, which could potentially be interpreted as GSCs [32]. By using tamoxifen-induced random labelling of cells, Lei and Spradling traced the numbers of follicles over time and argue that the follicle pool is in fact highly stable with a half-life of 10 – 11 months, which would make the follicle pool at birth large enough to support the ~500 ovulations required during the life time of a mouse [33]. They also failed to observe the generation of new follicles even after depletion with busulphan toxin, thus putting into question that there is in fact a need to explain regeneration of follicle numbers, a finding that is further supported by mathematical modelling [34]. Similarly other groups, failed to observe replenishment of the follicle pool by donor bone marrow-derived cells [35] or after chemical depletion [36].

It has been argued that some of these contradicting findings could be explained by the use of Ddx4 –which was previously thought to be a cytoplasmic germ cell marker, but was used by Zou et al [7] and White et al [8] to isolate cells based on their finding that the protein also contains a transmembrane domain in GSCs. They suggest that the reason many other laboratories have failed to identify GSCs in ovaries using Ddx4 [24] is due to the fact that cytoplasmic Ddx4 expression in oocytes masks the presence of rare GSCs. Similarly, SSEA-1, which was used by Johnson's group to identify ovarian GSCs [25], was reported by Bristol-Gould et al [34] to overlap only with cells from the HSC lineage. Widespread acceptance of adult ovarian GSCs will no doubt depend on the identification of a marker signature or more robust methods which allows identification of these cells by all laboratories with the relevant expertise.

The difficulty in pinpointing putative ovarian GSCs or their niche in vivo, while at the same time apparently being expandable and giving rise to the appropriate differentiated tissue in vitro, mirrors the experience for stem cells of somatic adult tissues, as we shall explore below. However, the unique properties and complex embryonic development of germ cells do in fact give rise to some important differences.

In the current model, primordial germ cells (PGCs) derive from a small number of epiblast cells which are specified before differentiation into the different germ layers begins. PGCs subsequently undergo a complex migration through the allantois, along the developing hindgut, finally entering the dorsal mesentery and the developing gonads. Despite the obvious importance of these early developmental processes for future fertility, they remain little understood. Mikedis and Downs have suggested that PGCs temporarily reside in an "allantoic core domain" (ACD) which they propose has similar functions to the Spemann organiser, consisting of a stem cell pool which extends the body axis in a posterior direction – contributing not only to the germ cell lineage but also the three germ layers – effectively creating a strong interface between the future umbilical cord and the developing embryo [37]. The stem cells in the ACD express Oct4, Blimp1, Stella and Fragilis – markers thought to be specific for PGCs –

but appear to contribute also to other tissues [38]. These observations, as well as the fact that hematopoietic stem cells also migrate from the proximal epiblast to the embryonic aorta-gonad-mesonephric region during the same period of development, imply that it is theoretically possible that there may be "intermixing" or indeed that there is a common precursor pool for PGCs and a subpopulation of bone marrow stem cells.

While a lineage tree analysis [39] based on somatic mutations accumulating in microsatellites found that oocytes form a cluster which is entirely distinct from other cell populations – suggesting that there is no intermixing of the germ cell precursor pool with that of any other cell type – it is conceivable that a very rare subpopulation of bone marrow stem cells would be missed in such an analysis. In fact their results also show that the number of mitotic divisions oocytes have undergone increases with age and following unilateral ovarectomy. This may be explained by recruitment of oocytes in the order in which they first differentiated during development – but is also consistent with the notion of continuous oocyte production from cycling stem cells. Many observers have suggested that if GSCs do exist, they are most likely to be derived from the normal developmental precursors of oocytes, i.e. PGCs or oogonia – which have not yet differentiated into oocytes and are still able to undergo mitosis [40-42]. The close relationship of PGCs to pluripotent cells is demonstrated by the fact that following isolation from the embryo, they can be converted back to a pluripotent phenotype termed embryonic germ cells in vitro without genetic manipulation [43-45].

Even if one accepts the existence of GSCs, there are a number of unanswered questions apart from their exact location. Firstly, it is not clear whether they contribute to oocyte production under normal physiological conditions, or only after injury. Secondly, if the follicle pool is replenished by GSCs, why does this replenishment eventually cease, leading to menopause? Niikura et al [27] have suggested that the aging niche environment itself may be responsible – however, the life-long production of spermatozoa in the testis indicates that this is not in itself a sufficient explanation. A large number of germ cells in neonatal mouse ovaries have not yet entered meiosis and can be induced to proliferate, increasing the follicular pool – but by the time animals enter reproductive age the numbers have returned to "normal" levels [46]. This suggests that mechanisms exist within the ovary to actively regulate the number of follicles. Together with the fact that a large proportion of oocytes are eliminated shortly after birth [47] this indicates a highly selective process to ensure the removal of oocytes with reduced meiotic fitness, which runs counter to the idea of continued oocyte replenishment from cycling precursors.

Whether or not GSCs exist in vivo, the presence of cells that can be expanded and differentiated to functioning oocytes in vitro – as suggested by Zou et al [7] and others [28] – would in itself be of huge potential benefit for the treatment of infertility. Nonetheless, the long time required to induce proliferation of these cells in vitro (around 10 weeks), compared to other adult stem cells, suggests that transformation of the cells in vitro may be responsible for the observed phenotype – similar to findings which describe the production of oocytes from other somatic stem cells in vitro – or that indeed the results may be explained by rare primordial oocytes that are carried over during the in vitro period.

4. Uterine endometrial stem cells

From the volume of work, it is fair to say that human endometrium has been the most intensively studied portion of the female genital tract in the field of adult stem cell research. This is partly due to the fact that the uterus is by far the most accessible portion of the tract, where sample collection and analysis is much less invasive compared to investigating ovary or fallopian tubes, and partly to the logic assumption that such intensely proliferating and renewing tissue should contain a stem cell pool. Since the stratum basalis of the endometrium is necessary for monthly renewal of the functional layer [48], characterized by intense proliferation under stimulation of rising estradiol levels, it is a prime candidate for harboring stem cells. Still, there is no common agreement yet where adult endometrial stem cells reside in vivo. LRCs were first identified by BrdU labeling in the epithelial layer and underlying stroma in postnatal mice [2]. However, by 12 weeks post labeling, no BrdU positive cells were left. Suboptimal timing of the pulse could mean that division of slow-cycling stem cells was missed such that only transitory amplifying progeny was labeled. Nevertheless, the study provided the first in vivo evidence for the existence of long-lived cells in the endometrium.

A significant advance over BrdU labeling for localizing stem cells in the mouse was recently achieved with the development of a histone2B-GFP (H2B-GFP) reporter in a Tet-inducible system, in which a doxyclicline pulse leads to fluorescent labeling of chromosomes, without mutagenic stress. This method allows induction of the construct during embryonic development, which increases the likelihood that stem cells will be labeled. With each cell division the GFP signal is reduced until after ~ 12 weeks it can only be detected in LRCs. In the first study using this system [49], the doxycyclin pulse was administred in adult animals and 12 weeks after pulse withdrawal LRCs were identified only in the distal oviduct segment of fimbrium and ampulla. Their presence was stable even after 47 weeks. In a follow-up study, labeling was extended to the embryonic period, confirming localization of LRCs to the distal oviduct, and identifying an additional population at the endocervical transition region but not within the endometrium. Only a pulse during the pre-pubertal period (post natal day 21-42) resulted in labeling rare cells in the glandular endometrial epithelium [4]. It is perhaps due to these experimental difficulties in identifying the adult stem cells in the uterus of a living organism, despite the undisputedly enormous capacity for endometrial regeneration and plasticity, that alternative models have been proposed to explain regenerative capacity of the endometrium.

An interesting hypothesis that has gained some traction within the scientific community is the possibility that bone marrow could be a source of endometrial stem cells. Previously, bone marrow stem cells were reported to be able to differentiate into a wide variety of cell types [50-52]. Analysis of bone marrow transplant recipients revealed the presence of donor cells in the opening of endometrial glands, raising the possibility that a stem cell pool outside of the reproductive system may contribute to the regeneration processes [53,54], mirroring the findings with female GSCs. Ikoma and colleagues used in situ hybridization to confirm the localization of Y chromosome-positive donor cells in the endometrium of a female patient who had undergone bone marrow transplant from a male donor [55]. However newer studies have questioned the functional importance of such findings, as bone marrow cells residing within the endometrium do not appear to contribute to the stem cell pool, which exhibits high clonogenic capacity in vitro, but are instead reminiscent of terminally differentiated cells [56], which could be explained by the capacity of bone marrow-derived cells to fuse with differentiated cells from a variety of organs [57].

Regardless of the difficulties in identifying and characterizing the adult stem cell niche in the intact epithelium in vivo, numerous studies in recent years have isolated distinct populations of putative adult stem or progenitor cells from uterine tissue and demonstrated their proliferative capacity and broad differentiation potential in vitro. Two different classes of adult stem cells are found in the endometrium: mesenchymal and epithelial. Mesenchymal stem cells have also been named Endometrial Regenerative Cells (ERC) due to their high proliferative potential. Although the ability to produce a clonal population of cells when seeded at very low dilution is much greater for the stromal mesenchymal subpopulation, the epithelium also contains cells capable of clonal expansion in vitro. Self renewal of endometrial stem cells in vitro was first demonstrated in pioneering work by the group of Caroline Gargett [2,58,59]. In the absence of a widely accepted surface marker for endometrial stem cells, they demonstrated that multipotent cells can be efficiently isolated based on the uptake of Hoechst 33342 [60]. The method of SP cells was originally established in procedures for adult stem cells isolation from the hematoopetic tissue [61]. It made use of the discovery that the small population of adult stem cells in the tissue differs from the differentiated cells by their capacity to efficiently eject Hoechst dye from the cytoplasm, presumably due to the high expression of the ABCG2 transporter [62]. Designated endometrial side population cells (ESP), consisting of both epithelial and mesenchymal SP cells, are able to generate endometrial glandular structures in vivo if transplanted under the kidney capsule of NOD/SCID immunocompromised mice [63]. Interestingly, a fraction of the ESPs migrated and generated blood vessels to supply the endometrium, showing potency to generate different types of tissue in the functional organ. Still, the overall efficiency of the procedure was rather low as endometrium formation occurred in only 2 out of 24 injected animals. Dramatic improvements in the outcome of this type of xenograft was achieved by expanding and cloning lines of SP cells from both epithelial and stromal compartments. Transplantation of individual lines supported by the administration of estrogen (E2) and progesterone (P) to mimic the menstrual cycle gave rise to endometrial tissue in all animals [56]. This is a particularly significant finding since endometrial stem cells do not express E2 and P receptors. Nevertheless, the hormonal environment greatly affects the outcome of the tissue regeneration process. This phenomenon was closely analyzed by Janzen and colleagues [64], who showed that hormonal withdrawal in the xenograft mouse model leads to a pronounced decrease in the size of endometrial tissue while the remaining cells are highly enriched in stem cells. This suggests that the stem cell niche is activated at the beginning of each cycle, when both estradiol and progesterone levels are low, while the subsequent hormonal stimulation drives regeneration and proliferation of transitory amplifying cells up to their final differentiation in response to progesterone stimulation from the corpus luteum in the second phase of menstrual cycle. On the molecular level, endometrial stem cells showed elevated Wnt signaling activity and increased expression of the Wnt target genes Axin2, Cyclin D1, ID2, CD44.

As well as studying the role of endometrial stem cells in regeneration of the functional endometrium, a number of methods have been developed to use endometrial progenitors in translational approaches for therapeutic purposes. In vitro differentiation assays suggest that mesenchymal stromal stem cells may be able to differentiate into diverse cell types [58,65,66], and can be efficiently isolated from menstrual blood. A clinical trial of endometrial stem cells as a source of autologous regenerative cardiomyocytes to treat ischemic cardiac injury is underway [67] and promising results were also reported for pancreatic island replacement in a xenograft diabetic mouse model [68].

In parallel to the translational approaches, we need to get a better understanding of the biology of endometrial stem cells in vivo to understand how these cells influence the development of pathologies in the uterus. There are strong indications that deregulation of the stem cell niche plays a role in endometriosis, a disease which affects up to 10% of all women and is present in nearly half of those experiencing fertility problems or pelvic pain after the age of 35 [69]. The hallmark of endometriosis is the presence of ectopic explants of endometrial tissue outside of the uterus, which can affect the fallopian tubes, the ovary but also more distant regions of the pelvic cavity, causing pain and discomfort. Endometriotic lesions have identical responses to hormonal stimuli as the endometrium of the uterus. The model of "retrograde menstruation" is accepted as the most plausible explanation for the spread of endometriosis tissue through the genital tract, but the molecular mechanisms which trigger and control the disease remain obscure. Analysis of menstrual tissue samples from affected women showed that endometriosis patients have significantly more fragments of the basalis layer in the flow than healthy controls [70]. Cells from menstrual blood of patients also have longer than average telomeres, which is in agreement with an enhanced stem cell presence [71]. According to this model, long-lived stem cells from the basalis are disseminated and give rise to endometriotic lesions. In healthy women, on the other hand, only the upper layer of the endometrium is ablated, leaving the basalis layer intact. These data offer hope that characterization of the endometrial stem cell niche and its regulatory mechanisms will lead to a breakthrough in prevention and treatment of endometriosis – a disease that poses an enormous burden for patients' quality of life as well as high costs for the healthcare system.

5. Fallopian tube stem cells

Our understanding of the biology of fallopian tube epithelium is still rudimentary. From the perspective of medical diagnostics, it is the least accessible portion of the female reproductive tract. The fallopian tube is barely visible by ultrasound and even exploratory laparoscopy offers only information about tubal patency, and no insight into the condition of the mucosa. Current knowledge about tubal histology and pathology comes solely from patients of salpingectomy procedures, where tubes are surgically removed, usually as part of a total hysterectomy. The importance of making progress in this area has been highlighted in recent years, as the potential significance of this tissue in disease initiation has become clear. The distal portion of the tubal fimbrium, and the ampulla with its abundant mucosal folds, are of critical importance for oocyte capture and fertilization. Intimate contact of the fimbrium with the surface of the ovary exposes the epithelial layer of the tube to the inflammatory signals associated with ovulation which are present in follicular fluid. Thus, there is an increased requirement for robust renewal mechanisms. Indeed, in vitro experiments with cell isolates from the fallopian tube have defined a population that is positive for CD44 and integrin $\alpha 6$ and has the capacity for clonal growth, self renewal and differentiation into the secretory and ciliated cells found in the tubal epithelial monolayer-specifically in the distal region of the tube [72]. Strong clonogenic potential of a H2B-GFP-retaining subset of epithelial cells in the distal oviduct and their capacity to form and maintain spheroids in vitro long-term has been demonstrated in a mouse model [49]. These spheroids are able to differentiate into more complex structures of the mucosa, strongly suggesting that they do indeed contain progenitor cells of the tube with the potential to proliferate and differentiate in vitro. However, the exact organization of the fallopian tube stem cell niche in vivo remains unknown apart from the positive identification of candidate cells in label-retaining experiments in vivo mentioned above. Moreover, the exact turnover rate of the epithelium and how it responds to different physiological and environmental stimuli or stresses is yet to be established.

Beyond the requirement for Wnt signaling for normal development of the genital tract, several studies have reported changes in Wnt signaling associated with pathology of the fallopian tube. For example, activation of β-catenin signaling is implicated as a contributing factor in ectopic pregnancy [73,74] and endometriosis [75]. Increased Wnt signaling is a mucosal response to infection [76], confirming that this paracrine pathway plays a role in homeostasis inside the fallopian tube. More studies are needed, however, to illuminate all aspects of paracrine signaling in the fallopian tube epithelium in health and disease. As mentioned above, Wnt signaling alterations are one of the hallmarks of ovarian cancer. Since there is now consensus in the medical community, based on cumulative molecular and clinical evidence, that a significant portion of high grade serous ovarian carcinoma originates from the fallopian tube fimbrium rather than the ovarian surface epithelium, it has become imperative to illuminate the regulatory mechanisms involved in fallopian tube epithelium homeostasis. Different models of ovarian carcinogenesis and the potential role of adult stem cells in this process will be reviewed fully later in this chapter.

6. Ovarian and cervical stem cells

In contrast to the uterus, direct evidence for adult stem cells in the ovary is sparse. However, the tissue remodeling processes involved in ovulation require a considerable regeneration potential within this organ that has to be maintained for several decades. Beyond the healing of the ovulatory wound created in the epithelial surface during follicle rupture, ovarian epithelial cells also undergo proliferation at the beginning of the menstrual cycle [77].

Label-retaining cells on the ovarian surface were identified in vivo by comparing a histone-GFP inducible transgenic model and BrdU labeling of animals [78]. It is important to note that both methods resulted in identification of positive cells in the OSE epithelium of the ovary but the two populations overlapped only in a minority of cases. This further underscores the complexity and difficulty of identifying adult stem cells solely on the basis of cell division rate. Recently, the ovarian surface was subjected to a more detailed analysis, and the highest proportion of LRCs was detected in the hilum region close to the fallopian tube. These cells expressed high levels of aldehyde dehydrogenase 1 (ALDH1), and were able to proliferate long-term in vitro [3], forming spheres. Long-term pulse experiments and staining with the proliferation marker Ki67 also suggested that slow-cycling cells are activated in a cyclical fashion during the estrus phase, which supports a repair mechanism for the damaged OSE monolayer. Finally, genetic lineage tracing of Lgr5 expression in the ovary was performed [3], by using the $Lgr5^{tm1(cre/ERT2)Cle}$/J mouse which harbors a genetic construct that enables GFP labeling of Lgr5 expressing cells and tamoxifen-inducible expression of Cre recombinase. When crossed with a strain harboring a stop codon in a dTomato sequence that can be excised by Cre recombinase, offspring mice are produced in which easy tracing of the progeny of Lgr5-expressing cells is possible by a simple tamoxifen pulse at the desired time point [79]. This experiment revealed an $Lgr5^{+}$ population at the hilum, which contributes to the whole OSE monolayer during renewal of the epithelium in the course of 1 month [3]. Notably, this study was focused on the ovary and no lineage tracing was analyzed in other portions of the genital tract. A similar approach, with the SOX2 gene promoter controlling expression of CRE recombinase in a tamoxifen-inducible fashion, revealed the existence of long-term lineage labeling in the cervical epithelium of mice (Arnold et al 2012). It remains unclear, however, if other parts of the female genital system contain epithelial cells originating from $SOX2^{+}$ progenitors and it will be interesting to find out how these populations of putative stem cell candidates identified by independent methods correlate with each other. It is of course feasible that different tissue compartments harbor stem cells which are defined by different molecular markers. Although these two studies represent a methodological breakthrough in the detection of adult stem cells, by demonstrating the in vivo capacity of these cells to give rise to differentiated progeny in the tissue through tracing cellular markers in physiological conditions, they are not entirely comprehensive. This approach is hypothesis-driven by selecting candidate genes; however, the list of potential stemness regulators that could be tested is much longer. For example, Lgr5 is only one member of the Lgr receptor family, and other members are also involved in regulating tissue regeneration, e.g. Lgr6 in the hair follicle [80] and Lgr4 in the prostate [81]. More detailed studies are needed to confirm whether the $LGR5^{+}$ cells detected in the hilum represent the pool of true stem cells or a more dynamic population of amplifying cells that have already undergone a degree of lineage commitment. The $SOX2^{+}$ cells identified in the cervix are a good starting point to define the molecular characteristics of the long-lived basal cells which have so far proven to be elusive using other tracing methods, and this genetic system will provide a valuable tool for a more detailed analysis of the adult stem cell niche in the future.

7. Adult stem cells between development and disease — Role of the Wnt pathway

The precise signaling pathways controlling the renewal processes of the fallopian tube epithelium, endometrium and ovary remain unknown, but there are several reasons to assume that they depend on paracrine Wnt, Notch and BMP signaling. Wnt signaling controls crucial developmental processes during all phases of embryogenesis and during adult life. Wnt ligands interact with a family of receptors, inducing a variety of responses depending on the cellular context. The main transducer of Wnt signalling within the cell is β-catenin, which translocates to the nucleus and induces expression of target genes via several transcription factors (Figure 2).

Although a detailed overview of Wnt signaling is beyond the scope of this chapter, we will briefly outline its involvement in the control of the cell behavior within a tissue, affecting, among other processes, proliferation, establishment of polarity, differentiation, morphogenetic movements and apoptosis. With respect to tissue homeostasis, Wnt signaling acts as a key cell-cell communication network during tissue formation in development, as well as for maintaining tissue function. The embryological development of the female genital tract relies on active Wnt signaling, as evidenced by Müllerian aplasia. This autosomal mutation in the Wnt4 gene leads to a severely underdeveloped or absent uterus. Mouse models have further revealed a strong dependence of developmental processes on Wnt 5a, 7a, and 9b, since mutant animals showed severe malformations of different parts of the genital tract raging from defective coiling of oviducts to absence of the upper vagina or uterine glands [82-84]. The importance of the Wnt pathway for maintenance of homeostasis is well-documented by molecular and genetic analysis of numerous human malignancies. Perhaps the most startling example of the tight relationship between Wnt signaling and control of proliferation in mucosal epithelium is provided by the relationship between Adenomatous polyposis coli (APC) mutations and colon cancer. APC protein in complex with Axin1 promotes degradation of β-catenin and thereby inhibits Wnt signaling transduction (see Fig 2). In familial adenomatous polyposis patients, who carry an APC functional deletion mutation, the risk of developing colon cancer is almost 100% and 5% of sporadic colon cancer patients harbor either APC loss-of-function or β-catenin activating mutations [85]. Mutations in components of the Wnt pathway are frequently found in numerous other malignancies as well, e.g. pancreas, liver, kidney, pituitary, and notably also in the ovary and endometrium [86].

Although the significance of Wnt signaling for tissue maintenance has been known for the last couple of decades, it is the discovery of adult tissue stem cells which led to a breakthrough in our understanding of the regulatory mechanisms at the molecular level. Lgr5-expressing stem cells of the intestine ensure renewal of the mucosa every 3-5 days. Although $Lgr5^{+}$ cells at the base of intestinal crypts are sufficient to recreate the complete epithelial layer in vitro, an alternative model attributes true "stemness" to rare, more quiescent cells in the wall of the crypt, which express Bmi1, HOPX and mTERT [87-89]. Ablation of $Lgr5^{+}$ cells does not disrupt homeostasis, as $Bmi1^{+}$ cells compensate for the loss [90]. This illustrates the complexity of the regulatory stem cell niche, where different cells can be recruited and even reprogrammed,

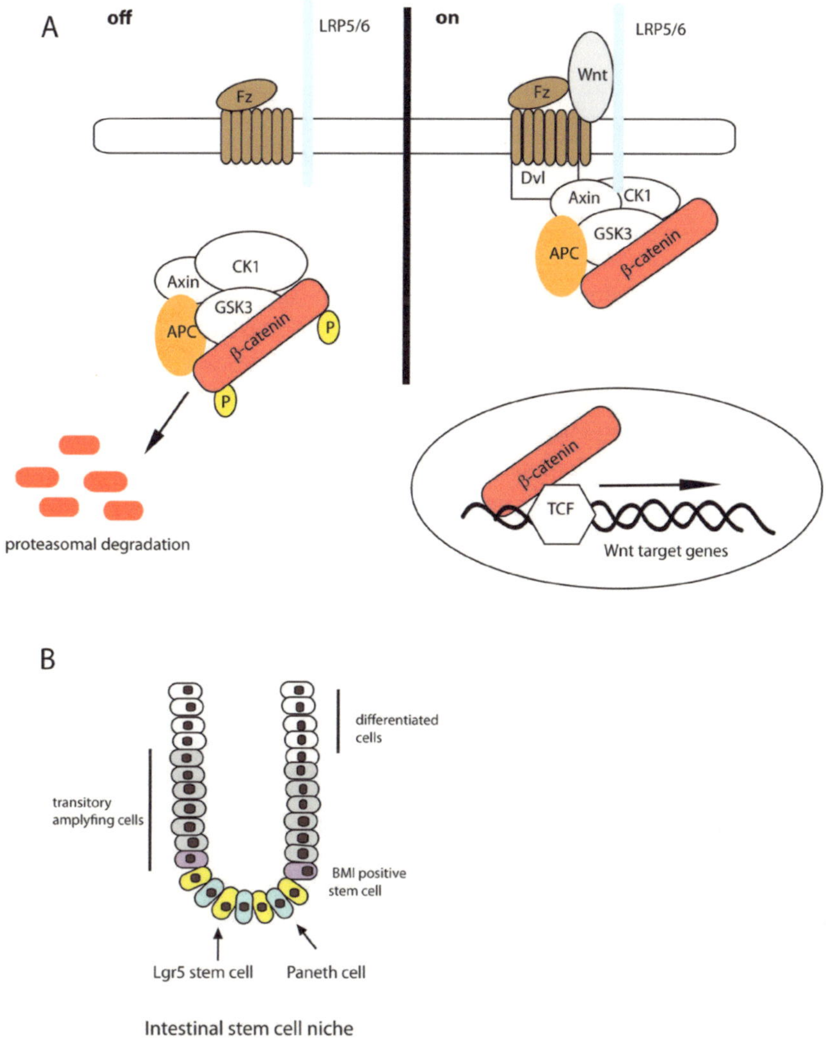

Figure 2. Paracrine regulation of the adult stem cell niche. A) Schematic representation of the cellular mechanism of β-catenin turnover in the cytoplasm. After the Wnt signal triggers dimerization of receptors, β-catenin is released from the degradation complex and translocates to the nucleus, where it activates Wnt target genes; B) Hierarchical organization of the intestinal crypt, with Lgr5+stem cells localizing to the bottom off the crypt between nurturing Paneth cells. Above the stem cells, there is a zone of intense proliferation (transitory amplifying cells), followed by terminally differentiated epithelium, which has short life span (1-2 days)

depending on the conditions during epithelial renewal. Although Lgr5 is a Wnt signaling target, crypt organization, cell fate determination and differentiation are also dependent on integration of signals from the Notch [91] and TGF-β pathways. Such paracrine signaling pathways have emerged as a common principle of functioning stem cell niches in other organs [92-95]. The inherent longevity of adult stem is a potent mechanism for dissemination of accumulated mutations within the tissue.

8. Cancer stem cells and the putative link to adult stem cells

Parallel to the discovery of adult stem cells in healthy tissues, cancer research has produced a bulk of evidence showing that most malignancies are not homogenous cell masses but rather heterogeneous tissues whose progression depends on the fitness of a distinct cellular fraction: cancer stem cells (CSCs). CSCs can confer resistance to chemotherapeutic agents or exhibit other characteristics that provide a competitive advantage to the cancer and ensure its progression [96-99]. In many cases, long-term prognosis and patient survival can be correlated with the frequency of cancer stem cells in the tumor at the time of diagnosis [100]. CSCs exhibit the characteristics of adult stem cells: self renewal and differentiation capacity. They can be distinguished from the bulk of the tumor tissue as the only fraction that is able to generate new tumors when transplanted into immunocompromised mice [101,102]. For this reason, they are sometimes referred to as tumor initiating cells. This property of cancer stem cells has been demonstrated for numerous malignancies, including ovarian and endometrial cancer. As with markers of adult stem cells, markers specifically associated with cancer stem cells are also proving difficult to identify. It is of course possible that there is no unique cancer stem cell for ovarian, endometrial or other genital malignancies and that stemness can be achieved by alternative routes in individual cases, thereby resulting in different combinations of surface markers such as CD44, EpCAM, ALDH1, CD117, CD133 etc, which have been identified by different studies [81,103,104]. Cancer stem cells are intermixed with the bulk of the tumor tissue, but are found enriched in advanced stages of metastasis, in spheres present in effusions recovered from ascites [105]. Spheres exhibit anchorage-independent growth but can efficiently adhere to mesothelial cells in the peritoneum via integrin 1, which may play a role in metastatic spread. It is not yet known what triggers formation of spheres, although the physiological implications for disease progression and response to treatment are immense. The compact organization of the cells in a cluster makes drug delivery ineffective, either due to the difficulty of drugs to penetrate, or induction of a quiescent state in cells at the core through tight cell-cell contacts, making them insensitive to agents targeting actively replicating cells. Either way, it is clear that 3D organization represents another layer of protection for cancer cells from chemotherapy. This effect has been demonstrated by comparing the response to the standard therapeutic drugs cisplatin and paclitaxel for 11 different ovarian cancer cell lines in 2D in 3D [106].

Importantly, however, it remains unknown if and how cancer stem cells are related to normal adult stem cells. Cancer stem cells may develop from adult stem cells by losing dependence on niche regulatory factors, or they may simply be the progeny of differentiated cells that have acquired "stemness" characteristics during the accumulation of mutations and cellular transformation. It is conceivable that both mechanisms could occur in all or some types of cancer. If there is indeed a causal relationship between dysregulation of the adult stem cell niche and carcinogenesis, this may open up new possibilities for early diagnosis and timely intervention. Since the research field of cancer stem cells is currently offering a variety of sometimes competing models of what defines this population in endometrial, cervical and serous ovarian cancer, we focus our attention on the current understanding of carcinogenesis in the genital tract. Understanding of the cellular events that lead to initial transformation

http://dx.doi.org/10.5772/58842

could prove to be of pivotal importance for resolving remaining questions about tumor spread and the role of cancer stem cells. In this light, we will address changes in tissue architecture and physiology of the female reproductive tract that could favor malignant transformation and bring it into context of therapeutic implications, particularly in light of recent evidence that dysregulation of tissue homeostasis could be the result of infection with certain sexually transmitted pathogens.

Malignant tumors of the genital tract are the third leading cause of cancer related deaths after breast cancer and lung cancer [107]. Based on the primary organ, affected cancers are classified as ovarian, uterine, cervical, vulval and vaginal. The high prevalence of these cancers is likely to be related to the great plasticity of these tissues, their regenerative potential and thus probably their high cell turnover as well. On top of this, openness to the external environment and exposure to pathogens makes them potentially vulnerable to the transformation. The latter has been demonstrated by the dramatic correlation between HPV infection cervical cancers, since nearly 100% of patients are HPV positive [108].

9. Serous ovarian cancer and stem cells of the fallopian tube

From the perspective of patient care and long-term prognosis, one of the biggest challenges for medicine represents high grade serous ovarian cancer, an aggressive form of epithelial ovarian cancer which has a survival rate of under 40 % after 10 years [109]. This deadly malignancy takes more than 14,000 lives per year in the US alone and no improvement can be expected in the foreseeable future, due to the absence of early screening methods and the aggressive nature of the cancer. The origin of high-grade serous ovarian cancer (HG-SOC) has puzzled medical doctors and pathologists for decades. Naturally, the site of carcinogenesis was initially attributed to OSE cells [110]. Numerous examples of developmental genetics show that relatively complex changes in phenotype are frequently induced by expression of only one master regulator gene. In the case of OSE cells, ectopic expression of homeobox transcription factor HOX9 in the xenograft mouse model causes Müllerean metaplasia-transformation of cuboidal to papillary columnar epithelium resembling fallopian tube mucosa and serous ovarian cancer [111]. However, there is as yet no clinical evidence that this conversion also occurs in vivo. The scarcity of precancerous lesions or early stage carcinoma in situ detected by pathologies [112], raises questions whether this hypothesis is supported by clinical data. In parallel, the hypothesis that the etiology of HG-SOC could be explained by malignant transformation in neighboring tubal epithelium has gained increasing support. The fallopian tube develops from the Müllerian duct tissue, encompassing a columnar epithelial monolayer, which continues to express PAX8, the main cellular marker of "Müllerian differentiation" – a phenotype which is reminiscent of cancer tissue from HG-SOC patients. A breakthrough came from a cohort of clinical studies in BRCA mutation carriers, who have a very high hereditary risk of developing ovarian and breast cancer later in life. Small malignancies within the tubal epithelium (so-called serous tubal intraepithelial carcinoma – STIC) were discovered in up to 10% of patients who underwent prophylactic surgery to remove both fallopian tubes and ovaries [113]. These apparently healthy patients thus already had cancer in their fallopian tubes, which had not yet metastasized to the ovary. Samples from HG-SOC patients subsequently confirmed that STICS were present in up to 60% of cases [114]. Although not definitive proof, these findings strongly support the hypothesis that the fallopian tube epithelium is the tissue of origin of serous ovarian cancer. Recently, malignant transformation of the fallopian epithelium into full-blown ovarian cancer has been successfully triggered in a transgenic mouse model [115]. Here, the Cre recombinase system was used under control of the PAX8 promoter to introduce mutations into Brca1/2, Tp53 and Pten, which are frequently found mutated in HG-SOC. The animals developed malignant disease with a HG-SOC phenotype, which metastasized to the ovary, peritoneum or liver – modes of spread that are also found in patients, and were for a long time considered as evidence for separate origins of the disease. Regardless of the important function that BRCA1/2-mediated DNA repair mechanisms have in fallopian tube cells, in order to solve the complex problem of HGSC etiology it is necessary to understand the role of the p53 tumor suppressor. p53 is mutated in nearly all HG-SOC patients, but somatic mutations of p53 on their own do not increase the likelihood of HGSC development [116]. Nevertheless, nuclear accumulation of p53 is much more frequent in tubes where STICs are also found, arguing that this phenotype is perhaps characteristic of a cellular state which is prone to transformation upon further "hits". However, a valid causal relationship between these cellular atypias and the appearance of fully transformed malignant serous tubal intraepithelial carcinoma (STICs) is not yet conclusively proven. Therefore, true premalignant lesions in the fallopian tube as precursors of HG-SOC are yet to be defined.

An understanding of the molecular mechanisms of epithelial homeostasis and renewal in healthy fallopian tube tissue is thus likely to be essential for successful resolution of the molecular events that lead to serous ovarian cancer. Of particular interest in this context is how slow-cycling adult stem cells in the distal fallopian tube respond to long-term changes in the tissue microenvironment from numerous different stimuli such as inflammation, genotoxic stress changes in extracellular matrix, or the presence of pathogens. The role of the microenviroment via cooperation of different cellular compartments and extracellular matrix not only provides conditions for physiologically regulated responses, such are repair and healing, but can also decisively influence the progression of disease. As discussed in the section on cancer stem cells, cancer tissue is heterogeneous and keeping in mind the monoclonal origin of cancers, it is unclear whether differentiated cells are subject to transformation followed by expansion, whether some of them are reprogrammed into cancer stem cells, or whether deregulated stem cells are the source of the malignancy after additional somatic mutations. The latter hypothesis is somewhat more likely, given the presence of cells with stem cell characteristics in tumors, and that the alternative would mean acquisition of pluripotency. However, at this stage reverse reprogramming of differentiated cells cannot be excluded as the underlying mechanism.

The same is true for endometrial cancer. In contrast to the complete absence of diagnostic tools for early detection of ovarian cancer, there is a detailed classification of neoplastic changes in the endometrium – known as endometrial intraepithelial neoplasia (EIN), although there is a lot of controversy among pathologists how reliable the existing methodology is as a prognostic factor. Moreover, there is still no consensus regarding the cellular origin of the malignant tissue, or indeed the role played by adult stem cells. Approximately 75-80% of endometrial cancers are classified as type 1-endometrioid endometrial cancer (EEC) – with the remaining cases belonging to papillary, mucinous or clear cell types. Histologically, they consist of malignant endometrial cells in the columnar monolayer, although squamous metaplasia is sometimes observed [117]. They frequently harbor alterations in PTEN and Wnt signaling pathways, and a prevalent staining pattern of nuclear catenin localization is found in up to 60% of endometrial hyperplasias and 30% of endometrial cancers [118]. In a mouse model with constitutively active Wnt signaling due to APC deletion in the genital tract, loss of PTEN function was found to be the rate-limiting step of carcinogenesis inside the uterus, while activation of β-catenin signaling increased the severity of the disease [119]. Strikingly, a follow-up study of the mice that were initially declared tumor free by inspection of the uterus revealed that 62% did get sendometrioid cancer, but not in the uterus, as would be expected, but rather in the neighboring distal oviduct [120]. Thus, this recent study argues that the fallopian tube may participate not only in the carcinogenesis HG-SOC but also in a subset of endometrial cancers. Notably, in human patients loss of PTEN associates with better prognosis and reduced risk of metastasis, since the tumor appears more differentiated [121]. These at first sight contradictory facts might be explained by the different requirements of cancer cells in the early and late stages of progression. While PTEN is a tumor suppressor, its major downstream effector, p-Akt, has dual functions in tumor cells. In mammary carcinoma, activated Akt enhances carcinogenesis via the ErbB-2 pathway, but inhibits invasion of the tumor and its metastasis [122]. Of course, mutations in the PTEN-PI3 kinase pathway components represent only one of the important signaling routes that have been found to be altered in endometrial cancer. Similarly, as in ovarian cancer, it remains to be seen how discrete cytological changes progress into full-blown cancer and which cells give rise to the cancer stem cells that are detected in the later stages of malignant disease. A recent study, based on a detailed analysis of 113 cases of endometrial cancer, postulated an important role for SALL4 protein which is known as a strong determinator of pluripotency in human embryonic cells [123]. The authors found SALL4 expressed in 47% of cancer samples while no expression was detected either in healthy controls or in the hyperplastic tissue, and increased levels of expression negatively influenced prognosis and patient survival. SALL4 increases c-Myc transcriptional activity and reduces the response of affected cells to carboplatin treatment. While it cannot be excluded that SALL4 expression in precancerous stages is limited to very low levels in sparse stem cells below detection limit of whole tissue sample, these results strongly indicate that pluripotent capacity of the tumor is acquired at later stages of disease development. The existence of cells with stem cell characteristics has recently also been demonstrated for cervical cancer. Sox2-expressing cells constitute a small percentage of cells in cervical cancer cell lines, but they show much greater tumor forming capacity in the xenograft mouse model than the remaining cells [124]. This correlates well with findings of tracing experiments from the Sox2 mouse model mentioned above, which identified stem cell lineages in the cervix [125]. It is becoming increasingly clear that the basal layer of cervical epithelium plays a decisive role in the initiation of cervical cancer, as these are the only dividing cells in squamous stratified epithelium. Active progression of the cell cycle is a prerequisite for HPV-driven cellular transformation, as host cells have to pass through the prophase of mitosis for transcription of viral genes to be initiated and infection established [126]. Cervical cancer develops from premalignant lesions called cervical intraepithelial neoplasia (CIN), which are routinely detected by regular Pap smears. Importantly, CINs occur almost exclusively within the transformation zone of the cervix, a region of metaplastic conversion of columnar to squamous epithelium. Regular screening has greatly improved the long-term prognosis and survival rates of cervical cancer [127].

The mounting evidence for an involvement of the fallopian tube in development of ovarian and potentially other cancers thus highlights the importance of developing methodologies to improve diagnostic sampling and visualization of this organ. In particular, there is an imperative to improve diagnostics of the tube in patients. Ideally, some kind of endoscopy would enable detailed exploration of the mucosa in vivo, including the taking of biopsies. This would improve our understanding of phenotypical changes that take place in the epithelium, better define and categorize alterations and discover real premalignant lesions or early malignancies. The result would hopefully be comparable to the advancement that routine colonoscopy brought to early diagnosis and proper management of colon cancer.

10. Conclusion

We have presented in detail the current "state of the art" in the field of adult stem cells of the female reproductive system, their role in maintaining healthy mucosal tissue and changes that occur during disease. Moreover, we have focused on the phenomenon of cancer stem cells, which appears to be very important for the progression of malignancies by giving tumor tissue a competitive advantage. Still, the question remains how adult stem cells relate to cancer stem cells, and the models which have been postulated so far, as in the case of cervical cancer, have yet to be experimentally proven.

Regardless of the final outcome, it is almost certain that adult stem cells play an important role in the process of cancer initiation. Carcinogenesis is considered to be a long, stepwise process of accumulation of mutations. Therefore, differentiated cells with short life spans are unlikely to pass on mutations unless the initial acquired change leads to immortalization. Adult stem cells on the other hand, are long-lived and thereby inevitably accumulate mutations over decades. With each asymmetric division, mutations are passed to the differentiating progeny. The final "transformation step" may occur afterwards in the differentiated cell, or the stem cell itself may reach the stage where it becomes cancerous. The latter model presumes that as tumor tissue grows, the stem cell continues to give rise to differentiated progeny, while becoming itself a cancer stem cell. In both cases, a definitive premalignant molecular fingerprint should be found in the original adult stem cell. Therefore, it will be of great importance to further characterize adult stem cells in the ovary, fallopian tube and uterus and elucidate the molecular mechanisms of epithelial renewal in healthy tissue, but also to determine the genomic changes which occur over time in response to altered hormonal stimulation, tissue injury or infection.

Recent studies with novel genetic lineage tracing tools in the mouse give a promising outlook that concrete molecular pathways controlling stemness will be defined in the near future. Only when this goal has been achieved is it conceivable that a major breakthrough in the early treatment and diagnosis of female reproductive cancers and endometriosis can be achieved. As we have outlined, the high activity of adult stem cells in the female reproductive tract that is required to maintain cyclical changes in tissue architecture put these tissues at a particularly high risk for accumulating mutations with the potential for transformation. This is likely to be exacerbated by the fact that the genital system is exposed to a variety of sexually transmitted pathogens. There is mounting evidence that these infections may play a role in initiation of malignancies in the ovary, uterus or as co-factors to HPV in the cervix [18-20]. Although deciphering the behavior of adult stem cells in disease remains a very challenging research area, a dynamic field of translational approaches has emerged, using stem cells as a source of healthy tissue in different models of in vitro differentiation and transplantation. There is justified optimism that such models will help to elucidate not only the events involved in deregulation of the adult stem cell niche, but also the interaction of human pathogens with this niche, as well as the somatic tissue derived from it. A more complete understanding of these events will provide a basis for the development of more effective preventative and therapeutic measures for diseases of the female reproductive tract.

Author details

Mirjana Kessler*, Rike Zietlow and Thomas F. Meyer

*Address all correspondence to: kessler@mpiib-berlin.mpg.de

Max Planck Institute for Infection Biology, Department of Molecular Biology, Berlin, Germany

References

[1] Kiel MJ, He S, Ashkenazi R, Gentry SN, Teta M, et al. (2007) Haematopoietic stem cells do not asymmetrically segregate chromosomes or retain BrdU. Nature 449: 238-242.

[2] Chan RW, Gargett CE (2006) Identification of label-retaining cells in mouse endometrium. Stem Cells 24: 1529-1538.

[3] Flesken-Nikitin A, Hwang CI, Cheng CY, Michurina TV, Enikolopov G, et al. (2013) Ovarian surface epithelium at the junction area contains a cancer-prone stem cell niche. Nature 495: 241-245.

[4] Patterson AL, Pru JK (2013) Long-term label retaining cells localize to distinct regions within the female reproductive epithelium. Cell Cycle 12: 2888-2898.

[5] Xie T, Spradling AC (2000) A niche maintaining germ line stem cells in the Drosophila ovary. Science 290: 328-330.

[6] Johnson J, Canning J, Kaneko T, Pru JK, Tilly JL (2004) Germline stem cells and follicular renewal in the postnatal mammalian ovary. Nature 428: 145-150.

[7] Zou K, Yuan Z, Yang Z, Luo H, Sun K, et al. (2009) Production of offspring from a germline stem cell line derived from neonatal ovaries. Nat Cell Biol 11: 631-636.

[8] White YA, Woods DC, Takai Y, Ishihara O, Seki H, et al. (2012) Oocyte formation by mitotically active germ cells purified from ovaries of reproductive-age women. Nat Med 18: 413-421.

[9] Behrman HR, Kodaman PH, Preston SL, Gao S (2001) Oxidative stress and the ovary. J Soc Gynecol Investig 8: S40-42.

[10] Murdoch WJ (2005) Carcinogenic potential of ovulatory genotoxicity. Biol Reprod 73: 586-590.

[11] Murdoch WJ, McDonnel AC (2002) Roles of the ovarian surface epithelium in ovulation and carcinogenesis. Reproduction 123: 743-750.

[12] Ahmed N, Maines-Bandiera S, Quinn MA, Unger WG, Dedhar S, et al. (2006) Molecular pathways regulating EGF-induced epithelio-mesenchymal transition in human ovarian surface epithelium. Am J Physiol Cell Physiol 290: C1532-1542.

[13] Zhu Y, Nilsson M, Sundfeldt K (2010) Phenotypic plasticity of the ovarian surface epithelium: TGF-beta 1 induction of epithelial to mesenchymal transition (EMT) in vitro. Endocrinology 151: 5497-5505.

[14] George SH, Milea A, Shaw PA (2012) Proliferation in the normal FTE is a hallmark of the follicular phase, not BRCA mutation status. Clin Cancer Res 18: 6199-6207.

[15] George SH, Greenaway J, Milea A, Clary V, Shaw S, et al. (2011) Identification of abrogated pathways in fallopian tube epithelium from BRCA1 mutation carriers. J Pathol 225: 106-117.

[16] King SM, Hilliard TS, Wu LY, Jaffe RC, Fazleabas AT, et al. (2011) The impact of ovulation on fallopian tube epithelial cells: evaluating three hypotheses connecting ovulation and serous ovarian cancer. Endocr Relat Cancer 18: 627-642.

[17] Bahar-Shany K, Brand H, Sapoznik S, Jacob-Hirsch J, Yung Y, et al. (2014) Exposure of fallopian tube epithelium to follicular fluid mimics carcinogenic changes in precursor lesions of serous papillary carcinoma. Gynecol Oncol 132: 322-327.

[18] Alibek K, Karatayeva N, Bekniyazov I (2012) The role of infectious agents in urogenital cancers. Infect Agent Cancer 7: 35.

[19] Madeleine MM, Anttila T, Schwartz SM, Saikku P, Leinonen M, et al. (2007) Risk of cervical cancer associated with Chlamydia trachomatis antibodies by histology, HPV type and HPV cofactors. Int J Cancer 120: 650-655.

[20] Shanmughapriya S, Senthilkumar G, Vinodhini K, Das BC, Vasanthi N, et al. (2012) Viral and bacterial aetiologies of epithelial ovarian cancer. Eur J Clin Microbiol Infect Dis 31: 2311-2317.

[21] Brosens JJ, Hayashi N, White JO (1999) Progesterone receptor regulates decidual prolactin expression in differentiating human endometrial stromal cells. Endocrinology 140: 4809-4820.

[22] Matsumoto H, Sakai K, Iwashita M (2008) Insulin-like growth factor binding protein-1 induces decidualization of human endometrial stromal cells via alpha5beta1 integrin. Mol Hum Reprod 14: 485-489.

[23] Zuckerman S (1951) The number of oocytes in the mature ovary. Recent Progress in Hormone Research 6: 63-109.

[24] Zhang H, Zheng W, Shen Y, Adhikari D, Ueno H, et al. (2012) Experimental evidence showing that no mitotically active female germline progenitors exist in postnatal mouse ovaries. Proc Natl Acad Sci U S A 109: 12580-12585.

[25] Johnson J, Bagley J, Skaznik-Wikiel M, Lee HJ, Adams GB, et al. (2005) Oocyte generation in adult mammalian ovaries by putative germ cells in bone marrow and peripheral blood. Cell 122: 303-315.

[26] Bukovsky A, Gupta SK, Virant-Klun I, Upadhyaya NB, Copas P, et al. (2008) Study origin of germ cells and formation of new primary follicles in adult human and rat ovaries. Methods Mol Biol 450: 233-265.

[27] Niikura Y, Niikura T, Tilly JL (2009) Aged mouse ovaries possess rare premeiotic germ cells that can generate oocytes following transplantation into a young host environment. Aging (Albany NY) 1: 971-978.

[28] Pacchiarotti J, Maki C, Ramos T, Marh J, Howerton K, et al. (2010) Differentiation potential of germ line stem cells derived from the postnatal mouse ovary. Differentiation 79: 159-170.

[29] Zhang D, Fouad H, Zoma WD, Salama SA, Wentz MJ, et al. (2008) Expression of stem and germ cell markers within nonfollicle structures in adult mouse ovary. Reprod Sci 15: 139-146.

[30] Virant-Klun I, Zech N, Rozman P, Vogler A, Cvjeticanin B, et al. (2008) Putative stem cells with an embryonic character isolated from the ovarian surface epithelium of women with no naturally present follicles and oocytes. Differentiation 76: 843-856.

[31] Byskov AG, Hoyer PE, Yding Andersen C, Kristensen SG, Jespersen A, et al. (2011) No evidence for the presence of oogonia in the human ovary after their final clearance during the first two years of life. Hum Reprod 26: 2129-2139.

[32] Kerr JB, Duckett R, Myers M, Britt KL, Mladenovska T, et al. (2006) Quantification of healthy follicles in the neonatal and adult mouse ovary: evidence for maintenance of primordial follicle supply. Reproduction 132: 95-109.

[33] Lei L, Spradling AC (2013) Female mice lack adult germ-line stem cells but sustain oogenesis using stable primordial follicles. Proc Natl Acad Sci U S A 110: 8585-8590.

[34] Bristol-Gould SK, Kreeger PK, Selkirk CG, Kilen SM, Mayo KE, et al. (2006) Fate of the initial follicle pool: empirical and mathematical evidence supporting its sufficiency for adult fertility. Dev Biol 298: 149-154.

[35] Eggan K, Jurga S, Gosden R, Min IM, Wagers AJ (2006) Ovulated oocytes in adult mice derive from non-circulating germ cells. Nature 441: 1109-1114.

[36] Kujjo LL, Chang EA, Pereira RJ, Dhar S, Marrero-Rosado B, et al. (2011) Chemotherapy-induced late transgenerational effects in mice. PLoS One 6: e17877.

[37] Mikedis MM, Downs KM (2014) Mouse primordial germ cells: a reappraisal. Int Rev Cell Mol Biol 309: 1-57.

[38] Mikedis MM, Downs KM (2012) STELLA-positive subregions of the primitive streak contribute to posterior tissues of the mouse gastrula. Dev Biol 363: 201-218.

[39] Reizel Y, Itzkovitz S, Adar R, Elbaz J, Jinich A, et al. (2012) Cell lineage analysis of the mammalian female germline. PLoS Genet 8: e1002477.

[40] Abban G, Johnson J (2009) Stem cell support of oogenesis in the human. Hum Reprod 24: 2974-2978.

[41] De Felici M (2010) Germ stem cells in the mammalian adult ovary: considerations by a fan of the primordial germ cells. Mol Hum Reprod 16: 632-636.

[42] De Felici M, Barrios F (2013) Seeking the origin of female germline stem cells in the mammalian ovary. Reproduction 146: R125-130.

[43] Labosky PA, Barlow DP, Hogan BL (1994) Embryonic germ cell lines and their derivation from mouse primordial germ cells. Ciba Found Symp 182: 157-168; discussion 168-178.

[44] Shamblott MJ, Axelman J, Wang S, Bugg EM, Littlefield JW, et al. (1998) Derivation of pluripotent stem cells from cultured human primordial germ cells. Proc Natl Acad Sci U S A 95: 13726-13731.

[45] Turnpenny L, Brickwood S, Spalluto CM, Piper K, Cameron IT, et al. (2003) Derivation of human embryonic germ cells: an alternative source of pluripotent stem cells. Stem Cells 21: 598-609.

[46] Bristol-Gould SK, Kreeger PK, Selkirk CG, Kilen SM, Cook RW, et al. (2006) Postnatal regulation of germ cells by activin: the establishment of the initial follicle pool. Dev Biol 298: 132-148.

[47] Kerr JB, Myers M, Anderson RA (2013) The dynamics of the primordial follicle reserve. Reproduction 146: R205-215.

[48] Padykula HA, Coles LG, Okulicz WC, Rapaport SI, McCracken JA, et al. (1989) The basalis of the primate endometrium: a bifunctional germinal compartment. Biol Reprod 40: 681-690.

[49] Wang Y, Sacchetti A, van Dijk MR, van der Zee M, van der Horst PH, et al. (2012) Identification of quiescent, stem-like cells in the distal female reproductive tract. PLoS One 7: e40691.

[50] Fukuda K, Yuasa S (2006) Stem cells as a source of regenerative cardiomyocytes. Circ Res 98: 1002-1013.

[51] Tao XR, Li WL, Su J, Jin CX, Wang XM, et al. (2009) Clonal mesenchymal stem cells derived from human bone marrow can differentiate into hepatocyte-like cells in injured livers of SCID mice. J Cell Biochem 108: 693-704.

[52] Wu R, Tang Y, Zang W, Wang Y, Li M, et al. (2014) MicroRNA-128 regulates the differentiation of rat bone mesenchymal stem cells into neuron-like cells by Wnt signaling. Mol Cell Biochem 387: 151-158.

[53] Du H, Taylor HS (2007) Contribution of bone marrow-derived stem cells to endometrium and endometriosis. Stem Cells 25: 2082-2086.

[54] Taylor HS (2004) Endometrial cells derived from donor stem cells in bone marrow transplant recipients. JAMA 292: 81-85.

[55] Ikoma T, Kyo S, Maida Y, Ozaki S, Takakura M, et al. (2009) Bone marrow-derived cells from male donors can compose endometrial glands in female transplant recipients. Am J Obstet Gynecol 201: 608.e601-608.

[56] Cervello I, Mas A, Gil-Sanchis C, Peris L, Faus A, et al. (2011) Reconstruction of endometrium from human endometrial side population cell lines. PLoS One 6: e21221.

[57] Alvarez-Dolado M, Pardal R, Garcia-Verdugo JM, Fike JR, Lee HO, et al. (2003) Fusion of bone-marrow-derived cells with Purkinje neurons, cardiomyocytes and hepatocytes. Nature 425: 968-973.

[58] Gargett CE, Schwab KE, Zillwood RM, Nguyen HP, Wu D (2009) Isolation and culture of epithelial progenitors and mesenchymal stem cells from human endometrium. Biol Reprod 80: 1136-1145.

[59] Schwab KE, Chan RW, Gargett CE (2005) Putative stem cell activity of human endometrial epithelial and stromal cells during the menstrual cycle. Fertil Steril 84 Suppl 2: 1124-1130.

[60] Kato K, Yoshimoto M, Kato K, Adachi S, Yamayoshi A, et al. (2007) Characterization of side-population cells in human normal endometrium. Hum Reprod 22: 1214-1223.

[61] Goodell MA, Brose K, Paradis G, Conner AS, Mulligan RC (1996) Isolation and functional properties of murine hematopoietic stem cells that are replicating in vivo. J Exp Med 183: 1797-1806.

[62] Scharenberg CW (2002) The ABCG2 transporter is an efficient Hoechst 33342 efflux pump and is preferentially expressed by immature human hematopoietic progenitors. Blood 99: 507-512.

[63] Masuda H, Matsuzaki Y, Hiratsu E, Ono M, Nagashima T, et al. (2010) Stem cell-like properties of the endometrial side population: implication in endometrial regeneration. PLoS One 5: e10387.

[64] Janzen DM, Cheng D, Schafenacker AM, Paik DY, Goldstein AS, et al. (2013) Estrogen and progesterone together expand murine endometrial epithelial progenitor cells. Stem Cells 31: 808-822.

[65] Wolff EF, Gao XB, Yao KV, Andrews ZB, Du H, et al. (2011) Endometrial stem cell transplantation restores dopamine production in a Parkinson's disease model. J Cell Mol Med 15: 747-755.

[66] Meng X, Ichim TE, Zhong J, Rogers A, Yin Z, et al. (2007) Endometrial regenerative cells: a novel stem cell population. J Transl Med 5: 57.

[67] Bockeria L, Bogin V, Bockeria O, Le T, Alekyan B, et al. (2013) Endometrial regenerative cells for treatment of heart failure: a new stem cell enters the clinic. J Transl Med 11: 56.

[68] Santamaria X, Massasa EE, Feng Y, Wolff E, Taylor HS (2011) Derivation of insulin producing cells from human endometrial stromal stem cells and use in the treatment of murine diabetes. Mol Ther 19: 2065-2071.

[69] Houston DE (1984) Evidence for the risk of pelvic endometriosis by age, race and socioeconomic status. Epidemiol Rev 6: 167-191.

[70] Leyendecker G, Herbertz M, Kunz G, Mall G (2002) Endometriosis results from the dislocation of basal endometrium. Hum Reprod 17: 2725-2736.

[71] Dracxler RC, Oh C, Kalmbach K, Wang F, Liu L, et al. (2014) Peripheral Blood Telomere Content Is Greater in Patients With Endometriosis Than in Controls. Reprod Sci. Epub ahead of print.

[72] Paik DY, Janzen DM, Schafenacker AM, Velasco VS, Shung MS, et al. (2012) Stem-like epithelial cells are concentrated in the distal end of the fallopian tube: a site for injury and serous cancer initiation. Stem Cells 30: 2487-2497.

[73] Kodithuwakku SP, Pang RT, Ng EH, Cheung AN, Horne AW, et al. (2012) Wnt activation downregulates olfactomedin-1 in Fallopian tubal epithelial cells: a microenvironment predisposed to tubal ectopic pregnancy. Lab Invest 92: 256-264.

[74] Li P, Zhu WJ, Ma ZL, Wang G, Peng H, et al. (2013) Enhanced beta-catenin expression and inflammation are associated with human ectopic tubal pregnancy. Hum Reprod 28: 2363-2371.

[75] Matsuzaki S, Darcha C (2013) In vitro effects of a small-molecule antagonist of the Tcf/ss-catenin complex on endometrial and endometriotic cells of patients with endometriosis. PLoS One 8: e61690.

[76] Kessler M, Zielecki J, Thieck O, Mollenkopf HJ, Fotopoulou C, et al. (2012) Chlamydia trachomatis disturbs epithelial tissue homeostasis in fallopian tubes via paracrine Wnt signaling. Am J Pathol 180: 186-198.

[77] Singavarapu R, Buchinsky N, Cheon DJ, Orsulic S (2010) Whole ovary immunohistochemistry for monitoring cell proliferation and ovulatory wound repair in the mouse. Reprod Biol Endocrinol 8: 98.

[78] Szotek PP, Chang HL, Brennand K, Fujino A, Pieretti-Vanmarcke R, et al. (2008) Normal ovarian surface epithelial label-retaining cells exhibit stem/progenitor cell characteristics. Proc Natl Acad Sci U S A 105: 12469-12473.

[79] Barker N, Clevers H (2010) Leucine-rich repeat-containing G-protein-coupled receptors as markers of adult stem cells. Gastroenterology 138: 1681-1696.

[80] Snippert HJ, Haegebarth A, Kasper M, Jaks V, van Es JH, et al. (2010) Lgr6 marks stem cells in the hair follicle that generate all cell lineages of the skin. Science 327: 1385-1389.

[81] Luo W, Rodriguez M, Valdez JM, Zhu X, Tan K, et al. (2013) Lgr4 is a key regulator of prostate development and prostate stem cell differentiation. Stem Cells 31: 2492-2505.

[82] Carroll TJ, Park JS, Hayashi S, Majumdar A, McMahon AP (2005) Wnt9b plays a central role in the regulation of mesenchymal to epithelial transitions underlying organogenesis of the mammalian urogenital system. Dev Cell 9: 283-292.

[83] Mericskay M, Kitajewski J, Sassoon D (2004) Wnt5a is required for proper epithelial-mesenchymal interactions in the uterus. Development 131: 2061-2072.

[84] Miller C, Sassoon DA (1998) Wnt-7a maintains appropriate uterine patterning during the development of the mouse female reproductive tract. Development 125: 3201-3211.

[85] Morin PJ (1997) Activation of beta-Catenin-Tcf Signaling in Colon Cancer by Mutations in beta-Catenin or APC. Science 275: 1787-1790.

[86] Anastas JN, Moon RT (2013) WNT signalling pathways as therapeutic targets in cancer. Nat Rev Cancer 13: 11-26.

[87] Montgomery RK, Carlone DL, Richmond CA, Farilla L, Kranendonk ME, et al. (2011) Mouse telomerase reverse transcriptase (mTert) expression marks slowly cycling intestinal stem cells. Proc Natl Acad Sci U S A 108: 179-184.

[88] Takeda N, Jain R, LeBoeuf MR, Wang Q, Lu MM, et al. (2011) Interconversion between intestinal stem cell populations in distinct niches. Science 334: 1420-1424.

[89] Tian H, Biehs B, Warming S, Leong KG, Rangell L, et al. (2011) A reserve stem cell population in small intestine renders Lgr5-positive cells dispensable. Nature 478: 255-259.

[90] Metcalfe C, Kljavin NM, Ybarra R, de Sauvage FJ (2014) Lgr5+stem cells are indispensable for radiation-induced intestinal regeneration. Cell Stem Cell 14: 149-159.

[91] VanDussen KL, Carulli AJ, Keeley TM, Patel SR, Puthoff BJ, et al. (2012) Notch signaling modulates proliferation and differentiation of intestinal crypt base columnar stem cells. Development 139: 488-497.

[92] Barker N, van Es JH, Kuipers J, Kujala P, van den Born M, et al. (2007) Identification of stem cells in small intestine and colon by marker gene Lgr5. Nature 449: 1003-1007.

[93] He XC, Zhang J, Tong WG, Tawfik O, Ross J, et al. (2004) BMP signaling inhibits intestinal stem cell self-renewal through suppression of Wnt-beta-catenin signaling. Nat Genet 36: 1117-1121.

[94] Reya T, Clevers H (2005) Wnt signalling in stem cells and cancer. Nature 434: 843-850.

[95] Walker L, Lynch M, Silverman S, Fraser J, Boulter J, et al. (1999) The Notch/Jagged pathway inhibits proliferation of human hematopoietic progenitors in vitro. Stem Cells 17: 162-171.

[96] Kato K, Takao T, Kuboyama A, Tanaka Y, Ohgami T, et al. (2010) Endometrial cancer side-population cells show prominent migration and have a potential to differentiate into the mesenchymal cell lineage. Am J Pathol 176: 381-392.

[97] Mirantes C, Eritja N, Dosil MA, Santacana M, Pallares J, et al. (2013) An inducible knockout mouse to model the cell-autonomous role of PTEN in initiating endometrial, prostate and thyroid neoplasias. Dis Model Mech 6: 710-720.

[98] Szotek PP, Pieretti-Vanmarcke R, Masiakos PT, Dinulescu DM, Connolly D, et al. (2006) Ovarian cancer side population defines cells with stem cell-like characteristics and Mullerian Inhibiting Substance responsiveness. Proc Natl Acad Sci U S A 103: 11154-11159.

[99] Yap TA, Carden CP, Kaye SB (2009) Beyond chemotherapy: targeted therapies in ovarian cancer. Nat Rev Cancer 9: 167-181.

[100] Silva IA, Bai S, McLean K, Yang K, Griffith K, et al. (2011) Aldehyde dehydrogenase in combination with CD133 defines angiogenic ovarian cancer stem cells that portend poor patient survival. Cancer Res 71: 3991-4001.

[101] Landen CN, Jr., Goodman B, Katre AA, Steg AD, Nick AM, et al. (2010) Targeting aldehyde dehydrogenase cancer stem cells in ovarian cancer. Mol Cancer Ther 9: 3186-3199.

[102] Rutella S, Bonanno G, Procoli A, Mariotti A, Corallo M, et al. (2009) Cells with characteristics of cancer stem/progenitor cells express the CD133 antigen in human endometrial tumors. Clin Cancer Res 15: 4299-4311.

[103] Gao MQ, Choi YP, Kang S, Youn JH, Cho NH (2010) CD24+cells from hierarchically organized ovarian cancer are enriched in cancer stem cells. Oncogene 29: 2672-2680.

[104] Meirelles K, Benedict LA, Dombkowski D, Pepin D, Preffer FI, et al. (2012) Human ovarian cancer stem/progenitor cells are stimulated by doxorubicin but inhibited by Mullerian inhibiting substance. Proc Natl Acad Sci U S A 109: 2358-2363.

[105] Burleson KM, Casey RC, Skubitz KM, Pambuccian SE, Oegema TR, Jr., et al. (2004) Ovarian carcinoma ascites spheroids adhere to extracellular matrix components and mesothelial cell monolayers. Gynecol Oncol 93: 170-181.

[106] Lee JM, Mhawech-Fauceglia P, Lee N, Parsanian LC, Lin YG, et al. (2013) A three-dimensional microenvironment alters protein expression and chemosensitivity of epithelial ovarian cancer cells in vitro. Lab Invest 93: 528-542.

[107] Ferlay J, Soerjomataram I, Ervik M, Dikshit R, Eser S, et al. (2013) GLOBOCAN 2012 v1.0, Cancer Incidence and Mortality Worldwide: IARC CancerBase No. 11 [Internet]. Lyon, France: International Agency for Research on Cancer; 2013. Available from: http://globocan.iarc.fr, (accessed on 30 April 2014)

[108] Bosch FX, Lorincz A, Munoz N, Meijer CJ, Shah KV (2002) The causal relation between human papillomavirus and cervical cancer. J Clin Pathol 55: 244-265.

[109] Baldwin LA, Huang B, Miller RW, Tucker T, Goodrich ST, et al. (2012) Ten-year relative survival for epithelial ovarian cancer. Obstet Gynecol 120: 612-618.

[110] Ahmed N, Thompson EW, Quinn MA (2007) Epithelial-mesenchymal interconversions in normal ovarian surface epithelium and ovarian carcinomas: an exception to the norm. J Cell Physiol 213: 581-588.

[111] Cheng W, Liu J, Yoshida H, Rosen D, Naora H (2005) Lineage infidelity of epithelial ovarian cancers is controlled by HOX genes that specify regional identity in the reproductive tract. Nat Med 11: 531-537.

[112] Aoki Y, Kawada N, Tanaka K (2000) Early form of ovarian cancer originating in inclusion cysts. A case report. J Reprod Med 45: 159-161.

[113] Kindelberger DW, Lee Y, Miron A, Hirsch MS, Feltmate C, et al. (2007) Intraepithelial carcinoma of the fimbria and pelvic serous carcinoma: Evidence for a causal relationship. Am J Surg Pathol 31: 161-169.

[114] Przybycin CG, Kurman RJ, Ronnett BM, Shih Ie M, Vang R (2010) Are all pelvic (nonuterine) serous carcinomas of tubal origin? Am J Surg Pathol 34: 1407-1416.

[115] Perets R, Wyant GA, Muto KW, Bijron JG, Poole BB, et al. (2013) Transformation of the fallopian tube secretory epithelium leads to high-grade serous ovarian cancer in Brca;Tp53;Pten models. Cancer Cell 24: 751-765.

[116] Xian W, Miron A, Roh M, Semmel DR, Yassin Y, et al. (2010) The Li-Fraumeni syndrome (LFS): a model for the initiation of p53 signatures in the distal Fallopian tube. J Pathol 220: 17-23.

[117] Zaino RJ, Kurman R, Herbold D, Gliedman J, Bundy BN, et al. (1991) The significance of squamous differentiation in endometrial carcinoma. Data from a Gynecologic Oncology Group study. Cancer 68: 2293-2302.

[118] Nei H, Saito T, Yamasaki H, Mizumoto H, Ito E, et al. (1999) Nuclear localization of beta-catenin in normal and carcinogenic endometrium. Mol Carcinog 25: 207-218.

[119] van der Zee M, Jia Y, Wang Y, Heijmans-Antonissen C, Ewing PC, et al. (2013) Alterations in Wnt-beta-catenin and Pten signalling play distinct roles in endometrial cancer initiation and progression. J Pathol 230: 48-58.

[120] van der Horst PH, van der Zee M, Heijmans-Antonissen C, Jia Y, Demayo FJ, et al. (2014) A mouse model for endometrioid ovarian cancer arising from the distal oviduct. Int J Cancer 135: 1028-1037.

[121] Akiyama-Abe A, Minaguchi T, Nakamura Y, Michikami H, Shikama A, et al. (2013) Loss of PTEN expression is an independent predictor of favourable survival in endometrial carcinomas. Br J Cancer 109: 1703-1710.

[122] Dillon RL, Marcotte R, Hennessy BT, Woodgett JR, Mills GB, et al. (2009) Akt1 and akt2 play distinct roles in the initiation and metastatic phases of mammary tumor progression. Cancer Res 69: 5057-5064.

[123] Li A, Jiao Y, Yong KJ, Wang F, Gao C, et al. (2013) SALL4 is a new target in endometrial cancer. Oncogene. doi: 10.1038/onc.2013.529.

[124] Liu XF, Yang WT, Xu R, Liu JT, Zheng PS (2014) Cervical cancer cells with positive Sox2 expression exhibit the properties of cancer stem cells. PLoS One 9: e87092.

[125] Arnold K, Sarkar A, Yram MA, Polo JM, Bronson R, et al. (2011) Sox2(+) adult stem and progenitor cells are important for tissue regeneration and survival of mice. Cell Stem Cell 9: 317-329.

[126] Pyeon D, Pearce SM, Lank SM, Ahlquist P, Lambert PF (2009) Establishment of human papillomavirus infection requires cell cycle progression. PLoS Pathog 5: e1000318.

[127] OECD (2011), "Screening, survival and mortality for breast cancer", in Health at a Glance 2011: OECD Indicators, OECD Publishing. http://dx.doi.org/10.1787/health_glance-2011-48-en (accessed 17 July 2014)

Section 3

Hepatic Stem Cell Niches

Chapter 7

Adult Hepatic Progenitor Cells

Peter Siig Vestentoft

Additional information is available at the end of the chapter

http://dx.doi.org/10.5772/58814

1. Introduction

The liver is the largest internal organ of the body only second in size compared to the skin. The liver not only functions as an endocrine and exocrine organ, but it also performs a multitude of vital functions including glycogen storage, detoxification and plasma protein synthesis [1–4]. The liver receives nutrients and environmental toxins from the digestive tract through the portal vein. This direct transport of potentially harmful agents is hypothesized to have exerted an evolutionary pressure on the liver to possess multiple pathways for regeneration [4,5]. In fact, the regenerative capacity of the liver is so enormous that this was renowned in ancient times and described in Mediterranean folklore. According to Greek mythology the Titan Prometheus stole fire from the Gods of Olympia and gave it to the mortals. As a consequence, Zeus, the king of Gods, chained Prometheus to a rock. An eagle would then appear each day and pecked out part of Prometheus' liver only to let it regenerate overnight [1]. This punishment was to be repeated for eternity, but according to one version of the story, Heracles (Hercules) eventually killed the eagle and freed Prometheus.

Despite of the famed renewal capacity of the liver, hepatic diseases constitute a worldwide problem. Hepatic diseases can broadly be divided into two major groups: acute and chronic liver diseases. Acute liver failure is characterized by the manifestation of sudden severe hepatic injury that can have several etiologies [6,7]. Frequent causes included viral hepatitis or drug intoxication, commonly paracetamol, leading to hepatic encephalopathy, coagulopathy and often progressive multiorgan failure [6,8–11]. In developed countries, acute liver failure is relatively rare with an incidence estimated between 1-6 cases per million people each year [6, 12]. In contrast, chronic liver diseases are caused by prolonged insults. Common causes include sustained alcohol consumption, non-alcoholic fatty liver disease and hepatitis B or C virus infection [2]. These insults can lead to hepatic fibrosis, a form of wound healing characterized by the presence of collagen-rich septae connecting the so-called portal areas. If untreated, this potentially reversible manifestation, can progress to end stage cirrhosis, where hepatic architecture is greatly disturbed and scar tissue encircles nodules of remaining hepatocytes [7,13–17]. Chronic liver diseases are estimated to affect 170 million patients worldwide. Those cases eventually progress to fibrosis and possibly cirrhosis in 25-30 % of these patients [2]. Where acute hepatic failure involves sudden massive cell death, chronic liver diseases are conversely characterized by continuous cell death [18–20].

When hepatic regeneration is hindered orthotopic liver transplantation is the only treatment that radically improves the outcome of hepatic failure [2,21]. However, the worldwide shortage of liver donors result in death of many patients waiting for transplantation [22]. Research into alternative methods of therapeutic treatment is therefore highly needed. The possibility of culturing hepatic stem cells holds the promise to treat certain liver diseases, even with autologous stem cells. This include correcting metabolic diseases characterized by inherited defects of hepatic enzymes or treating fulminant hepatic failure characterized by rapid onset of liver failure and death, when donor organs are unavailable [23]. The use of autologous stem cells would additionally prevent the lifespan administration of immunosuppressive agents currently employed to prevent allograft rejection. Therefore, there has been an increasing interest into using hepatic stem cell-based therapies as novel alternatives to traditional liver treatments. However, the stem cell biology of the liver is not well understood. In particular, the lack of specific markers for hepatic stem cell identification has hindered their characterization and isolation [24–28].

The present chapter will provide an overview of current knowledge of the rodent and human hepatic stem cell niche.

In particular, the chapter will go through the development of the hepatic stem cell niche, the associated extracellular matrix molecules and support cells. Attention will also be given to the various modes of hepatic regeneration and the involvement of hepatic stem cells in cancerous disease states.

2. Stem cells

Even though stem cells have been identified and characterized in several organs, no universally accepted definition of what constitutes a stem cell has been defined [29]. However, a broadly accepted view is that stem cells are cells that hold a capacity for unlimited or prolonged self-renewal and can also give rise to at least one type of highly differentiated progeny [30]. However, many classes of stem cell exist with different potentials. These range from the totipotent fertilized egg from which entire organisms develop over pluripotent embryonic stem cells that can give rise to the three germ layers to the unipotent tissue stem cells.

Typically, tissue or intra-organ stem cells are less differentiated cells that exist in a mitotically quiescent form [31]. This class of stem cells are so-called "determined", meaning that they lack markers of final differentiation, but are able to divide and differentiate into highly specialized effector cells [32,33]. When needed tissue stem cells are activated to divide and clonally regenerate the tissue in which they are located [32,34]. Upon activation, tissue stem cells would perform either symmetric or asymmetric cell division [30]. Symmetric stem cell division give rise to two daughter cells that themselves are stem cells, thereby maintaining the stem cell pool. Alternatively, this form for division may result in two daughter cells committed for differentiation. Asymmetric stem cell division, on the other hand, produces one stem cell and one differentiated daughter cell. Differentiated daughter cells are also known as progenitor cells or transit amplifying cells. They divide rapidly in order to generate a pool of continually more differentiated cells en route to replace senescent or damaged tissue cells [34]. Early progenitor cells hold multi-lineage potential and have characteristics similar to the parent stem cell whereas late progenitor cells are more differentiated and produce single-lineage progeny [32]. Therefore, even though stem cells have a high self-renewal capacity, they may divide relatively infrequently, whereas the transit amplifying cells greatly increase in number and differentiate into given tissue cells.

3. Stem cell niche

The potency of stem cells requires tight regulation of their behavior. Stem cell quiescence and activation must be regulated according to the needs of the organism. A critical actor in mediating the balanced response of stem cells to the needs of the organism is the stem cell niche.

The stem cell niche concept was first proposed by Schofield who conducted bone marrow studies [35]. It was suggested that stem cells reside in compartments that promote and maintain their characteristics [35]. It is believed that once postnatal tissues are formed, intra-organ stem cells reside in these special tissue microenvironments or niches. However, upon activation, the niche must change the composition of its microenvironment from favoring stem cell quiescence to induce stem cell activation and proliferation. Studies on *Drosophila spp.* gonads have helped understanding the factors constituting the stem cell niche and greatly expanded knowledge of stem cell activation and the generation of transit amplifying cells [36]. These studies have revealed a basal theme to reoccur. Structurally, the typical stem cell niche consists of stem cells resting on a scaffold of extracellular matrix components, having cell-cell interactions with differentiated neighboring cells [36–38]. In *Drosophila spp.* gonads, the extracellular matrix forms a repressing environment to stem cell differentiation, while promoting cellular adhesion [36]. Immediately outside this repressive zone, stem cell adherence is reduced while cellular differentiation is stimulated [36]. More specifically, integrins have been identified as key elements in this adhesion process. These transmembrane proteins that mediate adhesion to the extracellular matrix, are often highly expressed in stem cells and can suppress terminal differentiation in epidermal stem cells, for instance [39,40]. Conversely, the loss of integrins is associated with the epidermal stem cell niche disappearance, characterized by cellular differentiation [30].

The key factor to identify stem cell niches is the stem cell localization itself. For this purpose,label-retention assays may be applied, two common labels being ^{3}H-thymidine and the thymidine analog BrdU. Upon asymmetric cellular division, stem cells may incorporate either of these labels into their DNA thereby retaining 50% of the label with the resulting daughter stem cell and 50% with the transit amplifying cell. As transit amplifying cells are fast cycling the label is gradually diluted in the following chase period while the slow cycling daughter stem cell retain the marker. Use of these and similar label-retaining assays have been employed to identify the stem cell niche of the skin, hair follicle and peripheral cornea [41–45]. Though being an intriguing method for locating stem cells, label-retention techniques has certain disadvantages. Stem cells that did not enter the cell cycle during the labelling period will for instance remain unmarked, while progenitor cells that terminally differentiate and stop cell division can retain markers for longer periods of time [46]. Stem cells and their niches have, never the less, been identified in several organs. In vertebrates these include the bulge region of the hair follicle, the bone marrow and the lover region of the crypts in the small intestine [41,47–49]. The common denominator of these organs, however, is that they are characterized by a continuous supply of cells descending from the stem cells. Stem cells in tissues characterized by a lower cellular turnover are, on the contrary, more difficult to identify. One such organ is the liver, where mitotically quiescent hepatocytes have relatively long life spans and high proliferative capacities [50,51].

4. Hepatic anatomy

Although many cell types are present, the liver is characterized by two epithelial tissue components; cholangiocytes and hepatic cords containing hepatocytes, respectively. The hepatocytes secrete serum proteins, including albumin, and express monooxygenases from the cytochrome P450 family, the major enzymes involved in oxidative metabolism of xenobiotics [52]. Cholangiocytes, on the other hand, form biliary channels transporting bile from the liver towards the bile bladder.

Examination of hepatic tissue sections reveal an unvarying landscape of cords of hepatocytes with scattered central veins and so-called portal triads or portal tracts. The latter contain bile ducts and branches from the portal vein and portal artery, thereby forming a triad [53]. However, from a three-dimensional perspective, this dull landscape masks a highly complex organ [53,54]. Accurately defining the livers functional entities have historically been difficult, as multiple functions could be applied based on either enzymatic expression patterns or histological observations. A frequently used definition is the simple histological unit "lobule" (figure 1). The classic lobule is envisioned as a two-dimensional hexagonal structure centered around a central vein [55]. Each hexagonal corner contain a portal tract and cords of hepatocytes extend from the hepatocytic limiting plate at the periportal space, towards the central vein [55]. The terminal segments of the biliary system in the portal tracts connect directly with the hepatic cords through a specialized structure known as the Canal of Hering – thought to constitute the hepatic progenitor cell niche [56]. Canals of Hering are formed partly by biliary cells and partly by hepatocytes near the limiting plate [56]. Hepatic cords are separated from each other by a special form of blood vessels called sinusoids [55]. The sinusoids are lined by endothelial cells with open pores, or fenestrae, lacking a diaphragm and a basal lamina [57]. The resulting high endothelial permeability facilitate the exchange of macromolecules, solutes

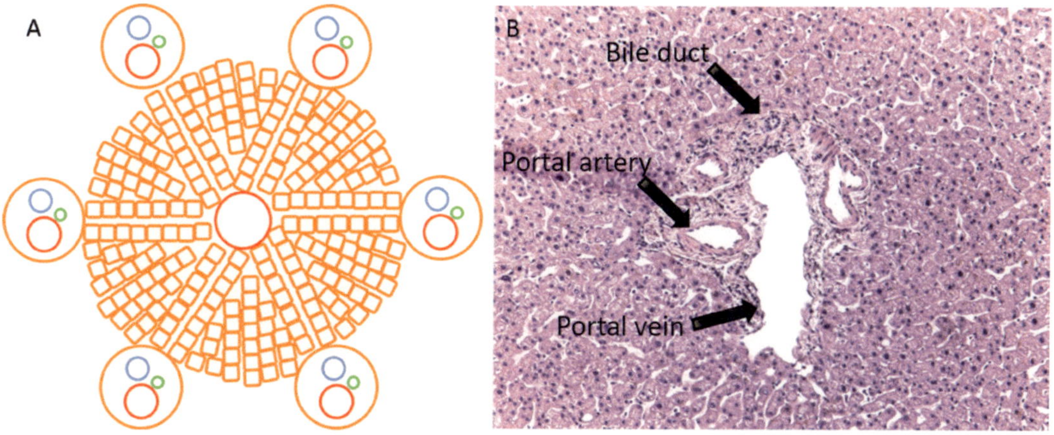

Figure 1. A. Cartoon of a stylized hepatic lobule. Each hexagonal corner of the hepatic lobule is marked by a portal area containing a portal vein, a portal artery and a bile duct. A central vein mark the center of the lobule. B. Hematoxylin and eosin staining of a tissue section from adult normal human liver. A portal area containing a portal vein, portal artery and bile duct is discernible. Magnification x100.

and water between sinusoidal blood and hepatocytes [57]. The sinusoidal wall is additionally separated from the hepatocytes by a lumen termed the space of Disse. The predominant view is that blood drain from the portal vein and portal artery branches and blends in the sinusoids from where it drains into the central vein [54,55]. Lymph, on the other hand, is thought to be generated by filtration of sinusoidal blood into the space of Disse from where it flows towards lymphatic vessels located in the portal tracts [54]. Bile canaliculi are narrow spaces formed from the apical membranes of adjacent hepatocytes in the hepatic cords [54]. Bile originating from the bile canaliculi is transported towards terminal bile ducts in the portal tracts through the canal of Hering [54,56].

Non-parenchymal cell types also present in the liver include stellate cells and Kupffer cells. Hepatic stellate cells, also known as Ito cells, are starshaped and contain lipid droplets with vast amounts of vitamin A [58]. In normal liver they are located to the space of Disse which is suggested to constitute the hepatic stellate cell niche [59]. Kupffer cells, on the other hand, are hepatic macrophages involved in the phagocytosis of cellular debris, extracellular matrix components and release of inflammatory factors [60].

5. Liver development

The liver in an endodermal derived organ with hepatocytes and cholangiocytes originating from a common progenitor termed "hepatoblast" or "primitive hepatocyte" [61]. Development of the liver goes through sequential stages including induction, specification, proliferation and maturation steps. The endoderm is important for inducing development of the neighboring cardiogenic mesoderm followed by maturation of the heart. Embryonic development of the liver is initiated in the ventral part of the anterior endoderm, whereas pancreas coordinately develops from the dorsal part. Within a short period of time, in a so-called "window of opportunity" around embryonic day (E) 8.5-11.5 in mouse endoderm, the anterior endoderm is competent for activation of a hepatic development gene program [62]. At the time of hepatic induction, the adjacent mesenchymal tissue, comprising the cardiogenic mesoderm and septum transversum, produces subtypes of growth factors: fibroblast growth factor (FGF) and bone morphogenetic protein (BMP), respectively [63]. Growth factors acid FGF, basic FGF, FGF4, BMP 2, and BMP4 initiate a hepatic gene expression program, while FGF or cardiogenic mesoderm suppresses the pancreatic gene expression program [62]. In the absence of BMPs or FGFs, the pancreatic gene expression program is initiated while the hepatic gene program is suppressed [62].

Following induction of hepatic gene expression, the endodermal cells adopt a columnar appearance at E8.5 and express albumin. At E9.5 a thickening of the endoderm is observed, interceded by primitive endothelial cells from the septum transversum [62,64]. This prospective liver, termed the hepatic diverticulum or "liver bud", is visible in the human embryo at the 17 somite stage, corresponding to 3 weeks and 5 days post conception [53,64]. Signaling molecules, including BMPs, hepatocyte growth factor (Hgf) and vascular endothelial growth factor receptor 2 (Vegfr-2) from the septum transversum and endothelial cells induce proliferation and migration of hepatoblast positive for cytokeratin (CK19), Hepatocyte Paraffin 1 (HepPar1), α-fetoprotein (AFP) and albumin, into the adjacent septum transversum [62,65–69]. At E11 the hepatoblasts addionationally stain for the intermediate filament proteins CK8, CK14 and CK18 [70–72]. Concurrent with the hepatoblast invasion the endothelial cells coalesce around spaces in the septum transversum thereby forming anastomosing primitive blood vessels around which hepatoblast are situated. The endodermal invasion displaces the septum transversum that eventually form the liver capsule, mesenchyme and possibly the hepatic stellate cells [62,68,69,73,74].

At E14 (in mouse) hematopoietic cells colonize the liver, making it a prenatal site for hematopoiesis. Concomitantly, hepatoblasts express markers of both the hepatocytic and cholangiocytic lineages and are capable of differentiating into either of the two epithelial cell types. The hepatoblasts, however, gradually commit to either the hepatocytic or cholangiocytic lineages. Three transcriptions factors, Hepatocyte Nuclear Factor (HNF)-4α, HNF-6 and HNF-1β, are found to be particularly important in this process. Microarray data have demonstrated that HNF-4α bind approximately half of the active genes in liver and is essential for determination toward a hepatocytic fate [75,76]. On the other hand, HNF-6 and HNF-1β are essential for development of the biliary lineage. Knockout mice for HNF6 and its downstream target HNF1β, develop no gallbladder and display abnormal development of the intrahepatic and extrahepatic bile ducts [77,78]. Around E16 (mouse) the hepatoblast are commited to either the hepatocytic or cholangiocytic lineages and are thereby no longer bipotential [62,79,80].

During development of the liver, morphogenesis of the biliary tree is also said to proceed through a series of developmental stages. These are categorized as the ductal plate, remodeling bile duct and remodeled bile duct stages [79]. The earliest indicator of biliary development comes from studies of the transcription factor SRY-related HMG box transcription factor 9 (SOX9). SOX9 is essential for the formation of certain stem cell niches, such as the hair follicle stem cell compartment [81]. SOX9 is expressed in the hepatic diverticulum but disappears during the endodermal invasion of the septum transversum. At E11.5, however, SOX9 is reexpressed in hepatoblasts located near the developing portal veins [82]. These prospective cholangiocytes lining the mesenchyme surrounding developing portal veins form a single-layered ring at E14.5 termed the "ductal plate". Studies on cells isolated from the ductal plate and from adult livers have shed important information on this structure. Cells from adult livers and the ductal plate, positive for epithelial cell adhesion molecule (EpCAM) and CK19 and negative for AFP can give rise to both the hepatic and biliary lineages, when injected into immunodeficient NOD/SCID mice [83,84]. The ductal plate is therefore not only suggested to constitute the pre-and perinatal hepatic progenitor cell niche, but also to be directly antecedent to the canal of Hering, the presumed adult hepatic progenitor cell niche [83–85].

The ductal plate, which can be envisioned as a biliary sleeve, increase expression of CK8, 18 and 19 relative to the remaining parenchymal cells [86,87]. Through a unique mode of tubulogenesis the cholangiocytes induce neighboring hepatoblast to differentiate into cholangiocytes themselves thereby developing a two-layered transiently asymmetric ductal plate around E16.5 [79,82]. Focal lumina appear between the mesenchymal and parenchymal ductal plate facing layers, thereby giving rise to early bile ducts at E16.5 [79]. In the following remodeling phase, these primitive bile ducts migrate into the portal mesenchyme in a complex process timely coordinated with the formation of hepatic portal arteries [65]. The parts of the ductal plate, which are not involved in bile duct formation, possibly regress as a result of apoptosis [88]. As of a result, the intrahepatic bile ducts loose contact with the ductal plate and become fully embedded in the portal mesenchyme in the remodeled stage. However, the intrahepatic bile duct system is still immature until several weeks after birth and remnants of the ductal plate can be identified, in particular, at the smaller vein branches [86,89]. As a final step in the maturation process, developing cholangiocytes initiate expression of CK7, a marker of adult bile duct cells [53,86]. The outlined development of the intrahepatic bile duct system is initiated at the hepatic hilum from where it gradually progresses towards the periphery of the liver, where the smaller portal branches reside [89].

6. Hepatic tissue homeostasis

The wide range of important metabolic functions performed by the liver and its proximity to ingested environmental toxins are hypothesized to have imparted the livers tremendous capacity for adaptation and regeneration [3,90].

Hepatocytes are the main component of liver and therefore, the most vulnerable to damage. The generation of adult hepatocytes, under non-pathogenic conditions, has been widely disputed. In normal liver, parenchymal turnover is slow with hepatocyte lifespans estimated 150 to 450 days in rat [50,51,91,92]. With a turnover rate of normal liver cells of approximately 1 in 20,000-40,000 at any given time the entire liver is estimated to be replaced by normal tissue at least once a year [93]. As hepatocytes supposedly are terminally differentiated cells, they were once hypothesized only to possess the capacity for one or two cell divisions. A number of studies of label-retaining markers of cells based on the incorporation of markers such as tritiated thymidine into DNA in rats or lack of markers such as cytochrome c in humans have located proliferative hepatocytes in the periportal region [94,95]. Cell tracking has illustrated a gradual invasion of these recognizable cells from the portal tract towards the terminal central vein. Based on these and similar experiments the "streaming liver" hypothesis was suggested in which mitotically active hepatocytes at the limiting plate in the periportal region continuously provided hepatocytic offspring. In a unidirectional fashion, these hepatocytes are hypothesized to stream along the sinusoids as they gradually change enzymatic expression and eventually replace dead hepatocytes in the perivenous region [94]. However, this model is still quite controversal. Long-term labelling of hepatocytes with beta-galactosidase, an enzyme capable of converting X-gal into an insoluble blue compound, found positive clusters of hepatocytes, ergo cells that had divided, throughout the liver lobule thereby contradicting the streaming liver hypothesis [51].

The relative mitotic quiescence of hepatocytes and cholangiocytes mask their huge proliferative potential. Resecting two-thirds of the liver in accordance with the partial hepatectomy protocol (PHx) leads to complete regrowth in approximately 10 days [1,3]. This regrowth is, however, not a true regeneration, given that it does not recreate original hepatic morphology but is compensatory hyperplasia in the residual liver lobes [55]. Even with this relatively harsh treatment of the liver, only 1-2 proliferative events of hepatic epithelial cells are needed to lead to complete compensatory regrowth with no or very little hepatic stem cell contribution [1,96]. Impressively, this procedure can be repeated at least 12 times in rats without regenerative failure or endangering liver function as hepatocytes maintain a fully differentiated state [97]. In an experimental animal model, mice deficient for the tyrosine catabolic enzyme, fumarylacetoacetate hydrolase (FAH), suffer from hepatocyte damage due to accumulation of fumarylacetoacetate and its precursor maleylacetoacetate [98]. However, wild-type hepatocytes are capable of rescuing this phenotype. In an elegant study serial transplantations of wild-type hepatocytes into FAH deficient mice repopulated 6 generations of livers corresponding to 69 cell doublings [98]. Therefore, during normal tissue homeostasis, hepatocytes could be regarded as the functional unipotent hepatic stem cell, capable of giving rise to more than 50 livers [1]. Furthermore, in some chronic biliary diseases such as primary biliary cirrhosis and primary sclerosing cholangitis, hepatocytes have even been observed differentiating into biliary cells [99,100]. Nonetheless, the replicative activity of even hepatocytes can apparently decrease in chronic hepatic injury in mice and advance cirrhosis in humans, possibly due to telomere shortening [101]. Regeneration through replication of hepatocytes and cholangiocytes is also known as the "first tier of defense" or a "level 1 response" [4]. This form of response was responsible was regenerating Prometheus' liver during night as the eagle essentially conducted partial hepatectomy during the day.

7. Localizing the hepatic progenitor cell niche

Locating stem cells is the first step into characterizing their niche. Stem cells and their niches have been defined in several tissues, including the hair follicle and skin, the hematopoietic system and in the intestinal crypts [42,44,49,102–105]. These organs are, unlike the liver, generally under constant renewal and require frequent stem cell division for tissue replenishment. Stem cells in these organs are therefore fulltime committed to perform stem cell function. However, stem cells in tissues with low turnover have been notoriously difficult to detect. As with arrangements in other stem cell niches the hepatic progenitor cell niche is thought to be structurally composed of a stem or progenitor cell population situated on a basal lamina and in contact with surrounding support cells [36,38]. As cellular turnover in the liver is already low and hepatic homeostasis and regeneration to a large extent is completed by differentiated parenchymal and non-parenchymal cells stem cells in this organ have been difficult to characterize.

While hepatocytes can conceptually be considered as the livers functional stem cells, the contribution of hepatic stem or progenitor cells to liver regeneration has been debated. Clues to a possible existence of stem cells in the liver came from early studies conducted by Farber, Wilson and Leduc [106,107]. Following dietary administration of DL-ethionine or carcinogenic 2-acetylaminofluorene (2-AAF) to rats, Farber observed the presence of pseudoductular structures consisting of small cells near the hepatic portal areas in rat liver (figure 2). These small cells are termed "oval cells" due to their oval shaped nucleus and scant cytoplasm [106]. In a following study Wilson and Leduc examined murine livers following dietary administration of ethionine and bentonite [107]. The presence of small cholangioles apparently giving rise to both bile-duct cells and parenchymal cells suggested the presence or a population of reserve cells or stem cells [107]. In acute liver failure and chronic liver diseases similar so-called "ductular reactions" may be noted at the portal triad interface [7,108–110]. The ductular response is thought to result from proliferating progenitor cells and represent the livers "second tier of defense" or "level 2 response" [4]. These cells termed oval cells in rodents are named progenitor cells in humans as rodent hepatic injury models and human diseases may not be directly comparable. However, we will collectively refer to them as hepatic progenitor cells (HPCs). The resulting arborizing network of ductular structures sprouting from the portal area is classified as an atypical ductular reaction due to a poorly defined lumen.

A number of rodent hepatic injury models have been developed to investigate various modes of regeneration and to mimic human hepatic diseases. Particularly notable models include partial hepatectomy, which induces proliferation of differentiated hepatocytes and cholangiocytes and thereby represent the first tier of defense [18,111]. Ligation of the common bile duct (Bile Duct Ligation) obstructs bile flow from the liver (figure 2). This surgical technique mimics cholestasis and induces proliferation of hepatocytes and the larger bile ducts without signs of differentiation towards the hepatocytic lineage in the latter [112,113]. Several injury models specifically induce HPC responses and thereby the second tier of defense. For example administration of the choline-deficient ethionine-supplemented (CDE) diet or carcinogenic agents such as 3,5-diethoxycarbonyl-1,4-dihydrocollidine (DDC) to rodents induces ductular reactions while carbon tetrachloride (CCl_4) administration additionally result in advanced hepatic fibrosis [114–116]. In the 2-AAF/PHx model administration of 2-acetylaminofluorene to rats is followed by two-thirds partial hepatectomy. This procedure block hepatocyte differentiation, ergo the first tier of defense, while at the same time providing a strong stimulus

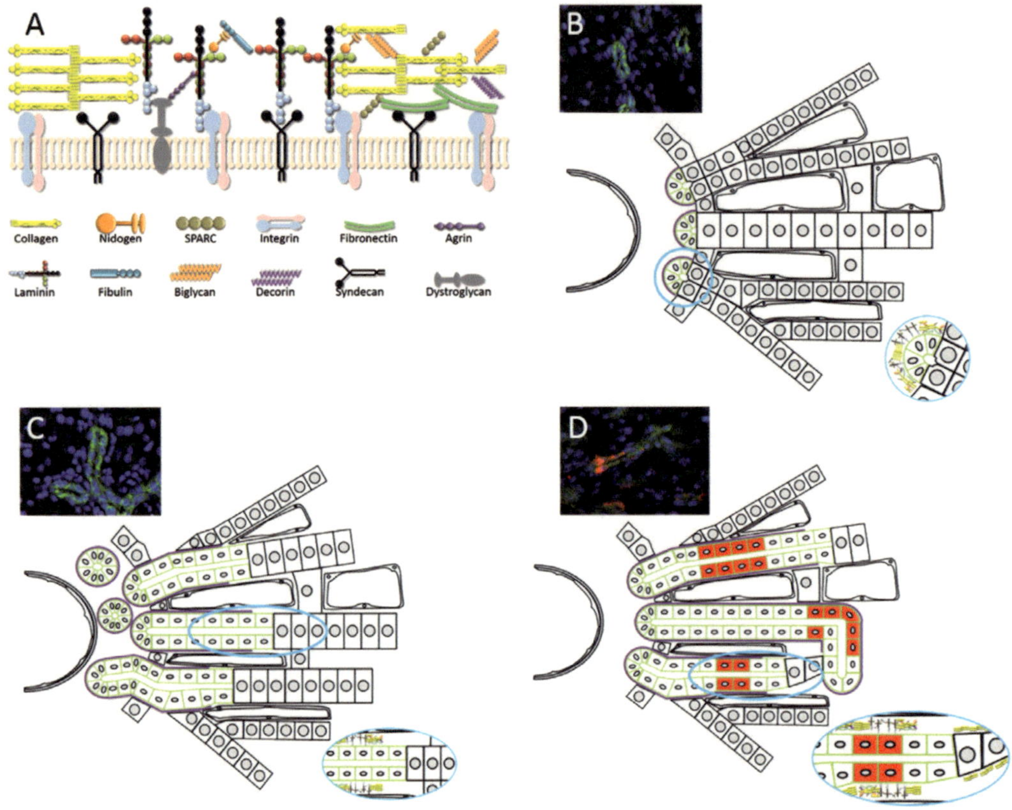

Figure 2. A. Cartoon of typical constituents of the extracellular matrix. B, C, D. Microphotographic images and cartoons of livers from rats subjected to B) sham operation, C) bile duct ligation and D) the 2-AAF-PHx model. Cholangiocytes and progenitor cells are stained for HAI-1 (green) and DLK1 (red). Cartoons in B, C and D portray part of portal areas with bile ducts (green) and their extracellular matrix in the portal mesenchyme bordering the limiting plate. B) In sham operated rat liver cholangiocytes are marked by HAI-1. C) In the bile duct ligation model in rats the larger bile ducts proliferate. D) In the 2-AAF-PHx model in rat liver a ductular reaction contain a subpopulation of hepatic progenitor cells positive for DLK1. Regardless of injury model extracellular matrix components escort the cholangiocytes. Upon exiting the hepatic progenitor cell niche the progenitor cells differentiate into hepatocytes. Microphotograph magnification x100. Adapted from Vestentoft et al. 2013 [129].

for growth. As a result, a ductular response is mounted (figure 2). Although these proliferating epithelial cells are collectively referred to as oval cells, it remains unclear if the oval cells resulting from different hepatic insults across different species have common characteristics as mice and rat respond differently to the same insults [117].

The ductular response is thought to represent proliferating progenitor cells. However, the origin of these progenitor cells is debated. Ductular reactions initiated, for example in the 2-AAF/PHx protocol, display both biliary and hepatocytic markers [26,118,119]. Moreover, destruction of the biliary tree through administration of 4,4'-methylenedianiline (MDA) inhibits progenitor cell proliferation, suggesting that progenitor cells originate from the biliary lineage [120]. However, administration of dexamethasone diminishes progenitor cell induction, but has no consequence on proliferation of larger bile ducts [113]. The anatomical location of the Canal of Hering, at the portal triad interface, makes this structure a prime candidate for the adult HPC compartment or "niche" [56,120–122]. Based on these experiments, the Canal of Hering represents, therefore, the HPC niche and the ductular reaction represents the activated HPC niche, respectively. However, the assumed stem cells located in the Canal of Hering may in fact not be "true" stem cells, but rather subpopulations of biliary or hepatocytic cells with increased stemness compared to other cells of their respective lineage [3]. Although progenitor cells morphologically resemble biliary cells, ductular reactions are phenotypically heterogenous [123]. In the ductular end connected to the biliary tree, the cells display cholangiocytic markers such as CK19, whereas the ductular end facing the parenchyma display hepatocytic markers including HepPar1 and the transcription factor HNF4 [119,123]. Between these extremes, hepatobiliary cells expressing cholangiocytic and hepatocytic markers to various degrees are found [123,124]. It is now clear that the ductular response can be divided into several distinct phases that are evident in the 2-AAF/PHx protocol [117]. In the activation phase, on day 1, few proliferating HPCs expressing CK19 are detectable in the biliary ductules. In the early proliferation and migration phase on day 5 multiple CK19-positive HPCs can be observed whereas progenitor cell expression of delta-like 1 homolog (DLK1/Pref1) and AFP is rare. In the late proliferation and migration phase on day 9, arborizing ductular structures expand from the portal area with HPCs expressing CK19, Dlk1 and AFP proteins.

Even though a number of HPC markers have been reported, none are specific for a pure population of hepatic stem cells [24,123,125,126]. What is more, only few of the reported HPC markers are expressed on the cellular surface and are therefore able to be employed for cellular isolation studies. CK19, OV-6 (an antibody recognizing a shared epitope between CK14 and CK19), EpCAM, CD24, hepatocyte growth factor activator inhibitor type 1 (HAI-1) and suppressor of tumorigenicity 14 (ST14) decorate both the intrahepatic bile ducts and the ductular reactions but only EpCAM, CD24, HAI-1 and ST14 are expressed on the surface [126–129]. AFP and DLK1, however, mark a subpopulation of HPCs suggesting the presence of an established hierarchy amongst the HPCs [129–131]. AFP and DLK1 are normally not expressed in the liver. However, both proteins are observed in hepatoblasts, the embryonic precursors to the cholangiocytic and hepatocytic lineages, suggesting that oval cells recapitulate a fetal phenotype when activated in hepatic injuries [72,132]. DLK1 is a transmembrane protein often described as an inhibitor of cellular differentiation and is expressed in less differentiated cells [133,134]. For example, forced expression of Dlk1 inhibits adipogenesis, whereas suppression promotes this process [133]. It is therefore conceivable that Dlk1 inhibits HPC differentiation thereby allowing transit amplifying cells to increase in numbers similar to the one observed in other stem cell niches [36]. With regard to AFP, elevated serum levels of this protein are associated with a favorable prognosis for patients with fulminant hepatic failure [135,136]. This observation supports the assumption that AFP marks cells capable of, at least, differentiating towards the hepatocytic lineage.

8. Support cells and the hepatic progenitor cell response

Proliferation and morphogenesis of cholangiocytes and HPCs is a complex interplay between the biliary cells, surrounding support cells and the extracellular matrix. All of these components contribute to the HPC niche. Cell-cell interactions and cell-matrix interplays are likely to be important for regulating stem cell behavior within niches [37].

Hepatic stellate cells possibly originate from the septum transversum-derived mesothelium lining the liver [137]. They are recognizable in their quiescent state by the expression of desmin and glial fibrillary acidic protein (GFAP), whereas they express alpha smooth muscle actin (α-SMA) when activated, often as a result of hepatic injury [138–141]. In the quiescent state they reside in the space of Disse, which constitute a laminin coated hepatic stellate cell niche, but when activated they give rise to contractile myofibroblast [59]. Both cell types are major producers of extracellular matrix components and activated hepatic stellate cells are the main source of matrix metalloproteinases and their inhibitors. However, so-called portal fibroblasts and vascular myofibroblasts can also transform into myofibroblasts thus giving rise to much confusion about the origins of the latter [142,143]. Additional confusion has been caused by misinterpretation of cellular markers. In particular, Thy-1, a cell surface protein initially suggested to mark oval cells, was later reclassified as a marker for hepatic myofibroblasts [144]

Hepatic stellate cells and myofibroblasts are greatly involved in the HPC response. In both the CDE model of HPC induction in mice, and the 2-AAF/PHx model in rat, hepatic stellate cell and myofibroblast response are invoked [121,145,146]. Hepatic stellate cells and myofibroblasts not only intimately escort the HPC invasion into the parenchyma, but cellular processes from the hepatic stellate cells disrupt the HPC basal lamina and form direct cellular contact [121]. Such direct cell-cell interactions between hepatic stellate cells and liver epithelial cells has been shown to induce differentiation of the latter into a hepatocytic fate *in vitro* [147]. The HPC response is a regulated process undergoing several stages. Both initiation and termination is under tight regulation and hepatic stellate cells may be involved in these processes. HGF is a potent mitogen for hepatocytes whereas TGF-β is a strong inhibitor of their proliferation [148]. TGF-β is additionally identified as a partaker in maintaining quiescence of stem cells in other niches, such as the melanocyte stem cells located to the hair follicle bulge region [149]. Not only are hepatic stellate cells activated and induced to transform into myofibroblasts by TGF-β, but hepatic stellate cells themselves are major producers of this cytokine [148]. Conditioned media harvested from hepatic stellate cells in the early HPC response is rich in hepatocyte growth factor (HGF). This media promote HPC proliferation, possibly due to an override of the antiproliferative effect of TGF-β [150]. In the terminal phases of liver regeneration hepatic stellate cells change cytokine expression profile and produce high levels of TGF-β which inhibits proliferation of hepatocytes [150]. Thus, hepatic stellate cells may be involved in both initiation and termination of the HPC response.

Other cells types involved in the HPC response are macrophages and Kupffer cells. Macrophages can remodel the extracellular matrix, partly through the production of matrix metalloproteinases [151]. As for hepatic stellate cells and myofibroblasts also bone marrow derived macrophages intimately associate with ductular reactions in rats [145]. Kupffer cells, on the other hand, are resident hepatic macrophages. Kupffer cells are greatly activated in the CDE model of HPC response in mice. Before onset of HPC proliferation and parenchymal invasion activated Kupffer cells gradually shift from a more periportal location towards a more centrilobular location. Depletion of Kupffer cells through clodronate injections result in greatly reduced invasion of HPCs into the hepatic parenchyma. However, HPC proliferation is unaltered. In conclusion these data suggest that hepatic stellate cells are involved in the initiation, proliferation and termination of the HPC response, whereas Kupffer cells are needed for HPC invasion into the hepatic parenchyma.

9. Extracellular matrix components and the hepatic progenitor cell response

Extracellular matrix can be defined as the complex molecular material surrounding cells and encompass both the basement membrane and the interstitial matrix [152]. Major components include the respective protein families of collagens, laminins, elastins, proteoglycans and glycosaminoglycans (figure 2) [152–155].

The extracellular matrix is a dynamic scaffold known to affect aspects of stem cell behavior such as morphology, growth and survival. A proportion of these responses are due to interactions between the extracellular matrix components and integrins, a family of dimeric extracellular matrix receptors that are linked to and transmit signals to the cytoskeleton [156,157]. The extracellular matrix may also contain growth factors which provide growth and morphogenic signals to nearby cells. Even physical features of the matrix, such as rigidity and geometry may influence cellular phenotype and behavior and has been shown to direct stem cell lineage specification [152,158–160]. Studies of *Drosophila spp.* stem cell niches have clarified that the microenvironment, as expected, may promote adherence to the niche and repress stem cell differentiation [36]. What is more, with age the molecular composition of the extracellular matrix change in an unfavorable direction for stem cell function and proliferation. This has been illustrated in experiments where stem cells transplanted from older mice, where stem cell self-renewal and differentiation has deteriorated, to extracellular matrix from younger mice rejuvenate stem cell function to levels comparable to that observed in younger mice [161,162].

Upon induction of the HPC response, the molecular composition of the HPC niche is thought to change in favor of promoting progenitor cell proliferation. Therefore, a key to understanding HPC biology and to characterize the HPC niche lies within unravelling the extracellular matrix composition of the niche. It is of particular interest to clarify which extracellular matrix molecules regulate the hepatic progenitor cell responses. A number of extracellular matrix molecules taking part in development of the intrahepatic bile ducts or in modulating the HPC response have been identified. Particularly, laminin and collagen I and IV are associated with these processes, but also other extracellular matrix components including tenascin, nidogen 1, agrin and fibronectin contribute.

The family of collagen fibrils comprises 28 members, all with at least one triple helical domain and arranged in a rope-like fashion [163,164]. Collagens are deposited in the extracellular space and particularly collagen I and collagen IV are implicated in hepatic development and regeneration [165]. However, their roles seem quite different. Collagen I is the main component of hepatic fibrosis, where it is laid down by the non-parenchymal hepatic stellate cells and myofibroblasts and contribute to the formation of scaring tissue [165,166]. Collagen IV, on the other hand, is part of the basement membrane of adult biliary cells and contributes to the ductal plate, the prenatal hepatic progenitor cell niche [167,168]. Collagen I and IV delineate expanding biliary cells not only in the HPC response, but also in the bile duct ligation model [129,146].

Members of the laminin family are trimeric proteins that, as for collagen IV, are part of the basal lamina [169]. In the HPC response laminin expression can be detected in hepatic stellate cells, myofibroblasts, endothelial cells and the progenitor cells themselves [170–172]. As for collagen IV, laminin contribute to the ductal plate during development and form the basal lamina escorting the HPC response in close apposition to stellate cells [121,168]. Several studies have highlighted the importance of remodeling the extracellular matrix in connection with the HPC response. In the CDE-induced murine model of HPC activation α-SMA positive cells and an extracellular matrix rich in collagen I are deposited in the periportal area prior to oval cell proliferation [146]. The ECM is laid down in a porto-veinous direction, thereby preforming a niche for the HPCs to invade. However, this invasion process is tightly correlated with ECM remodeling. Hepatic macrophages and stellate cells are sources of a variety of extracellular matrix degrading enzymes, such as matrix metalloproteinases (MMP) 2, 9, 12 and 13, and their inhibitor, tissue inhibitor of metalloproteinase type 1 (TIMP-1) [151,173,174]. Where the CCl_4 or CDE-models of HPC activation initiate a florid HPC response in wild-type mice this response is markedly attenuated in mice expressing a degradation resistant form of collagen I [175]. These mice also display a distinct paucity of laminin deposition suggesting that degradation of collagen is a prerequisite for HPC proliferation and parenchymal invasion.

Where ECM remodeling and collagen I degradation is necessary for the HPC response only laminin is important for the biliary phenotype. Primary murine HPCs cultured on laminin up-regulate expression of HPC and biliary associated genes, such as DLK1 and aquaporin 1, respectively, while hepatocytic gene expression, exemplified by C/EBPα is inhibited [145]. Collagen I and IV, on the other hand, inhibit or do not influence these biliary genes, whereas fibronectin promote C/EBPα expression. In support of these results, culturing HPCs with laminin support proliferation and expansion *in vitro* whereas culturing HPCs with collagen I result in growth arrest and differentiation [176,177]. The importance of the laminin-rich activated progenitor cell niche for maintaining the biliary/progenitor phenotype *in vivo* is also evident in the HPC response, as disappearance of the basement membrane induces differentiation [129,178]. Assuming that the canal of Hering truly constitutes the hepatic progenitor cell niche, this niche therefore appear to be sharply limited by the deposition of collagen I, collagen IV, laminin, nidogen 1 and agrin [129]. The niche support maintaining the biliary phenotype and proliferation of HPCs that will differentiate to a hepatocytic phenotype upon exit from the HPC niche, not unlike the scenario of other stem cell niches [36].

Stem cell niche component		Comment
Progenitor cell niche position	The Canal of Hering	Most distal part of the bile duct system. Composed of cholangiocytes and hepatocytes. Link bile ducts with canaliculi between hepatocytes.
Progenitor cell origin	Possibly the biliary lineage	Administration of dexamethasone selective diminishes the progenitor cell response.
Progenitor cell composition	Phenotypically heterogenous	Progenitor cells display biliary (CK19) and hepatocytic markers (HepPar1, HNF4α) to various degrees.
Progenitor cell markers	AFP, NCAM, DLK1/Pref1	NCAM and DLK1/Pref1 are cell surface markers. AFP and DLK1/Pref1 are expressed during hepatic development.
Associated cell types	Kupffer cells	Necessary for invasion of ductular reactions into the hepatic parenchyma.
	Hepatic stellate cells	Intimately associate with ductular reactions. Are necessary for initiation, proliferation and termination of ductular reactions.
	Macrophages	Intimately associate with ductular reactions.
Associated extracellular matrix components	Laminin Collagen I + IV Nidogen 1 Agrin	Laminin is essential for maintaining the biliary phenotype.

Table 1. Summary of components associated with the activated hepatic progenitor cell niche.

10. Activation and aberrant hepatic stem cell activation.

Animal studies have clarified that when regeneration through hepatocytic division fail HPCs from the canal of Hering contribute to liver regeneration. Despite several protein markers, such as Dlk1, EpCAM, CK19 and AFP, are associated with HPCs a pure population of hepatic stem cells or their niche have not been defined [83,179–182]. However, as elevated levels of AFP are associated with increased survival of patients suffering from acetaminophen-induced liver injury, hepatic stem cells may be activated in acute hepatic diseases [136]. OV6 and CK7 mark ductular reactions and intermediate hepatocytes, i.e. progenitor cells on route to a hepatocytic fate, suggesting stem cell involvement in a variety of human diseases and syndromes. These include hepatitis C virus infection, fatty liver disease and acute processes such as submassive liver cell necrosis, [183]. Generally, HPC activation seems correlated with the severity of inflammation and fibrosis [184]. In addition, the more aggressive the hepatocellular injury is, the larger a proportion of intermediate hepatocytes are observed [184]. Interestingly, the protein deleted in malignant brain tumor 1 (dbmt1) is specifically associated with ductular cell populations emerging after acetaminophen intoxication or infection with hepatitis B virus but not in primary biliary cirrhosis or large bile duct obstruction. Dbmt1 therefore may have a role in cellular fate decision [185].

Liver cancer is the second most frequent cause of cancer death in men and the sixth leading cause of cancer death in women [186]. Hepatocellular carcinoma (HCC) represent the major histological subtype accounting for 70-85 % of primary livers cancers, followed by an increase in incidents of intrahepatic cholangiocarcinomas (ICC) [186,187]. Given that cancer cells and stem cells share certain characteristics cancer is proposed to represent an abnormal stem cell disease [188,189]. Both categories of cells can self-renew, divide unlimited and give rise to heterogeneous progeny. Indeed, certain gliomas, intestinal adenomas and squamous skin tumours are now attributed to cancer stem cells (CSCs) [190–193]. Liver cancer most frequently arise in chronic liver diseases, such as chronic hepatitis, cirrhosis or both, where hepatocytic regeneration and continuous inflammation occur [194,195]. Hepatocarcinogenesis is considered as a slow process in which genomic changes progressively alter the hepatocellular phenotype [194]. In chronic liver diseases, the hepatic microenvironment is substantially altered in a fashion promoting cellular damage. Stellate cells are activated and infiltrating lymphocytes may cause inflammation through the release of free radicals and cytokines, resulting in DNA damage and cell proliferation, factors that may promote aberrant HPC activation [196,197]. However, as the hepatic stem cells are not fully defined, their involvements in these liver cancers have not been conclusively established. In addition to accumulation of genomic and epigenetic changes of genes and regulatory pathways the hepatic microenvironment is also involved in promoting liver cancer. For instance, activated hepatic stellate cells locate in the space between endothelial cells and trabeculae of cancer cells in HCC patients [198]. Conditioned media from such activated hepatic stellate cells both increase proliferation and migration of human HCC cells [199]. Thus, activated hepatic stellate cells may both drive fibrosis and proliferation of HCC cells.

A third form of primary liver cancer is the rare HCC-cholangiocarcinoma (HCC-CCA). In addition to the heterogenous cellular morphology also displayed by HCCs and ICCs, HCC-CCA's show signs of both hepatocellular and biliary epithelial differentiation [200]. Indeed, analysis of the expression pattern of hepatocytic marker HepPar1 and cholangiocytic markers CK7, CK19, EpCAM and CD133 in HCC-CCA's reveal subpopulations of cancer cells coexpressing both categories of markers [182,200–203]. These results seemingly confirm the hypothesis that HCC-CCA are of HPC origin and human hepatocarcinogenesis may originate from the transformation of HPCs [200]. The identification of bipotent CSCs possibly originating from HPCs is interesting, as stem cell like expression patterns in liver cancers reflect a particularly malignant nature and poor prognostic outcome [182,204–206]. However, the identification of bipotent cancer stem cells also opens for new therapeutic applications. Identification and elimination of CSCs could provide more effective treatment of certain tumors and prevent reoccurrence. Unfortunately, the niche controlling self-renewal, proliferation and differentiation of HPCs and CSCs is still not well described and the putative hepatic stem cells remain unidentified.

11. Conclusion and perspectives

The present chapter has attempted to provide a simplified overview of current knowledge of hepatic stem cells and their niches. Unfortunately, the putative hepatic stem cell has not been identified and therefore not been characterized. Knowledge of the hepatic stem cell therefore mainly originates from analysis of its progeny, the hepatic progenitor cells, in animal models where regeneration through hepatocyte division is impaired. As a result, the location of the HPC niche is unknown. However, evidences point to the canal of Hering as the HPC niche. Assuming that the canal of Hering truly represent the hepatic stem cell niche, and that HPCs are descendants of hepatic stem cell, animal studies and their corresponding human diseases has provided us with some knowledge of the constituents in the activated HPC niche:

- Activated hepatic progenitor cells are phenotypically heterogeneous and to various degrees display markers of the biliary and hepatocytic lineages.
- A cellular hierarchy is present with AFP, DLK1 and NCAM marking subpopulations of HPCs [123].
- Hepatic stellate cells and macrophages intimately associate with the ductular reactions.
- Hepatic stellate cells are necessary for initiation, proliferation and termination of ductular reactions.
- Kupffer cells are necessary for invasion of ductular reactions into the hepatic parenchyma.
- The extracellular matrix in the HPC micromillieu contain laminin, collagen I and IV, nidogen 1 and agrin.
- Laminin is necessary for maintaining the biliary phenotype of the HPCs.
- HPCs differentiate into hepatocytes upon exit from the activated HPC niche.

The establishment of HPC subpopulations suggests the presence of progenitor cells at different stages of differentiation. Identifying additional proteins expressed on the HPC surface could facilitate isolation and characterization of these subpopulations and evaluation of their potential for differentiation. Furthermore, as hepatic stem cells may be implicated in development of primary liver cancers, better characterization of the hepatic progenitor cells could provide new targets for treatment of cancerous diseases. Aberrant progenitor cell activation and proliferation is also dependent on the hepatic microenvironment. As stellate cells and Kupffer cells are involved in the proliferation and invasion of HPCs these cell types could also provide targets for alleviating HPC derived cancers. In addition, targeting stellate cells has the potential to reduce hepatic fibrosis. Therefore, more research is needed into characterizing the hepatic progenitor cell niche and obtaining a better understanding of the activation and differentiation of the hepatic progenitor cells.

12. Nomenclature

2-AAF: 2-Acetylaminofluorene.

AFP: α-Fetoprotein.

ASMA: alpha smooth muscle actin.

BMP: Bone morphogenetic protein.

BrdU: 5-Bromo-2′-Deoxyuridine.

CCl4: Carbon tetrachloride.

CDE: Choline-deficient, ethionine supplemented.

CK: Cytokeratin.

CSC: Cancer Stem Cell.

DDC: 3,5-diethoxycarbonyl-1,4-dihydrocollidine.

DLK1: Delta-Like 1 Homolog.

Dmbt1: Deleted in malignant brain tumors 1

FGF: Fibroblast growth factor.

GFAP: Glial fibrillary acidic protein.

HAI-1: Hepatocyte Growth Factor Activator Inhibitor Type 1.

HCC: Hepatocellular carcinoma.

HCC-CCA: HCC-cholangiocarcinoma.

HepPar1: Hepatocyte Paraffin 1.

HGF: Hepatocyte growth factor.

HNF: Hepatocyte nuclear factor.

HPC: Hepatic progenitor cell.

ICC: Intrahepatic cholangiocarcinoma.

MDA: 4,4'-Methylenedianiline.

MMP: Matrix metalloproteinase.

NCAM: Neural Cell Adhesion Molecule.

OV6: Oval Cell Marker Antibody 6.

PHx: Partial hepatectomy.

ST14: Suppressor of tumorigenicity 1.

SOX9: SRY-related HMG box transcription factor 9.

TIMP-1: tissue inhibitor of metalloproteinase 1.

Vegfr2: Vascular endothelial growth factor receptor-2.

Author details

Peter Siig Vestentoft*

Address all correspondence to: sqp608@alumni.ku.dk

Dako, Glostrup, Denmark

References

[1] Michalopoulos GK, DeFrances MC (1997) Liver regeneration. Science 276: 60-66.

[2] Russo FP, Parola M (2011) Stem and progenitor cells in liver regeneration and repair. Cytotherapy 13: 135-144. 10.3109/14653249.2010.545386 [doi].

[3] Michalopoulos GK (2007) Liver regeneration. J Cell Physiol 213: 286-300. 10.1002/jcp.21172 [doi].

[4] Riehle KJ, Dan YY, Campbell JS, Fausto N (2011) New concepts in liver regeneration. J Gastroenterol Hepatol 26 Suppl 1: 203-212. 10.1111/j.1440-1746.2010.06539.x [doi].

[5] Michalopoulos GK (2010) Liver regeneration after partial hepatectomy: critical analysis of mechanistic dilemmas. Am J Pathol 176: 2-13. S0002-9440(10)60317-3 [pii]; 10.2353/ajpath.2010.090675 [doi].

[6] Bernal W, Auzinger G, Dhawan A, Wendon J (2010) Acute liver failure. Lancet 376: 190-201. S0140-6736(10)60274-7 [pii];10.1016/S0140-6736(10)60274-7 [doi].

[7] Dechene A, Sowa JP, Gieseler RK, Jochum C, Bechmann LP, El FA, Schlattjan M, Saner F, Baba HA, Paul A, Dries V, Odenthal M, Gerken G, Friedman SL, Canbay A (2010) Acute liver failure is associated with elevated liver stiffness and hepatic stellate cell activation. Hepatology 52: 1008-1016. 10.1002/hep.23754 [doi].

[8] O'Grady JG (2005) Acute liver failure. Postgraduate Medical Journal 81: 148-154.

[9] Larsen FS, Kirkegaard P, Rasmussen A, Hansen BA (1995) The Danish liver transplantation program and patients with serious acetaminophen intoxication. Transplant Proc 27: 3519-3520.

[10] Ostapowicz G, Fontana RJ, Schiodt FV, Larson A, Davern TJ, Han SH, McCashland TM, Shakil AO, Hay JE, Hynan L, Crippin JS, Blei AT, Samuel G, Reisch J, Lee WM

(2002) Results of a prospective study of acute liver failure at 17 tertiary care centers in the United States. Ann Intern Med 137: 947-954. 200212170-00007 [pii].

[11] Rolando N, Wade J, Davalos M, Wendon J, Philpott-Howard J, Williams R (2000) The systemic inflammatory response syndrome in acute liver failure. Hepatology 32: 734-739. S0270913900079660 [pii];10.1053/jhep.2000.17687 [doi].

[12] Larson AM (2010) Diagnosis and management of acute liver failure. Curr Opin Gastroenterol 26: 214-221. 10.1097/MOG.0b013e32833847c5 [doi].

[13] Kisseleva T, Cong M, Paik Y, Scholten D, Jiang C, Benner C, Iwaisako K, Moore-Morris T, Scott B, Tsukamoto H, Evans SM, Dillmann W, Glass CK, Brenner DA (2012) Myofibroblasts revert to an inactive phenotype during regression of liver fibrosis. Proc Natl Acad Sci U S A 109: 9448-9453. 1201840109 [pii];10.1073/pnas.1201840109 [doi].

[14] Troeger JS, Mederacke I, Gwak GY, Dapito DH, Mu X, Hsu CC, Pradere JP, Friedman RA, Schwabe RF (2012) Deactivation of hepatic stellate cells during liver fibrosis resolution in mice. Gastroenterology 143: 1073-1083. S0016-5085(12)00925-0 [pii];10.1053/j.gastro.2012.06.036 [doi].

[15] Ellis EL, Mann DA (2012) Clinical evidence for the regression of liver fibrosis. J Hepatol 56: 1171-1180. S0168-8278(12)00044-X [pii];10.1016/j.jhep.2011.09.024 [doi].

[16] Guo J, Friedman SL (2007) Hepatic fibrogenesis. Semin Liver Dis 27: 413-426. 10.1055/s-2007-991517 [doi].

[17] Rojkind M, Giambrone MA, Biempica L (1979) Collagen types in normal and cirrhotic liver. Gastroenterology 76: 710-719.

[18] Rutherford A, Chung RT (2008) Acute liver failure: mechanisms of hepatocyte injury and regeneration. Semin Liver Dis 28: 167-174. 10.1055/s-2008-1073116 [doi].

[19] Riordan SM, Williams R (2003) Mechanisms of hepatocyte injury, multiorgan failure, and prognostic criteria in acute liver failure. Semin Liver Dis 23: 203-215. 10.1055/s-2003-42639 [doi].

[20] Canbay A, Friedman S, Gores GJ (2004) Apoptosis: the nexus of liver injury and fibrosis. Hepatology 39: 273-278. 10.1002/hep.20051 [doi].

[21] Riordan SM, Williams R (2000) Use and validation of selection criteria for liver transplantation in acute liver failure. Liver Transpl 6: 170-173. S1527646500459411 [pii];10.1002/lt.500060221 [doi].

[22] Lee WM (1993) Acute liver failure. N Engl J Med 329: 1862-1872. 10.1056/NEJM199312163292508 [doi].

[23] Azuma H, Paulk N, Ranade A, Dorrell C, Al-Dhalimy M, Ellis E, Strom S, Kay MA, Finegold M, Grompe M (2007) Robust expansion of human hepatocytes in Fah-/-/Rag2-/-/Il2rg-/-mice. Nat Biotechnol 25: 903-910. nbt1326 [pii];10.1038/nbt1326 [doi].

[24] Hixson DC, Allison JP (1985) Monoclonal antibodies recognizing oval cells induced in the liver of rats by N-2-fluorenylacetamide or ethionine in a choline-deficient diet. Cancer Res 45: 3750-3760.

[25] Evarts RP, Nagy P, Marsden E, Thorgeirsson SS (1987) A precursor-product relationship exists between oval cells and hepatocytes in rat liver. Carcinogenesis 8: 1737-1740.

[26] Evarts RP, Nagy P, Nakatsukasa H, Marsden E, Thorgeirsson SS (1989) In Vivo Differentiation of Rat Liver Oval Cells into Hepatocytes. Cancer Res 49: 1541-1547.

[27] Van Den Heuvel MC, Slooff MJ, Visser L, Muller M, de Jong KP, Poppema S, Gouw AS (2001) Expression of anti-OV6 antibody and anti-N-CAM antibody along the biliary line of normal and diseased human livers. Hepatology 33: 1387-1393. S0270913901415254 [pii];10.1053/jhep.2001.24453 [doi].

[28] Jensen CH, Jauho EI, Santoni-Rugiu E, Holmskov U, Teisner B, Tygstrup N, Bisgaard HC (2004) Transit-amplifying ductular (oval) cells and their hepatocytic progeny are characterized by a novel and distinctive expression of delta-like protein/preadipocyte factor 1/fetal antigen 1. Am J Pathol 164: 1347-1359.

[29] Jones DL, Wagers AJ (2008) No place like home: anatomy and function of the stem cell niche. Nat Rev Mol Cell Biol 9: 11-21. nrm2319 [pii];10.1038/nrm2319 [doi].

[30] Watt FM, Hogan BL (2000) Out of Eden: stem cells and their niches. Science 287: 1427-1430. 8287 [pii].

[31] Morrison SJ, Shah NM, Anderson DJ (1997) Regulatory mechanisms in stem cell biology. Cell 88: 287-298. S0092-8674(00)81867-X [pii].

[32] Potten CS, Loeffler M (1990) Stem cells: attributes, cycles, spirals, pitfalls and uncertainties. Lessons for and from the crypt. Development 110: 1001-1020.

[33] Weissman IL (2000) Stem cells: units of development, units of regeneration, and units in evolution. Cell 100: 157-168. S0092-8674(00)81692-X [pii].

[34] Lechler T, Fuchs E (2005) Asymmetric cell divisions promote stratification and differentiation of mammalian skin. Nature 437: 275-280. nature03922 [pii];10.1038/nature03922 [doi].

[35] Schofield R (1978) The relationship between the spleen colony-forming cell and the haemopoietic stem cell. Blood Cells 4: 7-25.

[36] Losick VP, Morris LX, Fox DT, Spradling A (2011) Drosophila stem cell niches: a decade of discovery suggests a unified view of stem cell regulation. Dev Cell 21: 159-171. S1534-5807(11)00252-8 [pii];10.1016/j.devcel.2011.06.018 [doi].

[37] Spradling A, Drummond-Barbosa D, Kai T (2001) Stem cells find their niche. Nature 414: 98-104. 10.1038/35102160 [doi];35102160 [pii].

[38] Fuchs E, Tumbar T, Guasch G (2004) Socializing with the neighbors: stem cells and their niche. Cell 116: 769-778. S0092867404002557 [pii].

[39] Jensen UB, Lowell S, Watt FM (1999) The spatial relationship between stem cells and their progeny in the basal layer of human epidermis: a new view based on whole-mount labelling and lineage analysis. Development 126: 2409-2418.

[40] Zhu AJ, Haase I, Watt FM (1999) Signaling via beta1 integrins and mitogen-activated protein kinase determines human epidermal stem cell fate in vitro. Proc Natl Acad Sci U S A 96: 6728-6733.

[41] Taylor G, Lehrer MS, Jensen PJ, Sun TT, Lavker RM (2000) Involvement of follicular stem cells in forming not only the follicle but also the epidermis. Cell 102: 451-461. S0092-8674(00)00050-7 [pii].

[42] Tumbar T, Guasch G, Greco V, Blanpain C, Lowry WE, Rendl M, Fuchs E (2004) Defining the epithelial stem cell niche in skin. Science 303: 359-363. 10.1126/science.1092436 [doi];1092436 [pii].

[43] Ohyama M, Terunuma A, Tock CL, Radonovich MF, Pise-Masison CA, Hopping SB, Brady JN, Udey MC, Vogel JC (2006) Characterization and isolation of stem cell-enriched human hair follicle bulge cells. J Clin Invest 116: 249-260.

[44] Cotsarelis G, Sun TT, Lavker RM (1990) Label-retaining cells reside in the bulge area of pilosebaceous unit: implications for follicular stem cells, hair cycle, and skin carcinogenesis. Cell 61: 1329-1337. 0092-8674(90)90696-C [pii].

[45] Cotsarelis G, Cheng SZ, Dong G, Sun TT, Lavker RM (1989) Existence of slow-cycling limbal epithelial basal cells that can be preferentially stimulated to proliferate: implications on epithelial stem cells. Cell 57: 201-209. 0092-8674(89)90958-6 [pii].

[46] Goldman B (2008) Magic marker myths. Nature Reports Stem Cells.

[47] Oshima H, Rochat A, Kedzia C, Kobayashi K, Barrandon Y (2001) Morphogenesis and renewal of hair follicles from adult multipotent stem cells. Cell 104: 233-245. S0092-8674(01)00208-2 [pii].

[48] Ehninger A, Trumpp A (2011) The bone marrow stem cell niche grows up: mesenchymal stem cells and macrophages move in. J Exp Med 208: 421-428. jem.20110132 [pii];10.1084/jem.20110132 [doi].

[49] Barker N, van Es JH, Kuipers J, Kujala P, van den Born M, Cozijnsen M, Haegebarth A, Korving J, Begthel H, Peters PJ, Clevers H (2007) Identification of stem cells in small intestine and colon by marker gene Lgr5. Nature 449: 1003-1007. nature06196 [pii];10.1038/nature06196 [doi].

[50] MACDONALD RA (1961) "Lifespan" of liver cells. Autoradio-graphic study using tritiated thymidine in normal, cirrhotic, and partially hepatectomized rats. Arch Intern Med 107: 335-343.

[51] Bralet MP, Branchereau S, Brechot C, Ferry N (1994) Cell lineage study in the liver using retroviral mediated gene transfer. Evidence against the streaming of hepatocytes in normal liver. Am J Pathol 144: 896-905.

[52] Bertz RJ, Granneman GR (1997) Use of in vitro and in vivo data to estimate the likelihood of metabolic pharmacokinetic interactions. Clin Pharmacokinet 32: 210-258. 10.2165/00003088-199732030-00004 [doi].

[53] Vestentoft PS, Jelnes P, Hopkinson BM, Vainer B, Mollgard K, Quistorff B, Bisgaard HC (2011) Three-dimensional reconstructions of intrahepatic bile duct tubulogenesis in human liver. BMC Dev Biol 11: 56. 1471-213X-11-56 [pii];10.1186/1471-213X-11-56 [doi].

[54] Saxena R, Theise ND, Crawford JM (1999) Microanatomy of the human liver-exploring the hidden interfaces. Hepatology 30: 1339-1346. S0270913999004991 [pii];10.1002/hep.510300607 [doi].

[55] Fausto N, Campbell JS (2003) The role of hepatocytes and oval cells in liver regeneration and repopulation. Mech Dev 120: 117-130.

[56] Roskams TA, Theise ND, Balabaud C, Bhagat G, Bhathal PS, Bioulac-Sage P, Brunt EM, Crawford JM, Crosby HA, Desmet V, Finegold MJ, Geller SA, Gouw AS, Hytiroglou P, Knisely AS, Kojiro M, Lefkowitch JH, Nakanuma Y, Olynyk JK, Park YN, Portmann B, Saxena R, Scheuer PJ, Strain AJ, Thung SN, Wanless IR, West AB (2004) Nomenclature of the finer branches of the biliary tree: canals, ductules, and ductular reactions in human livers. Hepatology 39: 1739-1745.

[57] Braet F, Wisse E (2002) Structural and functional aspects of liver sinusoidal endothelial cell fenestrae: a review. Comp Hepatol 1: 1.

[58] Blaner WS, O'Byrne SM, Wongsiriroj N, Kluwe J, D'Ambrosio DM, Jiang H, Schwabe RF, Hillman EM, Piantedosi R, Libien J (2009) Hepatic stellate cell lipid droplets: a specialized lipid droplet for retinoid storage. Biochim Biophys Acta 1791: 467-473. S1388-1981(08)00207-2 [pii];10.1016/j.bbalip.2008.11.001 [doi].

[59] Sawitza I, Kordes C, Reister S, Haussinger D (2009) The niche of stellate cells within rat liver. Hepatology 50: 1617-1624. 10.1002/hep.23184 [doi].

[60] Van HN, Lanthier N, Espanol SR, Abarca QJ, van RN, Leclercq I (2011) Kupffer cells influence parenchymal invasion and phenotypic orientation, but not the proliferation, of liver progenitor cells in a murine model of liver injury. Am J Pathol 179: 1839-1850. S0002-9440(11)00665-1 [pii];10.1016/j.ajpath.2011.06.042 [doi].

[61] Wilson JW, Groat CS, Leduc EH (1963) Histogenesis of the liver. Annals of the New York Academy of Sciences 111: 8-24. 10.1111/j.1749-6632.1963.tb36945.x.

[62] Zaret KS (2002) Regulatory phases of early liver development: paradigms of organogenesis. Nat Rev Genet 3: 499-512. 10.1038/nrg837 [doi];nrg837 [pii].

[63] Cascio S, Zaret KS (1991) Hepatocyte differentiation initiates during endodermal-mesenchymal interactions prior to liver formation. Development 113: 217-225.

[64] Severn CB (1971) A morphological study of the development of the human liver. I. Development of the hepatic diverticulum. Am J Anat 131: 133-158. 10.1002/aja.1001310202 [doi].

[65] Fabris L, Cadamuro M, Libbrecht L, Raynaud P, Spirli C, Fiorotto R, Okolicsanyi L, Lemaigre F, Strazzabosco M, Roskams T (2008) Epithelial expression of angiogenic growth factors modulate arterial vasculogenesis in human liver development. Hepatology 47: 719-728. 10.1002/hep.22015 [doi].

[66] Haruna Y, Saito K, Spaulding S, Nalesnik MA, Gerber MA (1996) Identification of bipotential progenitor cells in human liver development. Hepatology 23: 476-481. S0270913996000766 [pii];10.1002/hep.510230312 [doi].

[67] Terrace JD, Currie IS, Hay DC, Masson NM, Anderson RA, Forbes SJ, Parks RW, Ross JA (2007) Progenitor cell characterization and location in the developing human liver. Stem Cells Dev 16: 771-778. 10.1089/scd.2007.0016 [doi].

[68] Lemaigre FP (2009) Mechanisms of liver development: concepts for understanding liver disorders and design of novel therapies. Gastroenterology 137: 62-79. S0016-5085(09)00463-6 [pii];10.1053/j.gastro.2009.03.035 [doi].

[69] Zhao R, Duncan SA (2005) Embryonic development of the liver. Hepatology 41: 956-967. 10.1002/hep.20691 [doi].

[70] Brill S, Zvibel I, Reid LM (1995) Maturation-dependent changes in the regulation of liver-specific gene expression in embryonal versus adult primary liver cultures. Differentiation 59: 95-102. S0301-4681(09)60795-7 [pii];10.1046/j.1432-0436.1995.5920095.x [doi].

[71] Tee LB, Kirilak Y, Huang WH, Smith PG, Morgan RH, Yeoh GC (1996) Dual phenotypic expression of hepatocytes and bile ductular markers in developing and preneoplastic rat liver. Carcinogenesis 17: 251-259.

[72] Tanimizu N, Nishikawa M, Saito H, Tsujimura T, Miyajima A (2003) Isolation of hepatoblasts based on the expression of Dlk/Pref-1. J Cell Sci 116: 1775-1786.

[73] Severn CB (1972) A morphological study of the development of the human liver. II. Establishment of liver parenchyma, extrahepatic ducts and associated venous channels. Am J Anat 133: 85-107. 10.1002/aja.1001330106 [doi].

[74] Lemaigre F, Zaret KS (2004) Liver development update: new embryo models, cell lineage control, and morphogenesis. Curr Opin Genet Dev 14: 582-590. 10.1016/j.gde.2004.08.004 [doi];S0959-437X(04)00128-5 [pii].

[75] Odom DT, Zizlsperger N, Gordon DB, Bell GW, Rinaldi NJ, Murray HL, Volkert TL, Schreiber J, Rolfe PA, Gifford DK, Fraenkel E, Bell GI, Young RA (2004) Control of

pancreas and liver gene expression by HNF transcription factors. Science 303: 1378-1381. 10.1126/science.1089769 [doi];303/5662/1378 [pii].

[76] Li J, Ning G, Duncan SA (2000) Mammalian hepatocyte differentiation requires the transcription factor HNF-4alpha. Genes Dev 14: 464-474.

[77] Clotman F, Lannoy VJ, Reber M, Cereghini S, Cassiman D, Jacquemin P, Roskams T, Rousseau GG, Lemaigre FP (2002) The onecut transcription factor HNF6 is required for normal development of the biliary tract. Development 129: 1819-1828.

[78] Coffinier C, Gresh L, Fiette L, Tronche F, Schutz G, Babinet C, Pontoglio M, Yaniv M, Barra J (2002) Bile system morphogenesis defects and liver dysfunction upon targeted deletion of HNF1beta. Development 129: 1829-1838.

[79] Lemaigre FP (2003) Development of the biliary tract. Mech Dev 120: 81-87. S0925477302003349 [pii].

[80] Costa RH, Kalinichenko VV, Holterman AX, Wang X (2003) Transcription factors in liver development, differentiation, and regeneration. Hepatology 38: 1331-1347. 10.1016/j.hep.2003.09.034 [doi];S0270913903010048 [pii].

[81] Vidal VP, Chaboissier MC, Lutzkendorf S, Cotsarelis G, Mill P, Hui CC, Ortonne N, Ortonne JP, Schedl A (2005) Sox9 is essential for outer root sheath differentiation and the formation of the hair stem cell compartment. Curr Biol 15: 1340-1351. S0960-9822(05)00747-5 [pii];10.1016/j.cub.2005.06.064 [doi].

[82] Antoniou A, Raynaud P, Cordi S, Zong Y, Tronche F, Stanger BZ, Jacquemin P, Pierreux CE, Clotman F, Lemaigre FP (2009) Intrahepatic bile ducts develop according to a new mode of tubulogenesis regulated by the transcription factor SOX9. Gastroenterology 136: 2325-2333. S0016-5085(09)00300-X [pii];10.1053/j.gastro.2009.02.051 [doi].

[83] Schmelzer E, Zhang L, Bruce A, Wauthier E, Ludlow J, Yao HL, Moss N, Melhem A, McClelland R, Turner W, Kulik M, Sherwood S, Tallheden T, Cheng N, Furth ME, Reid LM (2007) Human hepatic stem cells from fetal and postnatal donors. J Exp Med 204: 1973-1987. jem.20061603 [pii];10.1084/jem.20061603 [doi].

[84] Schmelzer E, Wauthier E, Reid LM (2006) The phenotypes of pluripotent human hepatic progenitors. Stem Cells 24: 1852-1858. 2006-0036 [pii];10.1634/stemcells.2006-0036 [doi].

[85] Zhang L, Theise N, Chua M, Reid LM (2008) The stem cell niche of human livers: symmetry between development and regeneration. Hepatology 48: 1598-1607. 10.1002/hep.22516 [doi].

[86] Van EP, Sciot R, Callea F, van der Steen K, Moerman P, Desmet VJ (1988) The development of the intrahepatic bile ducts in man: a keratin-immunohistochemical study. Hepatology 8: 1586-1595. S0270913988001661 [pii].

[87] Shah KD, Gerber MA (1989) Development of intrahepatic bile ducts in humans. Immunohistochemical study using monoclonal cytokeratin antibodies. Arch Pathol Lab Med 113: 1135-1138.

[88] Terada T, Nakanuma Y (1995) Detection of apoptosis and expression of apoptosis-related proteins during human intrahepatic bile duct development. Am J Pathol 146: 67-74.

[89] Van EP, Sciot R, Desmet V (1988) Intrahepatic bile duct development in the rat: a cytokeratin-immunohistochemical study. Lab Invest 59: 52-59.

[90] Starzl TE, Fung J, Tzakis A, Todo S, Demetris AJ, Marino IR, Doyle H, Zeevi A, Warty V, Michaels M,. (1993) Baboon-to-human liver transplantation. Lancet 341: 65-71.

[91] Wood RL (1971) Regeneration of Liver and Kidney. JAMA: The Journal of the American Medical Association 218: 1306-1307.

[92] Zajicek G, Oren R, Weinreb M, Jr. (1985) The streaming liver. Liver 5: 293-300.

[93] Steiner JW, Perz ZM, Taichman LB (1966) Cell population dynamics in the liver. A review of quantitative morphological techniques applied to the study of physiological and pathological growth. Exp Mol Pathol 5: 146-181. 0014-4800(66)90012-8 [pii].

[94] Fellous TG, Islam S, Tadrous PJ, Elia G, Kocher HM, Bhattacharya S, Mears L, Turnbull DM, Taylor RW, Greaves LC, Chinnery PF, Taylor G, McDonald SA, Wright NA, Alison MR (2009) Locating the stem cell niche and tracing hepatocyte lineages in human liver. Hepatology 49: 1655-1663. 10.1002/hep.22791 [doi].

[95] Oren R, Zajicek G, Maaravi Y, Kenet G, Karmely F, Hubert A, Raanani P, Arber N (1997) Methimazole slows hepatocyte streaming in rats. Dig Dis Sci 42: 1433-1437.

[96] Grisham JW (1962) A morphologic study of deoxyribonucleic acid synthesis and cell proliferation in regenerating rat liver; autoradiography with thymidine-H3. Cancer Res 22: 842-849.

[97] Stocker E, Wullstein HK, Brau G (1973) [Capacity of regeneration in liver epithelia of juvenile, repeated partially hepatectomized rats. Autoradiographic studies after continous infusion of 3H-thymidine (author's transl)]. Virchows Arch B Cell Pathol 14: 93-103.

[98] Overturf K, Al-Dhalimy M, Ou CN, Finegold M, Grompe M (1997) Serial transplantation reveals the stem-cell-like regenerative potential of adult mouse hepatocytes. Am J Pathol 151: 1273-1280.

[99] Crosby HA, Hubscher S, Fabris L, Joplin R, Sell S, Kelly D, Strain AJ (1998) Immunolocalization of putative human liver progenitor cells in livers from patients with end-stage primary biliary cirrhosis and sclerosing cholangitis using the monoclonal antibody OV-6. Am J Pathol 152: 771-779.

[100] Laconi E, Oren R, Mukhopadhyay DK, Hurston E, Laconi S, Pani P, Dabeva MD, Shafritz DA (1998) Long-term, near-total liver replacement by transplantation of iso-

lated hepatocytes in rats treated with retrorsine. Am J Pathol 153: 319-329. S0002-9440(10)65574-5 [pii];10.1016/S0002-9440(10)65574-5 [doi].

[101] Rudolph KL, Chang S, Millard M, Schreiber-Agus N, DePinho RA (2000) Inhibition of experimental liver cirrhosis in mice by telomerase gene delivery. Science 287: 1253-1258. 8254 [pii].

[102] Morris RJ, Potten CS (1999) Highly persistent label-retaining cells in the hair follicles of mice and their fate following induction of anagen. J Invest Dermatol 112: 470-475. 10.1046/j.1523-1747.1999.00537.x [doi].

[103] Foudi A, Hochedlinger K, Van BD, Schindler JW, Jaenisch R, Carey V, Hock H (2009) Analysis of histone 2B-GFP retention reveals slowly cycling hematopoietic stem cells. Nat Biotechnol 27: 84-90. nbt.1517 [pii];10.1038/nbt.1517 [doi].

[104] Wilson A, Laurenti E, Oser G, van der Wath RC, Blanco-Bose W, Jaworski M, Offner S, Dunant CF, Eshkind L, Bockamp E, Lio P, Macdonald HR, Trumpp A (2008) Hematopoietic stem cells reversibly switch from dormancy to self-renewal during homeostasis and repair. Cell 135: 1118-1129. S0092-8674(08)01386-X [pii];10.1016/j.cell.2008.10.048 [doi].

[105] Sato T, Vries RG, Snippert HJ, van de Wetering M, Barker N, Stange DE, van Es JH, Abo A, Kujala P, Peters PJ, Clevers H (2009) Single Lgr5 stem cells build crypt-villus structures in vitro without a mesenchymal niche. Nature 459: 262-265. nature07935 [pii];10.1038/nature07935 [doi].

[106] FARBER E (1956) Similarities in the sequence of early histological changes induced in the liver of the rat by ethionine, 2-acetylamino-fluorene, and 3'-methyl-4-dimethylaminoazobenzene. Cancer Res 16: 142-148.

[107] Wilson JW, Leduc EH (1958) Role of cholangioles in restoration of the liver of the mouse after dietary injury. J Pathol 76: 441-449. 10.1002/path.1700760213.

[108] Xiao JC, Jin XL, Ruck P, Adam A, Kaiserling E (2004) Hepatic progenitor cells in human liver cirrhosis: immunohistochemical, electron microscopic and immunofluorencence confocal microscopic findings. World J Gastroenterol 10: 1208-1211.

[109] Richardson MM, Jonsson JR, Powell EE, Brunt EM, Neuschwander-Tetri BA, Bhathal PS, Dixon JB, Weltman MD, Tilg H, Moschen AR, Purdie DM, Demetris AJ, Clouston AD (2007) Progressive fibrosis in nonalcoholic steatohepatitis: association with altered regeneration and a ductular reaction. Gastroenterology 133: 80-90. S0016-5085(07)00994-8 [pii];10.1053/j.gastro.2007.05.012 [doi].

[110] Chen YK, Zhao XX, Li JG, Lang S, Wang YM (2006) Ductular proliferation in liver tissues with severe chronic hepatitis B: an immunohistochemical study. World J Gastroenterol 12: 1443-1446.

[111] Curado S, Stainier DY (2010) deLiver'in regeneration: injury response and development. Semin Liver Dis 30: 288-295. 10.1055/s-0030-1255357 [doi].

[112] Alatas FS, Masumoto K, Matsuura T, Hayashida M, Saeki I, Kohashi K, Oda Y, Taguchi T (2011) Synchronized expressions of hepatic stellate cells and their transactivation and liver regeneration during liver injury in an animal model of cholestasis. J Pediatr Surg 46: 2284-2290. S0022-3468(11)00755-X [pii];10.1016/j.jpedsurg.2011.09.016 [doi].

[113] Nagy P, Kiss A, Schnur J, Thorgeirsson SS (1998) Dexamethasone inhibits the proliferation of hepatocytes and oval cells but not bile duct cells in rat liver. Hepatology 28: 423-429. S0270913998003218 [pii];10.1002/hep.510280220 [doi].

[114] Preisegger KH, Factor VM, Fuchsbichler A, Stumptner C, Denk H, Thorgeirsson SS (1999) Atypical ductular proliferation and its inhibition by transforming growth factor beta1 in the 3,5-diethoxycarbonyl-1,4-dihydrocollidine mouse model for chronic alcoholic liver disease. Lab Invest 79: 103-109.

[115] Pritchard MT, Nagy LE (2010) Hepatic fibrosis is enhanced and accompanied by robust oval cell activation after chronic carbon tetrachloride administration to Egr-1-deficient mice. Am J Pathol 176: 2743-2752. S0002-9440(10)60796-1 [pii];10.2353/ajpath.2010.091186 [doi].

[116] Akhurst B, Croager EJ, Farley-Roche CA, Ong JK, Dumble ML, Knight B, Yeoh GC (2001) A modified choline-deficient, ethionine-supplemented diet protocol effectively induces oval cells in mouse liver. Hepatology 34: 519-522. S0270-9139(01)11774-X [pii];10.1053/jhep.2001.26751 [doi].

[117] Jelnes P, Santoni-Rugiu E, Rasmussen M, Friis SL, Nielsen JH, Tygstrup N, Bisgaard HC (2007) Remarkable heterogeneity displayed by oval cells in rat and mouse models of stem cell-mediated liver regeneration. Hepatology 45: 1462-1470. 10.1002/hep.21569 [doi].

[118] Evarts RP, Hu Z, Omori N, Omori M, Marsden ER, Thorgeirsson SS (1996) Precursor-product relationship between oval cells and hepatocytes: comparison between tritiated thymidine and bromodeoxyuridine as tracers. Carcinogenesis 17: 2143-2151.

[119] Nagy P, Bisgaard HC, Thorgeirsson SS (1994) Expression of hepatic transcription factors during liver development and oval cell differentiation. J Cell Biol 126: 223-233.

[120] Petersen BE, Zajac VF, Michalopoulos GK (1997) Bile ductular damage induced by methylene dianiline inhibits oval cell activation. Am J Pathol 151: 905-909.

[121] Paku S, Schnur J, Nagy P, Thorgeirsson SS (2001) Origin and structural evolution of the early proliferating oval cells in rat liver. Am J Pathol 158: 1313-1323.

[122] Theise ND, Saxena R, Portmann BC, Thung SN, Yee H, Chiriboga L, Kumar A, Crawford JM (1999) The canals of Hering and hepatic stem cells in humans. Hepatology 30: 1425-1433. S027091399900511X [pii];10.1002/hep.510300614 [doi].

[123] Zhou H, Rogler LE, Teperman L, Morgan G, Rogler CE (2007) Identification of hepatocytic and bile ductular cell lineages and candidate stem cells in bipolar ductular reactions in cirrhotic human liver. Hepatology 45: 716-724. 10.1002/hep.21557 [doi].

[124] Koenig S, Probst I, Becker H, Krause P (2006) Zonal hierarchy of differentiation markers and nestin expression during oval cell mediated rat liver regeneration. Histochem Cell Biol 126: 723-734. 10.1007/s00418-006-0204-3 [doi].

[125] Yovchev MI, Grozdanov PN, Zhou H, Racherla H, Guha C, Dabeva MD (2008) Identification of adult hepatic progenitor cells capable of repopulating injured rat liver. Hepatology 47: 636-647. 10.1002/hep.22047 [doi].

[126] Yovchev MI, Grozdanov PN, Joseph B, Gupta S, Dabeva MD (2007) Novel hepatic progenitor cell surface markers in the adult rat liver. Hepatology 45: 139-149. 10.1002/hep.21448 [doi].

[127] Okabe M, Tsukahara Y, Tanaka M, Suzuki K, Saito S, Kamiya Y, Tsujimura T, Nakamura K, Miyajima A (2009) Potential hepatic stem cells reside in EpCAM+cells of normal and injured mouse liver. Development 136: 1951-1960. 136/11/1951 [pii]; 10.1242/dev.031369 [doi].

[128] Bisgaard HC, Parmelee DC, Dunsford HA, Sechi S, Thorgeirsson SS (1993) Keratin 14 protein in cultured nonparenchymal rat hepatic epithelial cells: characterization of keratin 14 and keratin 19 as antigens for the commonly used mouse monoclonal antibody OV-6. Mol Carcinog 7: 60-66.

[129] Vestentoft PS, Jelnes P, Andersen JB, Tran TA, Jorgensen T, Rasmussen M, Bornholdt J, Grovdal LM, Jensen CH, Vogel LK, Thorgeirsson SS, Bisgaard HC (2013) Molecular constituents of the extracellular matrix in rat liver mounting a hepatic progenitor cell response for tissue repair. Fibrogenesis Tissue Repair 6: 21. 1755-1536-6-21 [pii]; 10.1186/1755-1536-6-21 [doi].

[130] Tanimizu N, Tsujimura T, Takahide K, Kodama T, Nakamura K, Miyajima A (2004) Expression of Dlk/Pref-1 defines a subpopulation in the oval cell compartment of rat liver. Gene Expr Patterns 5: 209-218. S1567-133X(04)00118-8 [pii];10.1016/j.modgep.2004.08.003 [doi].

[131] Bisgaard HC, Nagy P, Ton PT, Hu Z, Thorgeirsson SS (1994) Modulation of keratin 14 and alpha-fetoprotein expression during hepatic oval cell proliferation and liver regeneration. J Cell Physiol 159: 475-484. 10.1002/jcp.1041590312 [doi].

[132] Floridon C, Jensen CH, Thorsen P, Nielsen O, Sunde L, Westergaard JG, Thomsen SG, Teisner B (2000) Does fetal antigen 1 (FA1) identify cells with regenerative, endocrine and neuroendocrine potentials? A study of FA1 in embryonic, fetal, and placental tissue and in maternal circulation. Differentiation 66: 49-59. S0301-4681(09)60574-0 [pii];10.1046/j.1432-0436.2000.066001049.x [doi].

[133] Smas CM, Sul HS (1993) Pref-1, a protein containing EGF-like repeats, inhibits adipocyte differentiation. Cell 73: 725-734. 0092-8674(93)90252-L [pii].

[134] Moon YS, Smas CM, Lee K, Villena JA, Kim KH, Yun EJ, Sul HS (2002) Mice lacking paternally expressed Pref-1/Dlk1 display growth retardation and accelerated adiposity. Mol Cell Biol 22: 5585-5592.

[135] Taketa K (1990) Alpha-fetoprotein: reevaluation in hepatology. Hepatology 12: 1420-1432.

[136] Schmidt LE, Dalhoff K (2005) Alpha-fetoprotein is a predictor of outcome in acetaminophen-induced liver injury. Hepatology 41: 26-31. 10.1002/hep.20511 [doi].

[137] Asahina K, Zhou B, Pu WT, Tsukamoto H (2011) Septum transversum-derived mesothelium gives rise to hepatic stellate cells and perivascular mesenchymal cells in developing mouse liver. Hepatology 53: 983-995. 10.1002/hep.24119 [doi].

[138] Yokoi Y, Namihisa T, Kuroda H, Komatsu I, Miyazaki A, Watanabe S, Usui K (1984) Immunocytochemical detection of desmin in fat-storing cells (Ito cells). Hepatology 4: 709-714. S0270913984001051 [pii].

[139] Gard AL, White FP, Dutton GR (1985) Extra-neural glial fibrillary acidic protein (GFAP) immunoreactivity in perisinusoidal stellate cells of rat liver. J Neuroimmunol 8: 359-375.

[140] Ramadori G, Veit T, Schwogler S, Dienes HP, Knittel T, Rieder H, Meyer zum Buschenfelde KH (1990) Expression of the gene of the alpha-smooth muscle-actin isoform in rat liver and in rat fat-storing (ITO) cells. Virchows Arch B Cell Pathol Incl Mol Pathol 59: 349-357.

[141] Friedman SL (2008) Hepatic stellate cells: protean, multifunctional, and enigmatic cells of the liver. Physiol Rev 88: 125-172. 88/1/125 [pii];10.1152/physrev.00013.2007 [doi].

[142] Knittel T, Kobold D, Piscaglia F, Saile B, Neubauer K, Mehde M, Timpl R, Ramadori G (1999) Localization of liver myofibroblasts and hepatic stellate cells in normal and diseased rat livers: distinct roles of (myo-)fibroblast subpopulations in hepatic tissue repair. Histochem Cell Biol 112: 387-401. 91120387.418 [pii].

[143] Knittel T, Kobold D, Saile B, Grundmann A, Neubauer K, Piscaglia F, Ramadori G (1999) Rat liver myofibroblasts and hepatic stellate cells: different cell populations of the fibroblast lineage with fibrogenic potential. Gastroenterology 117: 1205-1221. S0016508599004655 [pii].

[144] Dezso K, Jelnes P, Laszlo V, Baghy K, Bodor C, Paku S, Tygstrup N, Bisgaard HC, Nagy P (2007) Thy-1 is expressed in hepatic myofibroblasts and not oval cells in stem cell-mediated liver regeneration. Am J Pathol 171: 1529-1537. ajpath.2007.070273 [pii]; 10.2353/ajpath.2007.070273 [doi].

[145] Lorenzini S, Bird TG, Boulter L, Bellamy C, Samuel K, Aucott R, Clayton E, Andreone P, Bernardi M, Golding M, Alison MR, Iredale JP, Forbes SJ (2010) Characterisation of a stereotypical cellular and extracellular adult liver progenitor cell niche in rodents and diseased human liver. Gut 59: 645-654. 59/5/645 [pii];10.1136/gut.2009.182345 [doi].

[146] Van Hul NK, Abarca-Quinones J, Sempoux C, Horsmans Y, Leclercq IA (2009) Relation between liver progenitor cell expansion and extracellular matrix deposition in a CDE-induced murine model of chronic liver injury. Hepatology 49: 1625-1635. 10.1002/hep.22820 [doi].

[147] Nagai H, Terada K, Watanabe G, Ueno Y, Aiba N, Shibuya T, Kawagoe M, Kameda T, Sato M, Senoo H, Sugiyama T (2002) Differentiation of liver epithelial (stem-like) cells into hepatocytes induced by coculture with hepatic stellate cells. Biochem Biophys Res Commun 293: 1420-1425. 10.1016/S0006-291X(02)00406-0 [doi];S0006-291X(02)00406-0 [pii].

[148] Karkampouna S, ten DP, Dooley S, Julio MK (2012) TGFbeta signaling in liver regeneration. Curr Pharm Des 18: 4103-4113. CPD-EPUB-20120522-28 [pii].

[149] Nishimura EK, Suzuki M, Igras V, Du J, Lonning S, Miyachi Y, Roes J, Beermann F, Fisher DE (2010) Key roles for transforming growth factor beta in melanocyte stem cell maintenance. Cell Stem Cell 6: 130-140. S1934-5909(09)00633-X [pii];10.1016/j.stem.2009.12.010 [doi].

[150] Chen L, Zhang W, Zhou QD, Yang HQ, Liang HF, Zhang BX, Long X, Chen XP (2012) HSCs play a distinct role in different phases of oval cell-mediated liver regeneration. Cell Biochem Funct 30: 588-596. 10.1002/cbf.2838 [doi].

[151] Fallowfield JA, Mizuno M, Kendall TJ, Constandinou CM, Benyon RC, Duffield JS, Iredale JP (2007) Scar-associated macrophages are a major source of hepatic matrix metalloproteinase-13 and facilitate the resolution of murine hepatic fibrosis. J Immunol 178: 5288-5295. 178/8/5288 [pii].

[152] Geiger B, Spatz JP, Bershadsky AD (2009) Environmental sensing through focal adhesions. Nat Rev Mol Cell Biol 10: 21-33. nrm2593 [pii];10.1038/nrm2593 [doi].

[153] Guilak F, Cohen DM, Estes BT, Gimble JM, Liedtke W, Chen CS (2009) Control of stem cell fate by physical interactions with the extracellular matrix. Cell Stem Cell 5: 17-26. S1934-5909(09)00293-8 [pii];10.1016/j.stem.2009.06.016 [doi].

[154] Hynes RO (2009) The extracellular matrix: not just pretty fibrils. Science 326: 1216-1219. 326/5957/1216 [pii];10.1126/science.1176009 [doi].

[155] Badylak SF, Freytes DO, Gilbert TW (2009) Extracellular matrix as a biological scaffold material: Structure and function. Acta Biomater 5: 1-13. S1742-7061(08)00282-1 [pii];10.1016/j.actbio.2008.09.013 [doi].

[156] van der Flier A, Sonnenberg A (2001) Function and interactions of integrins. Cell Tissue Res 305: 285-298.

[157] Daley WP, Peters SB, Larsen M (2008) Extracellular matrix dynamics in development and regenerative medicine. J Cell Sci 121: 255-264. 121/3/255 [pii];10.1242/jcs.006064 [doi].

[158] Brown BN, Barnes CA, Kasick RT, Michel R, Gilbert TW, Beer-Stolz D, Castner DG, Ratner BD, Badylak SF (2010) Surface characterization of extracellular matrix scaffolds. Biomaterials 31: 428-437. S0142-9612(09)00993-4 [pii];10.1016/j.biomaterials.2009.09.061 [doi].

[159] Discher DE, Mooney DJ, Zandstra PW (2009) Growth factors, matrices, and forces combine and control stem cells. Science 324: 1673-1677. 324/5935/1673 [pii];10.1126/science.1171643 [doi].

[160] Engler AJ, Sen S, Sweeney HL, Discher DE (2006) Matrix elasticity directs stem cell lineage specification. Cell 126: 677-689. S0092-8674(06)00961-5 [pii];10.1016/j.cell.2006.06.044 [doi].

[161] Sun Y, Li W, Lu Z, Chen R, Ling J, Ran Q, Jilka RL, Chen XD (2011) Rescuing replication and osteogenesis of aged mesenchymal stem cells by exposure to a young extracellular matrix. FASEB J 25: 1474-1485. fj.10-161497 [pii];10.1096/fj.10-161497 [doi].

[162] Ryu BY, Orwig KE, Oatley JM, Avarbock MR, Brinster RL (2006) Effects of aging and niche microenvironment on spermatogonial stem cell self-renewal. Stem Cells 24: 1505-1511. 2005-0580 [pii];10.1634/stemcells.2005-0580 [doi].

[163] Ricard-Blum S (2011) The collagen family. Cold Spring Harb Perspect Biol 3: a004978. cshperspect.a004978 [pii];10.1101/cshperspect.a004978 [doi].

[164] Bozec L, van der Heijden G, Horton M (2007) Collagen fibrils: nanoscale ropes. Biophys J 92: 70-75. S0006-3495(07)70805-6 [pii];10.1529/biophysj.106.085704 [doi].

[165] Duffield JS, Forbes SJ, Constandinou CM, Clay S, Partolina M, Vuthoori S, Wu S, Lang R, Iredale JP (2005) Selective depletion of macrophages reveals distinct, opposing roles during liver injury and repair. J Clin Invest 115: 56-65. 10.1172/JCI22675 [doi].

[166] Ramadori G, Saile B (2002) Mesenchymal cells in the liver--one cell type or two? Liver 22: 283-294. 1726 [pii].

[167] Yasoshima M, Tsuneyama K, Harada K, Sasaki M, Gershwin ME, Nakanuma Y (2000) Immunohistochemical analysis of cell-matrix adhesion molecules and their ligands in the portal tracts of primary biliary cirrhosis. J Pathol 190: 93-99. 10.1002/(SICI)1096-9896(200001)190:1<93::AID-PATH507>3.0.CO;2-A [pii];10.1002/(SICI)1096-9896(200001)190:1<93::AID-PATH507>3.0.CO;2-A [doi].

[168] Terada T, Nakanuma Y (1994) Expression of tenascin, type IV collagen and laminin during human intrahepatic bile duct development and in intrahepatic cholangiocarcinoma. Histopathology 25: 143-150.

[169] Timpl R, Rohde H, Robey PG, Rennard SI, Foidart JM, Martin GR (1979) Laminin--a glycoprotein from basement membranes. J Biol Chem 254: 9933-9937.

[170] Knittel T, Janneck T, Muller L, Fellmer P, Ramadori G (1996) Transforming growth factor beta 1-regulated gene expression of Ito cells. Hepatology 24: 352-360. S0270-9139(96)00330-8 [pii];10.1053/jhep.1996.v24.pm0008690404 [doi].

[171] Kikkawa Y, Mochizuki Y, Miner JH, Mitaka T (2005) Transient expression of laminin alpha1 chain in regenerating murine liver: restricted localization of laminin chains and nidogen-1. Exp Cell Res 305: 99-109. S0014-4827(05)00009-1 [pii];10.1016/j.yexcr.2005.01.005 [doi].

[172] Nikolova G, Strilic B, Lammert E (2007) The vascular niche and its basement membrane. Trends Cell Biol 17: 19-25. S0962-8924(06)00327-8 [pii];10.1016/j.tcb.2006.11.005 [doi].

[173] Pham VT, Couchie D, Martin-Garcia N, Laperche Y, Zafrani ES, Mavier P (2008) Expression of matrix metalloproteinase-2 and-9 and of tissue inhibitor of matrix metalloproteinase-1 in liver regeneration from oval cells in rat. Matrix Biol 27: 674-681. S0945-053X(08)00103-0 [pii];10.1016/j.matbio.2008.07.002 [doi].

[174] Pellicoro A, Aucott RL, Ramachandran P, Robson AJ, Fallowfield JA, Snowdon VK, Hartland SN, Vernon M, Duffield JS, Benyon RC, Forbes SJ, Iredale JP (2012) Elastin accumulation is regulated at the level of degradation by macrophage metalloelastase (MMP-12) during experimental liver fibrosis. Hepatology 55: 1965-1975. 10.1002/hep.25567 [doi].

[175] Kallis YN, Robson AJ, Fallowfield JA, Thomas HC, Alison MR, Wright NA, Goldin RD, Iredale JP, Forbes SJ (2011) Remodelling of extracellular matrix is a requirement for the hepatic progenitor cell response. Gut 60: 525-533. gut.2010.224436 [pii]; 10.1136/gut.2010.224436 [doi].

[176] Clayton E, Forbes SJ (2009) The isolation and in vitro expansion of hepatic Sca-1 progenitor cells. Biochem Biophys Res Commun 381: 549-553. S0006-291X(09)00350-7 [pii];10.1016/j.bbrc.2009.02.079 [doi].

[177] McClelland R, Wauthier E, Uronis J, Reid L (2008) Gradients in the liver's extracellular matrix chemistry from periportal to pericentral zones: influence on human hepatic progenitors. Tissue Eng Part A 14: 59-70. 10.1089/ten.a.2007.0058 [doi].

[178] Paku S, Nagy P, Kopper L, Thorgeirsson SS (2004) 2-acetylaminofluorene dose-dependent differentiation of rat oval cells into hepatocytes: confocal and electron microscopic studies. Hepatology 39: 1353-1361. 10.1002/hep.20178 [doi].

[179] Oertel M, Shafritz DA (2008) Stem cells, cell transplantation and liver repopulation. Biochim Biophys Acta 1782: 61-74. S0925-4439(07)00235-9 [pii];10.1016/j.bbadis.2007.12.004 [doi].

[180] Xu X, Liu RF, Zhang X, Huang LY, Chen F, Fei QL, Han ZG (2012) DLK1 as a potential target against cancer stem/progenitor cells of hepatocellular carcinoma. Mol Cancer Ther 11: 629-638. 1535-7163.MCT-11-0531 [pii];10.1158/1535-7163.MCT-11-0531 [doi].

[181] Yamashita T, Forgues M, Wang W, Kim JW, Ye Q, Jia H, Budhu A, Zanetti KA, Chen Y, Qin LX, Tang ZY, Wang XW (2008) EpCAM and alpha-fetoprotein expression defines novel prognostic subtypes of hepatocellular carcinoma. Cancer Res 68: 1451-1461. 68/5/1451 [pii];10.1158/0008-5472.CAN-07-6013 [doi].

[182] Yamashita T, Ji J, Budhu A, Forgues M, Yang W, Wang HY, Jia H, Ye Q, Qin LX, Wauthier E, Reid LM, Minato H, Honda M, Kaneko S, Tang ZY, Wang XW (2009) EpCAM-positive hepatocellular carcinoma cells are tumor-initiating cells with stem/progenitor cell features. Gastroenterology 136: 1012-1024. S0016-5085(08)02179-3 [pii];10.1053/j.gastro.2008.12.004 [doi].

[183] Roskams TA, Libbrecht L, Desmet VJ (2003) Progenitor cells in diseased human liver. Semin Liver Dis 23: 385-396. 10.1055/s-2004-815564 [doi].

[184] Bird TG, Lorenzini S, Forbes SJ (2008) Activation of stem cells in hepatic diseases. Cell Tissue Res 331: 283-300. 10.1007/s00441-007-0542-z [doi].

[185] Bisgaard HC, Holmskov U, Santoni-Rugiu E, Nagy P, Nielsen O, Ott P, Hage E, Dalhoff K, Rasmussen LJ, Tygstrup N (2002) Heterogeneity of Ductular Reactions in Adult Rat and Human Liver Revealed by Novel Expression of Deleted in Malignant Brain Tumor 1. Am J Pathol 161: 1187-1198.

[186] Jemal A, Bray F, Center MM, Ferlay J, Ward E, Forman D (2011) Global cancer statistics. CA Cancer J Clin 61: 69-90. caac.20107 [pii];10.3322/caac.20107 [doi].

[187] de Jong MC, Nathan H, Sotiropoulos GC, Paul A, Alexandrescu S, Marques H, Pulitano C, Barroso E, Clary BM, Aldrighetti L, Ferrone CR, Zhu AX, Bauer TW, Walters DM, Gamblin TC, Nguyen KT, Turley R, Popescu I, Hubert C, Meyer S, Schulick RD, Choti MA, Gigot JF, Mentha G, Pawlik TM (2011) Intrahepatic cholangiocarcinoma: an international multi-institutional analysis of prognostic factors and lymph node assessment. J Clin Oncol 29: 3140-3145. JCO.2011.35.6519 [pii];10.1200/JCO.2011.35.6519 [doi].

[188] Cohnheim J (1875) Congenitales, quergestreiftes Muskelsarkom der Nieren. Archiv f pathol Anat 65: 64-69. 10.1007/BF01978936.

[189] Wicha MS, Liu S, Dontu G (2006) Cancer stem cells: an old idea--a paradigm shift. Cancer Res 66: 1883-1890. 66/4/1883 [pii];10.1158/0008-5472.CAN-05-3153 [doi].

[190] Driessens G, Beck B, Caauwe A, Simons BD, Blanpain C (2012) Defining the mode of tumour growth by clonal analysis. Nature. nature11344 [pii];10.1038/nature11344 [doi].

[191] Chen J, Li Y, Yu TS, McKay RM, Burns DK, Kernie SG, Parada LF (2012) A restricted cell population propagates glioblastoma growth after chemotherapy. Nature. nature11287 [pii];10.1038/nature11287 [doi].

[192] Schepers AG, Snippert HJ, Stange DE, van den Born M, van Es JH, van de Wetering M, Clevers H (2012) Lineage tracing reveals Lgr5+stem cell activity in mouse intestinal adenomas. Science 337: 730-735. science.1224676 [pii];10.1126/science.1224676 [doi].

[193] Modrek AS, Bayin NS, Placantonakis DG (2014) Brain stem cells as the cell of origin in glioma. World J Stem Cells 6: 43-52. 10.4252/wjsc.v6.i1.43 [doi].

[194] Thorgeirsson SS, Grisham JW (2002) Molecular pathogenesis of human hepatocellular carcinoma. Nat Genet 31: 339-346. 10.1038/ng0802-339 [doi];ng0802-339 [pii].

[195] Seitz HK, Stickel F (2006) Risk factors and mechanisms of hepatocarcinogenesis with special emphasis on alcohol and oxidative stress. Biol Chem 387: 349-360. 10.1515/BC.2006.047 [doi].

[196] Hussain SP, Schwank J, Staib F, Wang XW, Harris CC (2007) TP53 mutations and hepatocellular carcinoma: insights into the etiology and pathogenesis of liver cancer. Oncogene 26: 2166-2176. 1210279 [pii];10.1038/sj.onc.1210279 [doi].

[197] Budhu A, Forgues M, Ye QH, Jia HL, He P, Zanetti KA, Kammula US, Chen Y, Qin LX, Tang ZY, Wang XW (2006) Prediction of venous metastases, recurrence, and prognosis in hepatocellular carcinoma based on a unique immune response signature of the liver microenvironment. Cancer Cell 10: 99-111. S1535-6108(06)00214-5 [pii];10.1016/j.ccr.2006.06.016 [doi].

[198] Enzan H, Himeno H, Iwamura S, Onishi S, Saibara T, Yamamoto Y, Hara H (1994) Alpha-smooth muscle actin-positive perisinusoidal stromal cells in human hepatocellular carcinoma. Hepatology 19: 895-903. S0270913994001187 [pii].

[199] Amann T, Bataille F, Spruss T, Muhlbauer M, Gabele E, Scholmerich J, Kiefer P, Bosserhoff AK, Hellerbrand C (2009) Activated hepatic stellate cells promote tumorigenicity of hepatocellular carcinoma. Cancer Sci 100: 646-653. CAS1087 [pii];10.1111/j.1349-7006.2009.01087.x [doi].

[200] Zhang F, Chen XP, Zhang W, Dong HH, Xiang S, Zhang WG, Zhang BX (2008) Combined hepatocellular cholangiocarcinoma originating from hepatic progenitor cells: immunohistochemical and double-fluorescence immunostaining evidence. Histopathology 52: 224-232. HIS2929 [pii];10.1111/j.1365-2559.2007.02929.x [doi].

[201] Komuta M, Spee B, Vander BS, De VR, Verslype C, Aerts R, Yano H, Suzuki T, Matsuda M, Fujii H, Desmet VJ, Kojiro M, Roskams T (2008) Clinicopathological study

on cholangiolocellular carcinoma suggesting hepatic progenitor cell origin. Hepatology 47: 1544-1556. 10.1002/hep.22238 [doi].

[202] Ma S, Chan KW, Hu L, Lee TK, Wo JY, Ng IO, Zheng BJ, Guan XY (2007) Identification and characterization of tumorigenic liver cancer stem/progenitor cells. Gastroenterology 132: 2542-2556. S0016-5085(07)00786-X [pii];10.1053/j.gastro.2007.04.025 [doi].

[203] Qiang L, Yang Y, Ma YJ, Chen FH, Zhang LB, Liu W, Qi Q, Lu N, Tao L, Wang XT, You QD, Guo QL (2009) Isolation and characterization of cancer stem like cells in human glioblastoma cell lines. Cancer Lett 279: 13-21. S0304-3835(09)00034-2 [pii]; 10.1016/j.canlet.2009.01.016 [doi].

[204] Lee JS, Heo J, Libbrecht L, Chu IS, Kaposi-Novak P, Calvisi DF, Mikaelyan A, Roberts LR, Demetris AJ, Sun Z, Nevens F, Roskams T, Thorgeirsson SS (2006) A novel prognostic subtype of human hepatocellular carcinoma derived from hepatic progenitor cells. Nat Med 12: 410-416. nm1377 [pii];10.1038/nm1377 [doi].

[205] Woo HG, Wang XW, Budhu A, Kim YH, Kwon SM, Tang ZY, Sun Z, Harris CC, Thorgeirsson SS (2011) Association of TP53 mutations with stem cell-like gene expression and survival of patients with hepatocellular carcinoma. Gastroenterology 140: 1063-1070. S0016-5085(10)01695-1 [pii];10.1053/j.gastro.2010.11.034 [doi].

[206] Cairo S, Wang Y, de RA, Duroure K, Dahan J, Redon MJ, Fabre M, McClelland M, Wang XW, Croce CM, Buendia MA (2010) Stem cell-like micro-RNA signature driven by Myc in aggressive liver cancer. Proc Natl Acad Sci U S A 107: 20471-20476. 1009009107 [pii];10.1073/pnas.1009009107 [doi].

Section 4

Stem Cell Niches in Adult Nentral Nervous System

Chapter 8

Typical and Atypical Stem Cell Niches of the Adult Nervous System in Health and Inflammatory Brain and Spinal Cord Diseases

Joshua Bernstock, Jeroen Verheyen, Bing Huang,
John Hallenbeck and Stefano Pluchino

Additional information is available at the end of the chapter

http://dx.doi.org/10.5772/58599

1. Introduction

"Once development was ended, the fonts of growth and regeneration of the axons and dendrites dried up irrevocably. In the adult centers, the nerve paths are something fixed, and immutable: everything may die, nothing may be regenerated."-Santiago Ramon y Cajal

The central nervous system (CNS) is inhabited by a heterogeneous population of cells (i.e. neurons and glia) and is marked by a highly complex anatomical structure [1]. In states of host homoeostasis the putative majority of cells in the CNS are long-lived and typically do not require replacement. Nonetheless, neurogenesis in the adult mammalian brain has been shown to occur in a myriad of locations, under a diverse set of physiologic/pathophysiologic conditions [2-10]. Neurogenesis is driven by stem cells which can be defined by their ability to produce both identical daughter cells (self-renewal) and progeny with more restricted fates (commitment and differentiation) [11]. To be classified as a neural stem cell (NSC), cells should be able to self-renew and give rise to a variety of mature progeny that make up the CNS, including neurons, astrocytes and oligodendrocytes [12-16]. However, fate-restricted precursor cells capable of self-renewal, but which concurrently display restricted differentiation potential, also reside in the CNS. These cells are often unipotent and are referred to as neural progenitor cells (NPC) [17, 18], for example, oligodendrocyte precursor cells (OPC) are able to self-renew, but typically produce only oligodendrocytes [18, 19].

Identification of NSC *in vivo* is clearly complicated and relies on the analysis of cell morphology, mitotic activity, and gene and protein expression. Commonly used NSC markers include nestin, glial fibrillary acidic protein (GFAP), Musashi 1/2, and the Shy-related high mobility group box transcription factor 2 (Sox2) [20-23]. Nestin is a class VI intermediate filament linked to mitotically active cells in the CNS [20, 24]. GFAP is expressed in multipotent ependymal cells, radial glia, and also in mature astrocytes [21]. Musashi 1 and 2 expression can be found in embryonic neuroepithelial cells [22] while Sox2 is found primarily in undifferentiated cells that possess self-renewal capabilities [23]. As noted above, NSC can exist in either a quiescent or mitotically active state. Quiescent cells have been shown to express Sox2 and FoxO3A, and are further demarcated by a prolonged retention of bromodeoxyuridine (BrdU) [24-28]. Dividing cells, on the other hand, show a rapid turnover of BrdU and simultaneously contain various markers of cell-cycle entry/progression: Mcm-2, Ki67, cyclin D1 and E (G1 phase), cyclin A (S phase), cytoplasmic cyclin B1 (G2 phase), and phosphohistone H3 (M phase) [10, 29]. Fate restricted precursor cells have traditionally been recognized via the expression of doublecortin (DCX) and the polysialylated-neural adhesion molecule (PSA-NCAM) [30, 31].

As stem cells (SC) continue to be identified, characterized and localized, the critical importance of specific signals from their microenvironment, or niche, have become apparent. Stem cell niches in the brain can be classified as either "typical" or "non-typical". The three typical NSC niches found in the CNS are the subventricular zone (SVZ), the subgranular zone (SGZ) and the central canal (CC) of the spinal cord [32-34]. Non-typical (germinal) niches have been identified in the hypothalamus, circumventricular organs (CVO), the meninges and the subpial layer of the cerebellum [32, 35-37]. Further, non-typical (non-germinal) niches can be found throughout parenchyma of the cerebral cortex, cerebellum and spinal cord, and are mainly comprised of restricted neuroglia precursors [10, 32, 38, 39]. Much of the aforementioned has recently been confirmed *in vitro* via an assortment of neurosphere assays, which are considered to represent the "gold-standard" technique for identifying the presence of NSC in the adult brain [33, 40]. Neurospheres have been obtained from many regions in the brain, including the olfactory bulb, cerebellum, various white matter tracts, spinal cord, substantia nigra, retina, hypothalamus, and hypophysis [41-45]. Finally, the concept of atypical niches has recently emerged and references the unique microenvironment formed upon exogenous stem cell transplantation. These niches are reported to evolve in close proximity to perivascular regions [32, 46].

The capacities of stem cells to contribute to growth and diversification during development and in so doing sustain homeostasis/repair processes throughout adult life is now clear. Elucidation of the mechanisms that govern stem cell behavior is therefore of fundamental significance in cell, developmental, and organismal biology. The capabilities arising from such knowledge are anticipated to have major biomedical and clinical translational applications [11, 47]. The remainder of this chapter will therefore offer an overview that will touch upon the distribution and relevant components (e.g. stem cells, support cells, signaling molecules) of stem cell niches in the CNS, in states of both homeostasis and various pathobiologies (e.g. ischemic, inflammatory, traumatic) and in the process will attempt to highlight potential therapeutic targets that may be manipulated in an effort to promote effective and translational repair and regeneration of the CNS after insult/injury.

2. Neural stem cells niches within the central nervous system

2.1. Definition/critical components of the "niche"

As stem cells in adult organs continue to be identified, characterized and localized, it has become clear that the vast majority of these cells depend on specific signals from the microenvironment of their niche to regulate their quiescence, activation, self-renewal and ultimate survival. Such a phenomenon was hypothesized by Schofield nearly 35 years ago and has been shown to hold true today [48]. The evolution of this concept has led to the definition of the niche as a microenvironment capable of integrating intrinsic and extrinsic factors and in so doing, influence stem cell proliferation, migration and fate specification [49, 50]. Intrinsic determinants are governed mainly by the genetic/epigenetic status of stem cells and their subsequent ability to decipher signals within the niche. Extrinsic determinants may be thought of as the processing of extracellular signals and include such events as cell-to-cell and cell-to-extracellular matrix (ECM) signaling [50, 51]. Generally, the cellular makeup of these niches has been shown to consist of a variety of cells, which typically include the immature progeny of NSC accompanied by endothelial, astroglial, and ependymal cells [50, 52]. Along with the ECM, they provide not only structural/trophic support, but have been shown to provide critical temporal and spatial information, thereby enabling stem cells to respond to both physiological and pathological stimuli [49]. Acting through these pathways, stem cell niches in the CNS have been shown to play essential roles in supporting active neurogenesis via the mobilization of endogenous neural stem/precursor cells and further serve to regulate different stages of adult neurogenesis in health/disease [52]. Clearly, an understanding of the detailed molecular, structural and functional properties of the niche may help to influence intractable neurological disease processes and/or yield novel clinically relevant NSC-based therapeutic approaches via the enhancement of endogenous regeneration and repair.

2.2. Subventricular zone (SVZ)

In the adult brain, NSC have traditionally been assumed to be restricted to certain regions, such as the SVZ of the lateral ventricles and the SGZ of the dentate gyrus (DG) of the hippocampus. Both of these niches have been shown to be capable of sustaining neurogenesis in the adult CNS [53-55]. The vast majority of adult neurogenesis in mammalian species has been demonstrated to occur within the SVZ niche as it retains many of its early embryonic features/primitive germinal layers. The SVZ also represents the largest neurogenic region and has by most accounts been the best characterized of the endogenous CNS niches [50]. Interestingly, recent work suggests that SVZ neural stem cells are not homogenous; rather they may represent a heterogeneous population capable of differentiating into restricted subsets/cells of differential fates [42]. Within the niche, a subset of GFAP-expressing astrocytes (type B/B1 cells) are thought to represent the NSC population (Figure 1) [56, 57]. These primary progenitors either slowly self-renew or differentiate and give rise to transit-amplifying cells (type C cells), which are capable of generating a substantial number of neuroblasts (type A cells) [58, 59]. SVZ neuronal precursors have been shown to migrate extensive distances in chains via the rostral migratory stream (RMS) [60] toward the olfactory bulb [61]. Upon arrival, they undergo the process of differentiation into mature neurons, and migrate into the granular and periglomerular layers [62, 63]. The type B cells mentioned above share morphologic features that are similar to astrocytes and strengthen the argument for a radial glial origin [64]. Uniquely, type B cells are in direct contact with blood vessels via their basal processes and concurrently interact with the ventricular lumen through apical processes [26, 65, 66].

Cells that eventually give rise to olfactory bulb neurons in the human brain have been identified via the expression of DCX in the SVZ [67]. Detailed studies have revealed a ribbon of SVZ astrocytes that line the lateral ventricles of adult human brain, and work has confirmed that these cells are in fact self-renewing and multipotent [68]. Interestingly humans do not display features characteristic of the RMS [68]. However, the migration of immature neurons away from the SVZ has been documented to occur [69, 70]. While some studies have indicated progressive decline in neuroblasts over the course of an adult life [70-72], recent work-utilizing carbon-14 has demonstrated that neurons continue to be generated and to integrate into host circuitry [73, 74]. Additionally, contemporary studies have begun to suggest a role for supraependymal 5-hydroxytryptamine (5-HT, serotonin) axons that directly contact NSC and therefore may serve in part to regulate neurogenesis via 5-HT2C receptors [75]. Such complex cytoarchitecture coupled with the emerging diversity of SVZ precursor cells leads to a unique microenvironment capable of supporting sustained neurogenesis throughout the life of an organism [25].

2.3. Subgranular Zone (SGZ) of the hippocampus

The second major region that produces new neurons in the adult mammalian brain is the SGZ of the hippocampus, which is located at the interface of the granule cell layer (GCL) and the hilus of the dentate gyrus [76, 77]. This has been shown to be true in a variety of mammalian species (e.g. rodents, primates, humans) [78-85]. In stark contrast to the new neurons born in the subventricular zone, newly formed hippocampal neurons transmigrate only a short distance into the granule cell layer before functionally integrating into existing hippocampal circuitry [77, 86-88]. While it has been suggested that neurogenesis in the adult hippocampus contributes to the processes involved in learning and memory, the definitive function of neuronal replacement in DG has yet to be elucidated [88, 89]. Similar to the SVZ, neurogenesis in the dentate gyrus has been demonstrated to occur throughout life [89, 90] and has been shown to be influenced/regulated by a multiplicity of physiological and environmental cues. These cues have not been fully characterized, but they include adrenal steroids, glutamate receptor activation, seizures, enriched environmental conditions, exercise, inflammation/brain injury, and antidepressant medication [59, 81, 83, 91, 92].

Given the presence of multiple precursor subtypes found within the adult hippocampus, a reliable method to distinguish molecular identities is needed in order to adequately reveal the degree to which primary precursors self-renew and/or differentiate into multiple progeny [93]. Briefly, a core tenant of the prevailing model of adult hippocampal neurogenesis is that GFAP/nestin/Sox2 expressing radial glia-like cells (RGL) [77, 86, 89, 93], or type-1 cells [94], represent a quiescent population which may be induced to generate the proliferative precursors known as intermediate progenitors, IPC1 (type-2a) and IPC2 (type-2b) cells. Via the use of anti-mitotic

http://dx.doi.org/10.5772/58599

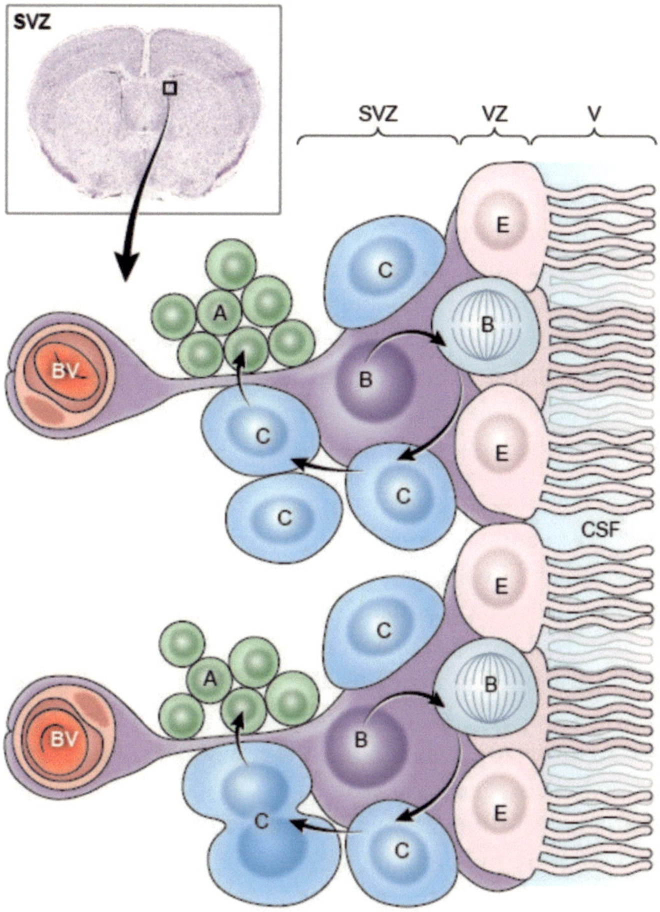

Figure 1. Subventricular Zone Niche. Coronal brain section (Allen Developing Mouse Brain Atlas) shows the location of the neurogenic subventricular zone (SVZ) niche. The SVZ can be found contacting the overlying ventricular zone (VZ), a pseudo-stratified epithelium layer that lines the cerebrospinal fluid (CSF) filled ventricles (V). NSC (type B cells, B) are found in a subependymal position, contacting both ependymal cells (E) and blood vessels (BV). Type B cells proliferate through asymmetric division, giving rise to transit-amplifying type C cells (C) that further differentiate to form neuroblasts (type A cells, A). Supported by type B cells, these neuroblasts proliferate, expand and migrate, allowing for adult neurogenesis. Adapted from Fuentealba et al. [100].

agents, genetic ablation, and transgenic fate mapping, a vast body of experimental evidence now exists in support of RGL as functional NSC [86, 95-98]. Of note, RGL seem to maintain both ultrastructural features and surface markers characteristic of astrocytes [59] and have been shown to be capable of undergoing several rounds of both self-renewal and differentiation over a prolonged period of time [99]. Importantly, RGL in the niche are polarized, a characteristic that provides a spatiotemporal nature to signals received within the niche. RGL zones within the niche can be subdivided into proximal, intermediate and distal domains along which RGL maintain their polarized structure (i.e. from apical-basal). They span from the hilar/

SGZ interface (proximal domain I) to the inner molecular layer (IML) (distal domain III) [100] (Figure 2). The proximal domain harbors a distinctive primary cilium which has been shown to be important for Sonic hedgehog (Shh) signaling, sensing/sampling of the hilus microenvironment, and contacting other RGL and blood vessels [100]. Here, endothelial cells provide access to critical factors, namely, vascular endothelial growth factor (VEGF), insulin-like growth factor (IGF) and brain-derived neurotrophic factor (BDNF), which together serve to coordinate the complex regulation between proliferation and differentiation [100]. RGL cell bodies/main shafts are located within the SGZ and GCL (domain II) and facilitate cell-cell based interactions of the RGL with progeny (feedback from which may serve to regulate RGL quiescence or transition via Notch signaling) and simultaneous sampling of local neural activity via resident granular cells [101, 102]. In the IML (domain III), RGL terminate and display an elaborate/branched structure. While the governing dynamics in this area have yet to be fully elucidated, it seems reasonable to deduce that inputs via interneurons and mossy cells have a role to play in the regulation of RGL/NSC [103].

Returning to the abovementioned progeny of the RGL, the IPC, it should be noted that they produce novel neuroblasts and eventually immature granule neurons (type-3 cells), which migrate into the inner granule cell layer, thereby differentiating into immature granule cells of the DG [88, 89, 100]. Retroviral mediated gene transduction has allowed such newborn neurons to be labeled and subsequently tracked. Using this technique, Zhao et al. demonstrated that these novel neurons extend dendrites toward the molecular layer and project axons through the hilus toward the CA3 region in a matter of days in an effort to become functionally integrated into host circuitry [104, 105]. Despite the complexity of events outlined above, and the relatively high rate of neurogenesis occurring in the SGZ, it is important to note that only a minority of newly born cells ultimately survive to mature and integrate within the granule cell layer of the hippocampus, highlighting the need for further exploration/characterization of the niche/neurogenic processes in the SGZ [106].

2.4. Central canal of the spinal cord

The spinal cord comprises the caudal part of CNS, extending from the medulla to the cauda equina. It contains 33 nerve segments, rostro-caudally grouped as the cervical, thoracic, lumbar, sacral, and coccygeal segments. At the center of the spinal cord lays the central canal, an ependymal region forming a round-shaped lumen, lined by epithelium, which contains cerebrospinal fluid (CSF). The spinal cord transmits signals between the brain and the rest of the body and contains complex circuitry thereby enabling reflexive and rhythmic motor patterns [107]. The inner region of the spinal cord surrounding the central canal is comprised of gray matter and contains neurons that are commonly arranged by function: motor neurons are clustered anteriorly, sensory projection neurons posteriorly, with a more mixed population in the intermediate areas, including the afferent and efferent neurons of autonomic nuclei. All regions are supported by and connected through a complex network of interneurons, which serve to modulate transmission and activity. The outer region is comprised of white matter and contains afferent and efferent axons arranged in tracts. Like the gray matter, white matter

http://dx.doi.org/10.5772/58599

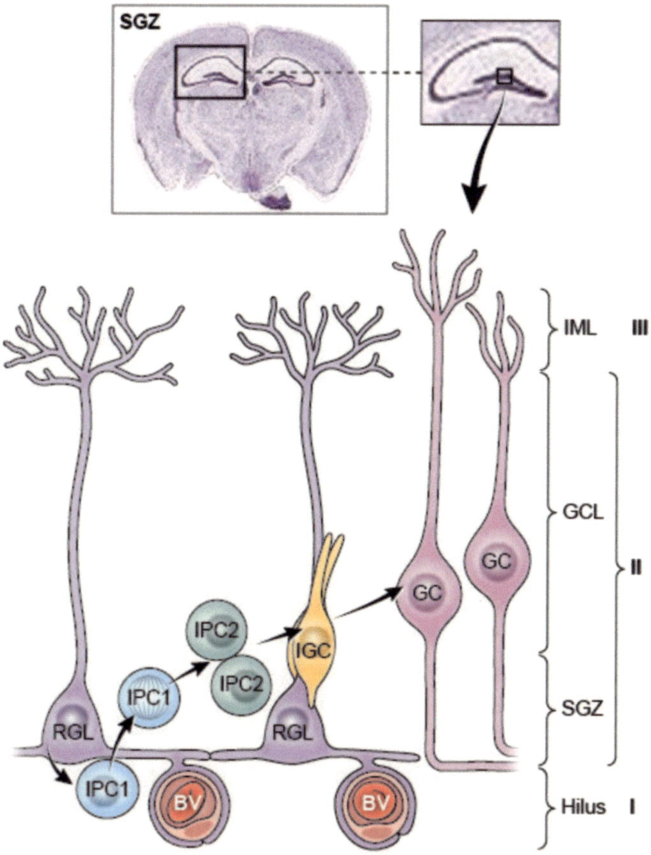

Figure 2. Subgranular Zone Niche. Coronal brain section (Allen Developing Mouse Brain Atlas) shows the location of the subgranular zone (SGZ) niche in the dentate gyrus of the hippocampus. Radial glial-like cells (RGL) are the type of NSC that make up the SGZ. In the proximal domain (I) or hilus, they contact blood vessels (BV) and their radial processes span the granule cell layer (GCL), in domain II, to reach the inner molecular layer (IML) in the distal domain (III). RGL divide asymmetrically to generate intermediate progenitor cells 1 (type-2a cells, IPC1) and 2 (type-2b cells, IPC2). These progenitors give rise to neuroblasts that differentiate to immature granule neurons (type-3 cells) that in turn migrate to the GCL and differentiate to immature granule cells (IGC). These cells further differentiate to form mature granule cells (GC), allowing for adult neurogenesis. Adapted from Fuentealba et al. [100].

exhibits functional organization, with afferent tracts clustered dorsally and at the lateral periphery, and efferent tracts clustered anteriorly and medially [107, 108].

The ependymal layer of the spinal cord is well known for its role in embryonic development and its function as neuroprogenitor niche. Ependymal cells divide symmetrically and migrate away from the central canal, giving rise to the different neural lineages [19, 109]. Postnatally, the spinal cord elongates and increases in size [110]. The proliferation required for such growth gradually declines, leaving adult rodents and humans with little to no ependymal proliferation [111, 112].

The presence of multipotent cells in the adult mammalian spinal cord was first discovered in the late 1990s. Rat and mouse NSC were isolated and characterized *in vitro*. Cultured cells were able to produce neurospheres capable of self-renewal, extended proliferation, passaging, and differentiation into the three major CNS cell types, i.e. neurons, oligodendrocytes, and astrocytes [12, 15, 113]. It was shown later that NSC reside at the central canal and in the parenchyma of the spinal cord [13, 14]. Although able to self-renew and generate mature oligodendrocytes, these parenchymal cells do not produce neurospheres, indicating that they are progenitors (i.e. restricted in fate) rather than NSC [114]. When spinal cord derived neurospheres are transplanted into the hippocampus they can give rise to neurons, a property that is not observed when transplanted back to the cord, and is suggestive of a non-conducive progenitor microenvironment [18].

The adult central canal is comprised of a pseudo-stratified epithelium with a myriad of cell types that contact the lumen or are present in a subependymal position (all $Sox2^{+}$) (Figure 3) [34, 112, 115]. The main constituents are ependymal cells, some of which are positive for GFAP [112, 116-118]. Although under physiological conditions most of these ependymal cells are quiescent, some proliferation has been observed at the dorsal tip of the central canal and ependymal cells from this region have enriched neurosphere-forming capabilities [112, 114, 119]. Dorsal ependymal cells show a radial morphology, much like radial glia, and their processes can reach up to the white matter or even the pial surface [112, 117, 119, 120]. They divide symmetrically, as they did during postnatal development [114]. Dorsal ependymal cells show enriched expression of GFAP, nestin, CD15 and/or brain lipid-binding protein (BLBP) [34, 112, 117, 119, 120]. A similar population and morphology has also been observed at the ventral part of the central canal, although to a lesser extent [112, 117, 119]. It has now been shown that ependymal cells are able to generate progeny of multiple fates under physiological and pathological conditions [114, 119, 121]. Other cells that make up the central canal are tanycytes and CSF-contacting neuron-like cells. Tanycytes, a specific subset of ependymal cells, contact blood vessels through their long basal processes and thus bridge the CSF and capillaries [119, 122]. Neuron-like cells that contact the CSF through dendrite-like processes are thought to be involved in CSF homeostasis (e.g. pressure sensing) and/or spinal cord extension/flexion sensing [112, 123, 124]. Surrounding the central canal, nerve fibers, neurons (NeuN+), oligodendrocytes (Olig2+) and blood vessels can also be found [34, 112]. Pericytes that are an active part of the blood brain barrier surrounding blood vessels have also been shown to be an important source of astrocytes, implicating stem cell-like properties for these cells. These astrocytes mainly contribute to astrogliosis during injury [125].

Central canal derived neurospheres tend to house a heterogeneous population of cells, much like neurospheres derived from other neurogenenic regions [34]. Neurosphere cells all express nestin but show variable expression levels of prominin-1 (CD133) (stem cell marker), GFAP, and aldehyde dehydrogenase 1 family member, L1 (ALDH1L1) (astrocytic markers), CD15, BLBP, glutamate aspartate transporter (GLAST), and radial glial cell marker-2 (RC2) (radial glial markers), and neuron-glial antigen 2 (NG2), A2B5 antigen (A2B5), and platelet-derived growth factor receptor α (PDGFRα) (oligodendrocytic markers) [34]. Only a small number of cells express neuronal markers such as microtubule-associated protein 2 (MAP2) and DCX, which correlates with the overall preference of the cord toward oligodendrocytic and astrocytic differentiation [34]. Expression of motor neuron development transcription factors (Islet1, lim1, HB9) has not been observed, reflecting the *in vivo* tendency towards production of GABAergic neurons [16, 112, 126]. Motor neuron differentiation can however be induced by certain morphogens, such as retinoic acid (RA) and Shh [126]. Notably, neurospheres preserve information related to their rostro-caudal location, namely the expression of certain combinations of developmental genes of the Hox family [112, 127].

In conclusion, the central canal of the spinal cord is mainly comprised of a heterogeneous population of ependymal cells. Stem cell properties have mainly been attributed to ependymal cells at the dorsal tip of the central canal and to pericytes. Further research is needed to fully unravel the neurogenic properties/potential of the central canal in states of both health and disease.

2.5. Non-typical neural stem cell niches

Beyond the typical NSC niches referenced above it should be noted that non-typical niches have now been identified and have begun to be characterized. These non-typical niches can be further divided into those areas that are germinal (neurogenic) and those that are not. Non-typical germinal regions include the hypothalamus, CVO, the meninges and the subpial layer of the cerebellum. Non-typical, non-germinal regions can be found throughout parenchyma of the cerebral cortex and spinal cord, and are mainly comprised of restricted neuroglia precursors [10, 32, 35-39, 128-131]. Accordingly, the following paragraphs will briefly discuss selected non-typical niches in neurogenic and non-neurogenic areas.

2.5.1. Non-typical germinal regions

As was the case with the typical niches, non-typical germinal regions are characterized by their inherent neurogenic capabilities, i.e. composed of a heterogeneous population of NSC able to self-renew and give rise to most of the neuronal and glial precursors [32, 132, 133]. To be characterized as neurogenic, isolated cells should be able to give rise to secondary neurospheres *in vitro* whilst being able to produce all three neuronal lineages [36, 37].

Constitutive adult neurogenesis has been identified in regions lining the third ventricle, including the hypothalamus and the CVO [131, 134-137]. Cells from these areas are not only positive for nestin, GFAP, Sox2 and Ki-67, but have also been shown to incorporate BrdU. Their ability to produce both proliferating and differentiating neurospheres *in vitro* strongly suggests that these areas represent germinal neurogenic NSC niches [137]. Furthermore, it should be noted that the ECM structure and composition of the aforesaid areas strongly resemble that of the SVZ [138].

Cells positive for nestin and DCX have also been found in the meninges of the brain and spinal cord [139-143]. These nestin^{+} cells are able to give rise to neurospheres *in vitro* and show highly efficient generation of excitable cells with neuronal phenotype and morphology [139], congruent with the meninges' important role during development, harboring neuroepithelial cells [144]. Further, within the adult meninges, neurogenic factors such as basic fibroblast

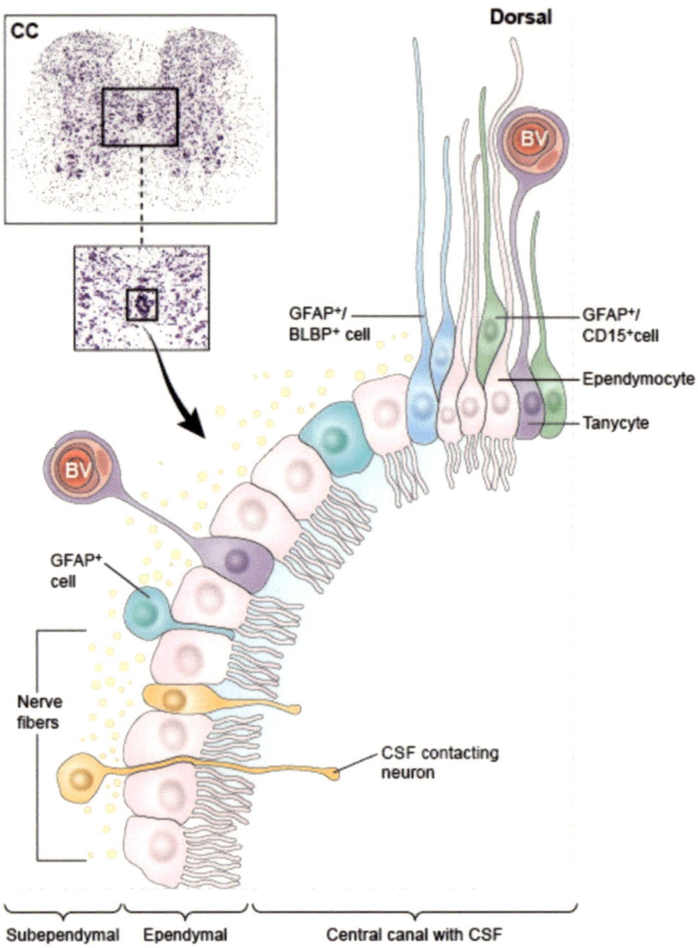

Figure 3. Central Canal Niche. Cross-section through the spinal cord at lumbar level 1 (Allen Developing Mouse Brain Atlas) shows the location of the central canal. Lining the lumen of the cerebrospinal fluid (CSF)-filled central canal is a pseudo-stratified epithelium with interspersed ependymal cells (ependymocytes). Ependymal cells GFAP can be found throughout the canal and are enriched in the dorsal and ventral part (latter not shown) where they have a radial morphology, much like that of radial glia. These radial GFAP+ cells are believed to be NSC since they proliferate and differentiate, allowing for (glial-restricted) neurogenesis. This is further supported by co-expression of stem cell markers, BLBP or CD15. Although mainly quiescent under physiological conditions, these cells become mitotically active under pathological conditions. After symmetrical division their progeny differentiate to astrocytes and oligodendrocytes. Other cells that make up the central canal are tanycytes that bridge the CSF and blood vessels (BV), and CSF-contacting neurons. Pericytes surrounding BV (not shown) have been found to also contribute to the generation of astrocytes under pathological conditions, and are thus considered another form of NSC around the central canal.

growth factor (bFGF), Chemokine (C-X-C motif) ligand 2 (CXCL2)/macrophage inflammatory protein 2-alpha (MIP2-alpha) and RA can still be observed [145-147].

Neurosphere-forming NSC have also been obtained from the cerebellum and are isolated based on their expression of the NSC marker prominin-1 (CD133) and their lack of markers of neuronal and glial lineage markers. Purified $CD133^+$ cells form self-renewing neurospheres and can differentiate into astrocytes, oligodendrocytes and neurons *in vitro* [148]. Although the exact location and composition of this niche remains unclear, proliferative elements have been putatively allocated to the subpial layer [149-151], with newly generated cells divided in two populations: DCX^+/PSA-$NCAM^+$/Pax^+ neuroblast neural precursors and microtubule-associated protein 5 ($MAP5^+$)/$Olig2^+$/$Sox2^+$ glial precursors [149, 150].

2.5.2. Non-typical non-germinal regions

Non-typical non-germinal regions are those that demonstrate proliferative properties, but are unable to induce comprehensive neurogenesis. Often these are areas within the parenchyma and consist of committed precursor cells that can self-renew and give rise only to a specific type of neuronal cell. The potential of cells in these areas to produce multipotent neurospheres is lost soon after birth [32, 38, 152, 153]. While there are non-typical regions that may be germinal in nature rather than non-germinal, proof is still lacking. These putative areas include the striatum, amygdala, substantia nigra, and vagal nucleus [35, 153].

In the cerebral cortex, $A2B5^+$ glial restricted precursors give rise to oligodendrocytes and astrocytes [154]. Oligodendrocyte precursor cells that express integral chondroitin sulfate proteoglycan 4 (CSPG4), also known as $NG2^+$ cells, can also be found through the cerebral cortex [155, 156]. These cells are restricted to producing oligodendrocytes and astrocytes. In the spinal cord, these $NG2^+$ cells can also be observed [152, 156]. $Olig2^+$ OPC are also widely found in the spinal cord. These cells are typically deemed to be more committed than $NG2^+$ cells, only able to give rise to oligodendrocytes [114, 152]. Furthermore, progenitors that produce immature DCX^+/GAD-65^+/GAD-67^+/$GABA^+$ neurons have been found enriched in the dorsal part of the spinal cord [19, 157].

The abovementioned progenitors are some of the more predominant cellular populations, yet it should be noted that parenchymal progenitors consist of an incredibly heterogeneous population, as evidenced by expression of stem cell markers. While crosstalk between cells populating non-typical niches under varied pathological conditions have also been begun to be highlighted [153], much work still needs to be done to fully elucidate the function and therapeutic potential of such regions [10, 35, 152, 153].

3. Neural stem cell niches in CNS disease

"I say all the most acute, most powerful, and most deadly diseases, and those which are most difficult to be understood by the inexperienced, fall upon the brain."-Hippocrates

Diseases of the central nervous system pose a massive societal burden and continue to be a leading cause of morbidity and mortality throughout the world; however, the medical community possesses few effective therapies that are able to modulate the pathogenesis of brain injury/illness. The paucity of viable therapeutic options stands in stark contrast to the intensity of research efforts and number of clinical trials that have been performed to date. As of yet, there are few, if any, treatments capable of markedly improving functional recovery to levels concordant with a pre-disease state (i.e. regenerative therapies). The restricted success of such a massive research investment demands a reevaluation of the pathobiology of the injured and/or dysfunctional brain.

Beyond homeostasis, it has been clearly established that the basic biological descriptors of neural stem cells-which include self-renewal, proliferation/differentiation, and migration-are affected by certain pathogenic stimuli e.g. excitotoxicity, mechanical trauma, ischemic and/or inflammatory) [158-167]. It follows that a greater knowledge of the factors involved in the dynamic regulation of adult neurogenesis may pave the way for the development of suitable treatments and preventative strategies that would delay the onset and/or mitigate the symptoms of a number of devastating brain disorders. Therefore, the remainder of this section will seek to highlight core components of the response of adult neurogenic regions in the face of the distinctly relevant clinical entities: ischemic stroke, multiple sclerosis (MS) and spinal cord injury (SCI).

3.1. Effects of ischemic stroke on the neurogenic process/niche

Stroke is the one of the most common causes of death and disability worldwide. Due to an aging population, the burden will markedly increase in the coming decades and will be particularly pronounced in developing countries [168, 169]. Of strokes that occur in the United States, 87% are ischemic and 10% are intracerebral hemorrhagic strokes, whereas 3% are subarachnoid hemorrhage strokes [169]. Based on this distribution, the remainder of this discussion will focus on ischemic stroke. Cerebral ischemia triggers the pathological pathways of the "ischemic cascade" that if untreated causes irreversible neuronal injury in the ischemic core within mere minutes of the onset [170-172]. Cerebral ischemia and, if applicable, reperfusion cause extreme changes in the parenchymal microenvironment to include variations in oxygen (O_2) concentrations, depletion of cellular energy stores e.g. adenosine triphosphate (ATP), perturbation of ion homeostasis, inflammation and aberrant neurotransmitter release [173]. The primary drivers of this pathogenic process stem from a crisis in energy availability and result from a reduction in O_2 and glucose [173]. Clearly such a vast array of pathology would suggest that the incidence/activity of endogenous neurogenic niches would be affected and this has proven correct.

Numerous studies have now demonstrated that ischemic stroke is in fact capable of increasing neural stem cell proliferation [158, 159, 174-184]. In the SGZ, ischemia seems to act preferentially on proliferation of type 1 and 2 progenitor cells, and to a lesser extent neuroblasts [167, 185]. Within the SVZ, stroke selectively increases the number of type A and C cells [186], yet there is also data to suggest that type B cells undergo a period of transient symmetric division after stroke [187]. Ependymal cells bordering the SVZ have also been noted to proliferate transiently after ischemic stroke [188]. Mitotic activity appears to peak during between 7-10 days post ischemia then returns to baseline levels between the 3-5th week [160, 175, 187, 189-191]. While maximal cell proliferation occurs on the order of days-weeks it should be noted that neuroblasts have been documented to exist for at least one year after an ischemic insult [176]. Signals that stimulate the stroke-induced neurogenesis have yet to be fully elucidated but likely involve the interplay of many non-dominant effectors, namely cytokines and growth factors/neurotrophins that have been shown to be upregulated during brain ischemia, the putative majority of which have established links to the neurogenic process [189, 192]. bFGF, BDNF, epidermal growth factor (EGF), glial cell-derived neurotrophic factor (GDNF), bone morphogenic protein (BMP) and erythropoietin (EPO), ciliary neurotrophic factor (CNTF), transforming growth factor (TGF)-α, VEGF and erythropoietin (EPO) have all been proposed to play prominent roles in neurogenesis [191, 193-210]. Insulin-like factor-1 (IGF-1) and granulocyte-colony stimulating factor (G-CSF) have also been shown to be inextricably involved in the abovementioned stroke-induced neurogenic process [211, 212]. It is also important to note that the physiologic stressors of ischemia directly affect other components of the neurogenic niche and in so doing may influence neurogenesis as highlighted by studies of cerebral endothelial cells [27, 213, 214].

Of particular note, inflammation also accompanies ischemic insults/injuries and is predominantly driven in the CNS by the activation of resident microglia, astrocytes and infiltrating immune cells, which go on to release a plethora of inflammatory cytokines/chemokines and reactive oxygen species [189, 215]. Inflammatory mediators have been shown to have varying effects on neural progenitor cell proliferation, migration, differentiation, survival and incorporation of newly born neurons into the CNS circuitry [216-222]. These studies suggest that additional work is warranted and will be needed to clarify the precise effects/outcomes as influenced by inflammation post-stroke. Further complicating the picture, evidence has emerged to suggest that neurotransmitters and associated excitotoxicity also mediate stroke-induced neurogenesis [223, 224].

In the post-ischemic brain newly generated cells from DG and SVZ have been shown to be capable of replacing dying neurons via directed migration toward areas of damage [225]. Studies have indicated that newly arrived neuroblasts in the ischemic boundary zones display phenotypes that are indeed characteristic of mature/functional neurons [160, 176, 181, 190, 191, 226-228]. The neural precursors that develop, transmigrate and integrate display an innate form of pathotropism [229, 230]. Work has come to suggest that EPO may promote neuroblast migration via the secretion of matrix metalloproteinases, MMP2 and MMP9, by EPO-activated endothelial cells [231]. Additional factors presumed to be involved in the progenitor cell migration to sites of injury are C-X-C motif chemokine 12/stromal cell-derived factor 1 (SDF-1)/ its receptor CXCR4: stroke has been shown to upregulate penumbral SDF-1 and NSC/ neuroblast CXCR4 expression [232-236]. Lastly, chemokine (C-C motif) ligand 2 (CCL2)/ monocyte chemotactic protein-1 (MCP-1) has also been shown to regulate migration of neuroblasts to the areas of damage as the expression of MCP-1 has been localized to the activated microglia/astrocytes present in ischemic areas post reperfusion [237]; correspondingly, ischemia-induced migrating neuroblasts express the MCP-1 receptor CCR2 [237, 238].

The experimental evidence that has been put forth hitherto clearly suggests that ischemia stimulates neurogenesis in the adult brain. Recently reports have emerged which demonstrate that the endogenous neurogenic response following experimental stroke influences the course of recovery in both short and long-term settings [239, 240]. Although this evidence indicates that cerebral ischemia-induced neurogenesis may affect neurological recovery after stroke, it is clear that such an endogenous repair response is far from ideal as patients continue to experience various levels of physical/cognitive morbidities post-injury [241-243]. In order to become a clinically valuable tool, the stroke induced neurogenic response will need to be markedly enhanced which requires consideration of ways to support/supplement the process. Understanding that the process of generating new neurons essentially consists of four phases: proliferation, migration, differentiation, and survival [89, 244] one might begin to design interventions that rationally target one or more of the aforementioned (e.g. therapeutics to prevent the death of the vast majority of neuroblasts) [176, 245]. Specifically, Kokaia et al. note "of particular importance for the promotion of neurogenesis and its functional benefit [will] be to increase the survival of stroke-induced neuroblasts and mature neurons [as the] the majority of new neuroblasts die soon after formation" [176, 245, 246].

3.2. Effects of multiple sclerosis on CNS neurogenic processes/niches

Multiple sclerosis is one of the most common causes of chronic neurologic disability beginning in early to middle adult life (median age of onset being 29 years of age) and is characterized by a triad of inflammation, demyelination and gliosis [247-249]. MS is idiopathic in nature yet is presumed to be driven by the complex interaction of autoimmunity, genetic predisposition, and environmental associations [248, 250]. MS affects approximately 400,000 people in the United States and 2.5 million worldwide [251]. Symptoms of MS have primarily been shown to result from a disruption in the integrity of myelinated tracts in the CNS [247, 252]. More recently research has also highlighted the underappreciated involvement of gray matter in MS pathogenesis, which may be especially relevant when one considers the development of irreversible disability [253, 254]. As such, the need to understand mechanisms governing endogenous stem cell/stem cell niches in MS is clearly justified.

Contrasting reports have emerged with regard to the activation of the SVZ and its cellular components in MS, in both the human disease state and in animal models. SVZ activation has been shown to be especially dependent on the temporal nature of the disease (i.e. acute vs. chronic inflammation) [163, 255, 256]. Such findings suggest that inflammation may be either advantageous or deleterious depending on the pathophysiologic context (see Table 2). In experimental autoimmune encephalomyelitis (EAE), the most widely used/accepted animal model of MS [257], alterations in SVZ NSC proliferation and mobilization have been demonstrated throughout the disease process [87, 163, 255, 258]. Such changes are concordant with other models of CNS injury (e.g. stroke) in which surviving cells that activate locally or infiltrate post-damage, secrete mediators that alter the neurogenic process [256, 259, 260]. Beyond the preclinical animal models, increases in SVZ activity have also been noted in humans with MS [261]. Further, enhanced proliferation has been found at the level of the hippocampal neurogenic niche in animal models of MS. However, the downstream network dynamics of these progenitors appears to be altered, leading to aberrant differentiation i.e. these EAE animals exhibited a significantly higher percentage of newborn radial-glia-like NSC yet the mean percentage of newborn/mature neurons was decreased [262, 263]. Such findings align with the clinical phenotypes/histopathology [264, 265] displayed by many human patients and correlate with findings on magnetic resonance imaging (MRI), which highlight the existence of focal hippocampal hyperintensities [266] and hippocampal atrophy [267]. Of note, neurogenesis/gliogenesis in the spinal cord in various murine models of MS [46, 166, 268] and in human patients with MS [269] have also been demonstrated to occur. Although accumulating evidence indicates that endogenous neurogenesis/gliogenesis do occur as part of an intrinsic attempt at self-repair (i.e. oligodendrocyte precursors in the MS lesions of human patients) [270-272], it has become clear that the endogenous stem cell compartment's capacity for mobilization is unable to achieve meaningful restoration of impaired CNS function in the face of a chronic inflammatory disorder [46]. Data now suggest that inflammatory components, such as infiltrating blood-born mononuclear cells, reactive CNS-resident cells (i.e. astrocytes, endothelial cells and microglia), and humoral mediators such as cytokines/chemokines may be partially responsible for such an inadequate response as they can and do affect proliferation/differentiation of NSC [32, 46, 87, 163, 222, 256, 259, 273]. It is clear then, that the molecular mechanisms capable of inducing and/or inhibiting neurogenesis in the CNS of MS patients under defined spatiotemporal conditions warrant further investigation.

3.3. Effects of spinal cord injury on CNS neurogenic processes/niches

SCI is often induced by trauma and subsequently leads to both motor and sensory deficiencies [274]. Typically, such injuries manifest clinically in presentations of pain, anesthesia/paresthesia, fasciculations, and/or weakness [275]. In severe cases, SCI can lead to complete paralysis and/or result immediately in life threatening impairments to respiration, heart rate, and blood pressure [107, 108]. SCI pathophysiology is marked by a pathophysiology with a complex temporospatial profile, and is characterized by three phases: acute (seconds to minutes after injury), subacute (hours to weeks post-injury), and chronic (weeks to years post-injury) [276, 277]. During these phases the injured environment undergoes distinct biochemical and anatomical alterations, involving a diverse group of molecules and cells (i.e. nervous, immune, vascular) [276]. The acute phase is initiated by mechanical disruption which results in such insults as ischemia, edema, vasospasm, ionic/neurotransmitter imbalance and ultimately cell death [276]. Factors released during the acute phase result in secondary inflammatory degeneration, the hallmark of the subacute phase. During the subacute period, progressive neurodegeneration occurs as a result of the pro-inflammatory neurotoxic environment (driven by neutrophils, monocytes, microglia, T-cells) and results in the continued demyelination/Wallerian degeneration of damaged axons [276, 278-281]. Over the course of the same period, astrocytes become reactive in a process called astrogliosis which ultimately facilitates the formation of glial scar. This scar tissue poses a physical and chemical barrier to axonal re-growth, thus inhibiting regeneration [282-284]. On the other hand, this scar tissue aids in regeneration and repair by regulating the immune response, preventing the spread of neurotoxic factors, enabling partial reestablishment of homeostasis, and by providing neurotrophic support through enrichment of IGF, nerve growth factor (NGF), BDNF and neurotrophins (NT-3) [282-295]. The provision of these neuroprotective, neurogenic and regenerative cues (and others) is continued during the chronic phase in an effort to repair damaged axons. This effect is however limited, due to an inhibitory microenvironment created by the glial scar and the persistence of other secondary degeneration mechanisms referenced earlier [281, 296].

Interestingly, ependymal stem cells which are quiescent under physiological conditions become activated following SCI [39]. Evidence suggests a proliferative and pathotrophic NSC response. Such mitotic activation has also been observed *in vitro* through enhancement of neurosphere-formation capabilities post-injury[13, 114, 119, 152]. These proliferating ependymal cells show a transient increase in GFAP, S100b, nestin, and Pax6 expression [16, 297, 298]. The lineage potential of these transiently activated progenitors *in vivo* seems to be predominantly restricted to glial cells, namely astrocytes and oligodendrocytes [114, 119, 121]. As mentioned before, pericytes are another source of astrocytes during spinal cord injury [125]. Newly produced astrocytes function mainly in aiding the establishment of the glial scar [114, 119, 121]. Parenchymal $NG2^+$ OPC are also activated and lead to oligodendrocyte differentiation. Newly produced oligodendrocytes participate in attempts to remyelinate injured axons [114]. Unfortunately, neuronal production has not yet been reported, and may be explained by the host of powerful pro-glial cues that emanate from the spinal cord [114, 119, 121, 299]; as a result functional recovery post-injury is modest at best.

4. Molecular characteristics of neural stem cell niches

"Look deep into nature, and then you will understand everything better." – Albert Einstein

A wealth of molecular signals have been shown to influence NSC maintenance and neurogenesis via control of survival, self-renewal, activation of quiescent NSC and regulation of their proliferative expansion/differentiation. Cues that influence the behavior of NSC within the niche include autocrine, paracrine and endocrine factors, as well as direct cell-cell and cell-ECM contact [10, 300, 301]. A summarized overview of molecular signaling influencing NSC maintenance and neurogenesis is given in Table 1.

4.1. Growth factors

4.1.1. Fibroblast Growth Factors (bFGF) and Epidermal Growth Factors (EGF)

EGF and bFGF are factors necessary for *in vitro* growth and expansion of NSC [40]. They are produced by cells in the SVZ and induce proliferation in cells that reside in the subependymal layer lining the lateral ventricles of the forebrain [302, 303].

4.1.2. Hepatocyte Growth Factor (HGF)

HGF is also expressed in SVZ cells and has been shown to function as a survival factor for neuroblasts and cortical neurons while also increasing proliferation of SVZ cells [304, 305]. Furthermore, it has been shown that HGF has neuroprotective properties as it can reduce apoptosis in stress conditions, probably mediated by PI3K/Akt signaling [306, 307].

4.1.3. Vascular Endothelial Growth Factor (VEGF)

VEGF is important for angiogenesis and hematopoiesis [308-310]. However, VEGF receptors have also been found in the subependymal zone of the SVZ, the SGZ, and on NSC [311, 312]. It is secreted by endothelial cells, NSC, and astrocytes [313]. VEGF exerts indirect effects on NSC and neurogenesis by inducing angiogenesis thereby providing structural and trophic support [313]. It also operates directly via the promotion of proliferation and maintenance of NSC and neurogenesis [314, 315]. Furthermore, VEGF has been shown to be neuroprotective during disease and injury [316, 317].

4.1.4. Insulin-like Growth Factors (IGF)

IGF activate the PI3K/Akt signaling pathway, activating the target of rapamycin (TOR) kinase and FoxO transcription factors [318]. IGF-1 is expressed in various areas of the CNS, including hippocampus, olfactory bulbs, and cerebellum [319, 320]. Multiple knockout studies have indicated that IGF-1 is needed for maintaining proliferation and stem cell characteristics [321, 322].

4.1.5. Pigment-Epithelium Derived Growth Factor (PEGF)

PEGF was first identified as a factor that induces differentiation of retinoblastoma cells into a neuronal phenotype [323, 324]. It has been found to be expressed by retinal cells, adipocytes and hepatocytes, and also endothelial and ependymal cells in the adult brain [325]. Although NSC do not express these factors themselves, they are responsive to them. It has no effect on survival, but increases NSC self-renewal and activates quiescent subependymal cells [325]. It is believed that PEGF function is dependent on Notch signaling and keeps cells undifferentiated through upregulation of Hes1, Hes5, and Sox2 [325, 326].

4.1.6. Platelet-Derived Growth Factors (PDGF)

PDGF is produced by endothelial cells and binds PDGF receptor α (PDGFRα) on NSC whereby it regulates neurogenesis [327]. PDGF receptor β (PDGFRβ) is expressed in brain pericytes, neurons and astrocytes and is implicated in neuroprotection after ischemic stroke [328].

4.2. Developmental factors and morphogens

4.2.1. Wingless-related integration site (Wnt) signaling

Wnt signaling pathways are major regulators of stem cell activity in the developing and adult brain, where it functions in both NSC maintenance and neurogenesis [300, 329-333]. These diverse and opposing functions are enabled by heterogeneous group of Wnt proteins that modulate canonical (involving β-catentin) and non-canonical signaling pathways with further regulation by a wide range of interaction partners and regulators [300, 334, 335]. Wnt3, for instance, is secreted by astrocytes and induces NSC proliferation and neurogenesis [333]. Wbt7b is regulated by retinoic acid and can expand the number of proliferating cells [336, 337]. The canonical pathway normally allows for an increase in cytoplasmic β-catentin, which induces proliferation and inhibits differentiation. However, when factors such as homeodomain interacting protein kinase 1 (HipK1) are upregulated in the SVZ, the same pathway can induce differentiation [338]. Furthermore, in pathological conditions such as stroke and hypoxia, Wnt signaling has been shown to drive neurogenesis through NSC proliferation and differentiation. Interestingly, these activated cells divide symmetrically leading to NSC expansion, as opposed to the asymmetrical division that normally takes place in the subendymal zone [303, 339].

4.2.2. Bone Morphogenic Proteins (BMP)

BMP and their receptors are expressed by cells adjacent to the SVZ. They inhibit proliferation of neuroblasts while blocking neurogenesis and favoring gliogenesis [340]. Noggin is secreted by ependymal cells of the SVZ and SGZ and opposes the effect of BMP by binding and inactivating them thereby maintaining cell proliferation [340-342].

4.2.3. Sonic Hedgehog (Shh)

Activation of the Shh can increase proliferation of NSC. Shh receptors (Patched (Ptc)) can for instance be found in hippocampal regions such as the hilus and pyramidal cells in CA1-CA3 [343]. Shh also plays a role in maintenance of NSC pools in telencephalic niches [344].

4.3. Hormones

4.3.1. Erythropoietin (EPO)

Although mainly produced by the kidney, EPO and its receptor were found to be expressed in adult neurogenic regions, such as the SVZ and SGZ [210, 345]. Under hypoxic stress EPO expression is upregulated in the adult brain [346]. EPO affects NSC by increasing proliferation, increasing neurogenesis, and enhancing survival [202, 347-352]. Conditional knockouts of EPO have shown that it is a critical factor for proliferation [202]. Its promotion of survival operates by reducing apoptosis of NSC and their progeny [350, 352].

4.3.2. Insulin

Insulin is produced by beta cells of the pancreas. Controversial evidence now suggests that it is also produced by cultured neuronal and glial cells and in the hippocampus [353]. In general, it allows for survival, self-renewal and proliferation of NSC [353-357]. Insulin is able to replace EGF and bFGF *in vitro*, allowing for self-renewal and long-term passaging [354].

4.3.3. Adipocyte-derived leptin and adiponectin

Leptin and adiponectin enhance the survival of NSC *in vivo* and *in vitro* [358-362]. They activate the glycogen synthase kinase β (GSKβ) signaling pathway in hippocampal NSC, allowing for accumulation of β-catenin and consequent promotion of proliferation of NSC [358, 359].

http://dx.doi.org/10.5772/58599

4.4. Cytokines

4.4.1. Leukemia Inhibitory Factor (LIF)

LIF is highly expressed in the adult injured brain, mediating inflammation and inducing NSC proliferation [363, 364]. LIF leads to an expansion of astrocytes while depleting neurons. Furthermore, it promotes NSC self-renewal rather than the generation of committed progenitors [364, 365]. Treatment of neurospheres with LIF *in vitro* increases the generation of secondary neurospheres [364].

4.4.2. Ciliary Neurotrophic Factor (CNTF)

CNTF receptor (CNTFR) expression is restricted to periventricular regions [365]. CNTF binding activates the LIF receptor/gp130 complex, enhancing maintenance, survival and self-renewal of NSC, while restricting differentiation of the glial lineage [366, 367]. Endogenous CNTF expression is upregulated after stroke and leads to increased proliferation of SVZ cells [204].

4.4.3. Stem Cell-Derived Neural Stem/Progenitor Cell Supporting Factor (SDNSF)

SDNSF is expressed in the DG of the hippocampus and is upregulated after ischemia. It has been shown to allow NSC to survive *in vitro* when bFGF is removed. Although cells maintain their self-renewal and differentiation potential, SDNSF alone does not promote proliferation [368].

4.4.4. C-X-C Motif Chemokine 12 (CXCL12)/Stromal Cell-Derived Factor 1 (SDF-1)

SDF-1 also known as CXCL12 is a chemokine produced by endothelial cells. It binds C-X-C motif receptor 4 (CXCR4) on NSC. It favors neurogenesis by driving survival and migration of neuronal and oligodendrocytic progenitors [369, 370]. After stroke, SDF-1 promotes migration and integration of new neurons, participating in functional recovery [371].

4.4.5. Macrophage Migration Inhibitory Factor (MIF)

Dendritic cells secrete MIF which mediates NSC expansion through the MIF receptor CD74, both *in vivo* and *in vitro* [372, 373].

4.4.6. Interleukin 1 (IL-1)

IL-1α and IL-1β have both been found to positively regulate neurogenesis [374, 375]. Interestingly, the effect of IL-1β depends on its concentration. Under physiological conditions, it increases differentiation of neural progenitors, whereas it inhibits neurogenesis under high inflammatory concentrations [376-378].

4.4.7. Interleukin 6 (IL-6)

At low concentrations, IL-6 promotes differentiation of NSC to neurons, astrocytes and oligodendrocytes [379-381]. However, at high concentrations IL-6 has been shown to reduce neurogenesis [161].

4.4.8. Cytokines during inflammation

Inflammatory cytokines (pro/anti) are produced by activated immune cells (including leukocytes, lymphocytes, astrocytes, microglia, and endothelial cells) after disturbance of homeostasis or during pathology. These cytokines influences NSC maintenance and neurogenesis in a very heterogeneous and context dependent manner, summarized in Table 2 [382, 383].

4.5. Neurotransmitters

4.5.1. Glutamate

Glutamate acts on NSC through metabotropic glutamate receptors (mGluR). Although excitotoxic for neurons, high levels of glutamate have been shown to promote survival and proliferation of NSC in the SVZ and DG [384-389].

4.5.2. Gamma-aminobutyric acid (GABA)

GABA is non-synaptically released by neuroblasts after spontaneous depolarization. It has been shown to reduce proliferation of GFAP+ NSC, suggestive of a feedback system regulating the NSC population [390, 391].

4.5.3. Serotonin

Serotonin has been shown to positively influence survival, proliferation, and neurogenesis [392-396]. Serotonin receptors have been found in the SVZ and the DG [392]. Their activation increases neurogenesis and affects symmetric division of a specific population of NSC [395].

4.5.4. Dopamine

Adult NSC in the SVZ and the DG have receptors for dopamine. Activation of certain dopamine receptors can indirectly promote NSC survival and differentiation due to the activation of A-disintegrin and metalloproteinases (ADAM) and consequent release of membrane bound EGF [397, 398]. By contrast, activation of the dopamine D2 receptor on NSC inhibits their proliferation and neurogenesis in a CNTF-dependent manner [399, 400].

4.5.5. $_{D}$-Serine

Although mainly produced by astrocytes, $_{D}$-Serine has recently found to be expressed by neurons and NSC [401-408]. Although it does not enhance NSC expansion and neurogenesis, $_{D}$-Serine is associated with NSC self-renewal and maintenance [409, 410].

4.5.6. Nitric oxide (NO)

NO is produced by neurons and inflammatory cells, but not NSC. Conflicting evidence exists on whether they stimulate or reduce SVZ and hippocampal NSC proliferation. It has been postulated that high NO concentrations promote proliferation, whereas low NO concentrations inhibit proliferation [411-413].

4.6. Extracellular matrix (ECM) components

Chondroitin sulfate proteoglycans (CSPG) are major constituents of the NSC niche ECM and play pivotal roles in the development, regeneration and plasticity of neuronal networks [414-416]. Enzymatic degradation of CSPG reduces self-renewal of NSC in the SVZ, as well as of neurospheres *in vitro* [417]. In other studies, however, degradation of CSPG resulted in increased NSC proliferation, differentiation and migration via an integrin-dependent mechanism [418]. These different outcomes may be the result of differences in the cell types being analyzed and further studies are needed to unravel the exact role of CSPG on adult NSC [418]. Heparan sulfate proteoglycans (HSPG) have also been implicated in the survival and proliferation of NSC, probably by interaction with bFGF [419, 420]. Sulfotransferases are expressed in adult neurogenic regions and in neurospheres and have been shown to be important for preserving the functional activity of CSPG and HSPG in NSC survival [421].

Laminins are other ECM components that can be found in NSC niches such as those in the SVZ [422]. Laminin receptors such as integrins, syndecans and dystroglycans can all be found expressed on NSC [423]. Notably, $\alpha6\beta1$ integrins are expressed in high levels on proliferating NSC and progenitors [65, 424, 425]. Quiescent NSC do not express $\beta1$ integrins; activation of NSC through daughter cell depletion or administration of CXCL12/SDF-1, however, leads to upregulation of $\beta1$ integrins, showing the pivotal role of $\beta1$ integrins in neurogenesis and NSC proliferation [425].

4.7. Direct cell-to-cell signals

4.7.1. Notch signaling

Notch is a membrane bound developmental factor and its signaling is of major importance in maintaining and expanding embryonic and adult NSC [426, 427]. Notch ligands such as Jagged and delta like ligand 4 (Dll4) are also membrane bound and regulate neurogenesis by stimulating NSC proliferation [428, 429]. Interestingly, NSC but not fate-restricted progenitors express Notch, a characteristic which has been used to distinguish between both populations [430, 431]. Progenitors communicate with NSC through Notch-epidermal growth factor receptor (EGFR) interactions, whereby regulating the balance between both cell populations in the SVZ. Enhanced EGFR signaling results in the expansion of the progenitor pool and reduces NSC numbers and their self-renewal [431]. Recent work also suggests that there is a strong interplay between Notch and Shh in regulating neurogenesis [432].

4.7.2. Ephrin signaling

Ephrin ligands and receptors are also membrane bound developmental factors. Ephrin A and B ligands and their receptors are expressed by NSC in the SVZ [433]. Ephrin signaling has been implicated in both proliferative and anti-proliferative effects on NSC [433-437]. They have been linked to NSC maintenance, survival, and inhibition of differentiation [438-441].

4.8. Neurotrophic Factors (NTF)

The NTF family includes BDNF, NGF, GDNF and NT-3, NT-4. They are important for differentiation, survival, and functioning of neurons in both the developing and adult brain [300]. NTF and their tropomyosin-related kinase (Trk) receptors are expressed in NSC. They have been shown to protect NSC against excitotoxicity and apoptosis during injury and to promote NSC differentiation [442-445].

4.9. Other factors

4.9.1. Apolipoprotein E (ApoE)

ApoE is a constituent of plasma lipoprotein particles. It has been found to be secreted by astrocytes *in vivo* and by neurospheres *in vitro,* contributing to neuritogenesis and maintenance of NSC in the DG [446-450].

Signaling factors	Source	Effect on NSC	References
Growth factors			
FGF	EC, A, CSF	Renewal, proliferation, differentiation, migration	[40, 302, 303]
EGF	EC, A, CSF	Renewal, proliferation, differentiation, migration	[40, 302, 303]
HGF	NSC	Survival, proliferation	[304-307]
VEGF	EC, NSC, A	Survival, renewal, migration	[308-317]
IGF	CSF	Renewal	[318-322]
PEGF	EC, NSC	Renewal	[323-326]
PDGF	EC	Survival, renewal	[327, 328]
Developmental factors			
Wnt signaling	A	Renewal, proliferation, differentiation*	[329-331, 333, 338, 339]
BMP	EC, A, CSF	Differentiation	[340-342]
Noggin	NSC	Renewal, proliferation	[340-342]
Shh	A, CSF	Renewal, proliferation, migration	[343, 344]
Hormones			
EPO	B, A, N	Survival, proliferation	[347-349, 351, 352,
Insulin	B	Survival, renewal, proliferation	451, 452]
Leptin/adiponectin	B	Proliferation	[353-357]

Signaling factors	Source	Effect on NSC	References
			[358-362]
Cytokines			
LIF	IC	Renewal, differentiation	[363, 364]
CNTF	IC	Survival, renewal, differentiation	[204, 366, 367]
SDNSF	IC	Survival, renewal	[368]
SDF-1	IC, EC, NSC	Survival, migration	[369-371]
MIF	IC	Proliferation	[372, 373]
IL-1	IC	Differentiation	[374-378]
IL-6	IC	Differentiation	[161, 379, 380]
Neurotransmitters			
Glutamate	N	Survival, proliferation, differentiation	[384, 385, 387-389]
GABA	N, NB	Proliferation**, differentiation, migration	[391, 453]
Serotonin	N	Survival, proliferation	[392-396]
Dopamine	N	Survival, proliferation, differentiation	[397, 399, 400]
D-Serine	A, N, NSC	Renewal	[401-405, 407, 410]
NO	N, IC	Proliferation*	[411-413]
Extracellular matrix			
CSPG		Survival, renewal	[415-418, 454, 455]
HSPG		Survival, renewal	[419, 420]
Laminins		Survival, proliferation*	[65, 424, 425]
Direct cell-to-cell signals			
Notch	NSC	Renewal*, proliferation*	[426, 429-432]
Ephrin	NSC	Renewal*, proliferation*	[433-439, 441]
Neurotrophic factors			
BDNF, NGF, GDNF, NT-3, NT-4	NSC, A, EC	Survival, renewal, proliferation differentiation	[442-444]
Others			
ApoE	A	Renewal, differentiation	[446, 447, 449, 450]

Abbreviations: bFGF, basic fibroblast growth factor; EGF, epidermal growth factor; HGF, hepatocyte growth factor; VEGF, vascular endothelial growth factor; IGF, insulin-like growth factor; PEGF, pigment-epitheliun derived growth factor; PDGF, platelet-derived growth factors; Wnt, wingless-related integration site; BMP, bone morphogenic proteinss; Shh, sonic hedgehog; EPO, erythropoietin; LIF, leukemia inhibitory factor; CNTF, ciliary neurotrophic factor; SDNSF, stem cell-derived neural stem/progenitor cell supporting factor; SDF-1, stromal cell-derived factor 1; MIF, macrophage migration inhibitory factor; IL-1, interleukin 1; IL-6, interleukin 6; TNF-α, tumor necrosis factor α; GABA, gamma-Aminobutyric acid; NO, nitric oxide; CSPG, chondroitin sulfate proteoglycans; HSPG, heparan sulfate proteoglycans; NTF, neurotrophic factors.; BDNF, bone-derived neurotrophic factor; NGF, nerve growth factor; GDNF, glial cell-line derived neurotrophic factor; NT, neurotrophin; A, astrocytes; B, blood; CSF, cerebrospinal fluid; EC, endothelial cells; IC, immune cells; N, neurons; NB, neuroblasts; NSC: neuronal stem cells. * context dependent; ** of progenitors, not stem cells.

Table 1. Molecular Components of the Niche Environment

Soluble factors	Role in NSC biology	Cell sources	Pathological models	Ref
CCL5	NSC proliferation↑	Reactive astrocytes, activated lymphocytes, microglia/macrophages	Entorhinodentate lesions; axonal degeneration (*in vivo*).	[456-458]
CXCL12/SDF1α	NSC migration↑	Reactivated astrocytes, activated endothelial cells, meningeal cells	Hypoxic–Ischemic (HI) Cerebral Injury; multiple sclerosis; stroke	[459-462]
CX3CL1	NSC proliferation↑	Reactivated astrocytes, activated lymphocytes, microglia/macrophages	Neurospheres, hippocampal slice cultures (*in vitro*)	[458]
CCL11	NSC proliferation↓ differentiation ↓	Reactivated astrocytes, activated lymphocytes, microglia/macrophages	Aging model	[463]
IFN-α	NSC proliferation↓	Plasmacytoid dendritic cells, activated macrophages, endothelial cells, neurons	Young and old Cr2(-/-) mice	[464]
IL-1β	Neuronal fate (dopaminergic neurons)	Reactivated astrocytes, activated lymphocytes, microglia/macrophages	Tyrosine hydroxylase (TH)-induced immunoreactivity (*in vitro)*	[465-467]
IFN-γ	NSC proliferation↓	T cells (Th1), natural killer cells	Experimental allergic encephalomyelitis (EAE)	[468, 469]
IL-6 family of neurotrophic cytokines (LIF, CNTF, CT-1)	(Astro)glial differentiation	Reactivated astrocytes, activated lymphocytes, microglia/macrophages	Cortical precursor culture (*in vitro*)	[465, 470]
IL-4	NSC migration↑ differentiation↑	T cells (Th2), through effect on microglia/macrophages	EAE related chemokines treatment (*in vitro*)	[471, 472]
IL-10	NSC migration↑	Reactivated astrocytes, activated lymphocytes, microglia/macrophages	EAE related chemokines treatment (*in vitro*)	[472]
IL-15	NSC proliferation↑	Activated microglia	IL-15–/– mice	[473]
TNF-α	NSC proliferation↓	Activated microglia/ macrophages	EAE; TNF-R1(-/-),TNF-R2(-/-) and TNF-R1/R2(-/-) mice. Lipopolysaccaride (LPS)-stimulation (*in vitro*).	[468, 474, 475]

Abbreviations: CC/CXC, chemokines; SDF1α, stromal cell-derived factor 1α; IFN, interferon; IL, interleukin; CNTF, ciliary neurotrophic factor; CT-1, cardiotrophin-1; LIF, leukaemia inhibitory factor; TNF-α, tumor necrosis factor α.

Table 2. The Influence of Inflammatory Mediators on NSC [382, 383]

http://dx.doi.org/10.5772/58599

5. Therapeutic modulation of the neural stem cell niche

Due to the indispensable role of the niche microenvironment in regulating NSC (e.g. control of the maintenance, expansion and differentiation), different molecular strategies have been investigated in an effort to modulate the NSC response and in so doing enhance neurogenesis. Such work has the potential to benefit a myriad of degenerative neurological disorders by facilitating repair and aiding in functional recovery. Most prominent are approaches using novel pharmacological targets within NSC niches [50]. Rational engineering of the niche must also be considered as an approach for CNS homeostasis and repair [476]. This section will therefore focus both on selected drugs that have been shown capable of modulating the niche and on current efforts geared toward the engineering of microenvironments to support enhanced/sustained niche homeostasis.

5.1. Molecular therapies

Various endogenous regulators of NSC have been investigated for their therapeutic value with regard to neurogenesis. Intraventricular administration of exogenous EGF, PEDF, HGF and CNTF in mice has been shown to enhance NSC proliferation [305, 325, 366, 477]. Additionally, the peripheral administration of human recombinant EPO (hrEPO) has been shown to enhance neurogenesis and improve functional outcome in models of both ischemic stroke and traumatic injury. It is unlikely, however, that such effects can be solely attributed to the enhancement of neurogenesis, being that hrEPO has also been demonstrated to suppress inflammation and induce angiogenesis [478]. Administration of other factors such as RA, bFGF, EGF, BDNF and VEGF have also been shown to enhance neurogenesis in similar disease models ultimately leading to enhanced recovery [177, 213, 311, 479-483]. Despite the plethora of positive effects demonstrated in animal models, many of these endogenous factors have been difficult to translate into clinical use due to invasive routes of administration, off target physiologic effects, cost of recombinant factors, etc.

5.2. FDA approved small molecules

Certain small molecules have been shown to exert similar effects via the direct or indirect modification of endogenous cues. Briefly, certain antidepressants have been shown capable of increasing the neurogenic response [484, 485]. As an example, *Fluoxetine* (a selective serotonin reuptake inhibitor) has been shown to give rise to maturation of immature neurons and enhanced neurogenesis [486]. Whether this function is mediated through an increase in 5-HT receptor activation on NSC remains unclear [395]. However, it is prudent to note that the clinical benefits of such typical antidepressant drugs are only partly dependent on neurogenesis [487]. Antipsychotic drugs have also been associated with neurogenesis, yet the precise mechanisms of action remain unclear. The antipsychotic drug *Haloperidol* (D2 receptor antagonist) has been shown to reverse dopamine-induced inhibition of NSC proliferation [399]. Similar effects have also been observed for other antipsychotics including *Clonazepam* and *Risperidone* [409, 488-490]. GABA has been observed to have a negative influence on NSC proliferation and migration [491-493] and so it should not be surprising that GABA-based treatments, such as *Phenobarbital* and *Clonazepam* have been shown to inhibit cell proliferation in the DG of the hippocampus [494, 495]. In contrast, pharmacological inhibition of GABA receptors via such agents as *Bicuculline* (i.e. GABA antagonists) can enhance NSC proliferation and differentiation, thereby positively influencing neurogenesis [489, 490].

As discussed above, behavior of NSC is largely regulated by signals from the niche under physiological and pathological conditions. Small molecules capable of altering NSC niche function may provide a tool for modulation of NSC and neurogenesis in disease states and concurrently open up novel experimental routes for the investigation of mechanisms of niche activation.

5.3. Therapeutic stem cell transplantation in CNS diseases and the development of atypical neural stem cell niches

The therapeutic benefits of stem cell transplantation in modulating CNS disease processes have been supported by a multitude of reports. Yet, the therapeutic efficacy appears to be most pronounced in disorders that display key components of inflammation (i.e. multiple sclerosis, stroke and spinal cord injury) [87, 301, 496]. It is relevant to note that this effect is not limited to direct delivery (i.e. focal), but has also been reported after systemic or subcutaneous injection of stem cells [87, 496, 497]. While NSC have the potential to integrate into the host system and may contribute to replacement of damaged cells, other somatic stem cells such as hematopoietic stem cells (HSC), mesenchymal stem cells (MSC), and umbilical cord cells also allow for functional recovery in mouse models of inflammatory degeneration [87, 496, 498-502]. This suggests that the therapeutic effect of stem cells goes beyond mere cell integration, differentiation, and replacement and involves a "shared stemness-related" functional signature.

Transplanted NSC migrate toward well-defined areas in the inflamed perivascular microenvironment [503, 504]. This leads to the establishment of ectopic stem cell niches, also called atypical niches, which are molecularly reminiscent of prototypical germinal niches and regulate the long-term survival and the behavior of NSC [503, 505, 506]. The term "therapeutic plasticity" has been suggested to describe the remarkable inherent flexibility of NSC to migrate to inflamed CNS areas and establish atypical ectopic stem cell niches through which they modulate their environment in support of a therapeutically beneficial outcome [496, 507]. This modulatory capacity is exerted through regulated cross-talk of NSC with other components of the atypical niche, including endothelial cells, blood-born inflammatory cells, activated macrophages and microglia, and reactive astrocytes [301, 496]. A myriad of cell-to-cell signaling pathways allows for this NSC-driven pathophysiologic modulation and enhanced clinical recovery [301, 496, 499, 508, 509].

The preferential migration of NSC toward CNS lesions is referred to as pathotropism. During an insult (e.g. hypoxia or injury) cytokines cause a subsequent activation of microglia, astrocytes and endothelial cells [46, 510]. As a result, reactive astrocytes and activated endothelial cells produce chemokines such as SDF-1, MCP-1, and VEGF that function collectively as a homing beacon, not only for inflammatory cells, but also for NSC [301, 510-514]. Much like leukocytes, NSC express adhesion molecules (CD44), integrins (α4 β1) and chemokine receptors (CCR1, CCR2, CCR5, CXCR3, CXCR4). This enables NSC to follow the concentration gradient of these chemokines toward the inflamed parenchyma and extravasate in a process of tethering, rolling and adhering to endothelial cells followed by transendothelial migration [183, 503, 515-517]. Factors such as bFGF and IGF-1 are also produced by activated astrocytes and support NSC proliferation, survival and differentiation [510, 511, 518]. Conversely, hypertrophic GFAP-enriched astrocytes of the glial scar produce factors such as slit homologue 2 (SLIT2), TNF-α and hyaluronan that repel NSC and limit the regenerative potential of their progeny [510, 511, 519, 520].

Once an atypical niche is established, undifferentiated NSC survive in the perilesional region in close proximity to activated microglia (expressing ionized calcium-binding adapter molecule 1 (IBA)) and to blood vessels [502, 521, 522]. The mechanisms by which transplanted NSC remodel the injured nervous system is irrespective of the experimental disease characteristics (e.g. focal vs. multifocal) and only a small number of cells undergo final differentiation [522-524]. When migrating to the lesional parenchyma, NSC contribute to cell replacement, mainly by differentiating into astrocytes, but also into neurons [522, 525, 526]. More striking, however, are the "bystander" capacities of undifferentiated NSC, which include the provision of trophic support and the modulation of the immune response. These beneficial effects lead to the establishment of a homeostatic environment [382, 496, 497, 524, 527]. In models for MS and stroke this has been shown to mediate efficient myelin repair and axon rescue [515, 525, 526, 528-531].

Trophic and neuroprotective effects are exerted by providing neurotrophins, growth factors, developmental stem cell regulators, and immune modulators through modulation of the microenvironment [301, 382, 496]. In models for MS, systemically administered NSC have shown to stimulate OPC proliferation and differentiation, and consequent remyelination through secretion of PDGF-A and bFGF [515, 528]. In models for stroke, focally injected NSC have been shown to enhance expression of BDNF, GDNF, CNTF, bFGF, VEGF, HGF, and IGF in the perilesional region [525, 526]. Finally, focal grafting of NSC in SCI models has been shown to support growth of motor and sensory axons due to upregulation of NGF, BDNF, and GDNF [532].

Transplantation of stem cells enables the switch to a more conservative and anti-inflammatory lesional environment [87, 301, 498]. In models for MS, NSC drive the reduction of perivascular infiltrates and $CD3^+$ T-cells and the increase of regulatory $CD25^+$ or $CD25^+/CD62L^+$ T-cells, accompanied by a downregulation of inflammatory markers, intercellular adhesion molecule 1 (ICAM-1), and lymphocyte function-associated antigen 1 (LFA-1) [503, 533]. *In vitro* studies have shown that NSC can 1) induce apoptosis of Th1 and Th17, but not Th2 lymphocytes through Fas ligand (FasL), TNF-related apoptosis-inducing ligand (TRAIL) and Apo-3 ligand (APO3L), 2) reduce T-cell proliferation through nitric oxide and prostaglandin E2 (PGE2), 3) reduce T-cell receptor (TCR) dependent T-cell activation, 4) inhibit interleukin 2 (IL-2) (T-cell) and IL-6 (B-cell) signaling, and 5) reduce local populations of monocytes and macrophages through cytotoxic TNF-α secretion [503, 505, 506, 533-539]. Immune-modulating capabilities have also been shown in models for stroke, and include an increase of VEGF, SDF-1 and TGF-β, as well as a reduced expression of pro-inflammatory genes *Ifng*, *TNF-α*, *Il1b* and *Lepr* [502, 529]. Furthermore, NSC-induced increases in activated microglia ($CD11b^+$) have been shown to lead to IGF-1, VEGF, TGF-β, and BDNF production, yielding better motor function and axonal sprouting, highlighting the beneficial role of microglia [529-531]. However, other studies have shown that NSC transplantation reduced microglia/macrophage presence with improvement of both neuronal survival and locomotor functions [502, 540]. Models for SCI also show a skewing of microglia/macrophage infiltrates. Here, focally transplanted NSC have been shown to make cellular junctions (Connexin 43) with phagocytic cells and astrocytes, and to reduce the presence of classically-activated pro-inflammatory M1 macrophages [522].

Grafted stem cells do not only home to the the inflamed CNS, but also to the secondary lymphoid organs where they modulate inflammation [505, 506, 540, 541]. NSC hinder the activation of myeloid dendritic cells (DC), limiting the expansion of antigen-specific encephalogenic T-cells. DC maturation is hindered, partially due to secretion of BMP-4. Furthermore, induced secretion of BMP-4/7, Shh and Noggin by transplanted NSC and immune cells, promoted survival of endogenous NSC [505, 506, 541]. An increase in the presence of LIF leads to a reduction of Th17 differentiation, further ameliorating the functional outcome of MS. In stroke models, a reduction of both neutrophil infiltration and activation of macrophages in lymphoid organs can be observed after NSC transplantation [540].

In an effort to translate these therapeutic approaches to clinic, human embryonic stem cells (hESCs) and induced pluripotent stem cells (iPSC) created from human fibroblasts have been studied for their neurogenic and neuroprotective properties after MS, stroke and SCI. Although some differences can be observed, e.g. higher cytotoxic potential against monocytes and lower cytotoxic potential against T-cells, human-derived cell functions are largely similar to those of animal-derived cells and they also increase clinical recovery. The therapeutic use of these cells is however limited by ethical constraints, genetic instability, and tumorgenicity [505, 538, 539, 542-545].

The therapeutic value of stem cell grafts, especially NSC, in inflammatory neurodegenerative disorders has become increasingly evident. Transplanted stem cells are able to home to the lesion areas where they take part in the establishment of an atypical perivascular niche, allowing stem cells to survive undifferentiated and to provide neurotropic support, modulate the inflammation, and allow for further migration into the lesional parenchyma to take part in neuronal differentiation and cell replacement. This has been shown to modulate the pathophysiology of disease, enhancing axonal conservation and regeneration, leading to increased functional recovery in animal models of MS, stroke and SCI.

5.4. Engineering the NSC niche

Approaches for niche engineering are centered around efforts to mimic multiple aspects of the niche microenvironment, which include architectural, mechanical, bioactive and growth factor cues [476]. ECM mimicking scaffolds support the survival and differentiation of transplanted NSC [546, 547]. Clearly, an understanding of ECM architecture is important in the designing of these scaffolds and to this extent studies have shown a correlation between scaffold fiber diameter and NSC behavior. For example, fibers with a 283nm diameter promote proliferation and differentiation to oligodendrocytes while fibers within 749-1452nm diameter range promote neuronal differentiation [548]. Apart from the 3D structure, the mechanical properties

http://dx.doi.org/10.5772/58599

of scaffolds have been shown to modulate morphology, proliferation, and differentiation of stem cells [549]. Polyethylene glycol (PEG) – poly-L-lysine (PLL) hydrogels allow for good NSC migration when their elastic modulus mimics that of brain tissue. Gels with a higher elastic modulus, on the other hand, limit migration [550]. Other studies have demonstrated that softer substrates promote neuronal differentiation whereas more rigid substrates induce glial differentiation [551]. Bioactive polymers such as those made from the laminin-1-derived IKVAV peptide further promote neuronal differentiation [552]. When seeded with NSC and transplanted into animal models of spinal cord injury, these structures have stimulated a marked enhancement in functional recovery [553]. Bioactive polymers which include tripeptide Arg-Gly-Asp (RGD) motifs showed promotion of cell attachment, self-renewal and differentiation [554]. Additionally, incorporation of signaling molecules relevant to NSC regulation can also positively influence the behavior of cells within these scaffolds [555]. Wnt and Notch ligands keep cells in a proliferative, undifferentiated state while the addition of BMP-4 enhances glial and neuronal differentiation [551]. Altogether, niche engineering represents a promising approach for regenerative medicine, as it enables control over the behavior of transplanted NSC, and may soon come to have vast therapeutic value.

6. Concluding remarks and future directions

"As long as our brain is a mystery, the universe, the reflection of the structure of the brain will also be a mystery."-Santiago Ramón y Cajal

The presence of neural stem cells/neurogenic niches in the adult mammalian central nervous system has been clearly established by a body of rigorous scientific work. The functional significance of adult neurogenesis continues to grow as new studies describe its critical roles in states of both health and disease. Despite this growing body of information and improvements in our understanding of NSC and niche functions in both the physiologic/pathologic conditions, several critical questions remain. Chief among them is the relevance of the basic biology that has so far been described in animal models to the ultimate goal of translating adult neurogenesis into clinical trials. Further work with regard to the definitive nature/location of NSC needs also to be carried out. Finally the definitive molecular mechanisms that influence endogenous stem cell migration/pathotropism will also be key in helping to develop suitable treatments and strategies to prevent, mitigate, and treat varied CNS injuries and disease.

Abbreviations

SVZ-subventricular zone

SGZ-subgranular zone

NSC – neural stem/precursor cells

CNS – central nervous system

NPC – neural progenitor cells

OPC-Ooigodendrocyte precursor cells

GFAP – glial fibrillary acidic protein

Sox2 – SRY (sex determining regionY) – box2

Oct4 – octamer-binding transcription factor 4

FoxO – Forkhead box

BrdU – bromodeoxyuridine

DCX – doublecortin

PSA-NCAM-polysialylated-neural adhesion molecule

SCs – stem cells

CC – central canal

CVO-circumventricular organs

ECM-extracellular matrix

DG-dentate gyrus

RMS-rostral migratory stream

5-HT – 5-hydroxytryptamine

GCL-granule cell layer

RGL-glial-like cells

IML-inner molecular layer

Shh-Sonic hedgehog signaling

VEGF-vascular endothelial growth factor

IGF-insulin-like growth factor

BDNF-brain-derived neurotrophic factor

CA – cornu ammonis region

CSF – cerebrospinal fluid

BLBP-brain lipid-binding protein

NeuN – neuronal nuclear antigen

Olig2+-oligodendrocytes

CD133 – prominin 1

ALDH1L1-aldehyde dehydrogenase 1 family member, L1

http://dx.doi.org/10.5772/58599

GLAST-glutamate aspartate transporter

RC2-radial glial cell marker-2

NG2-neuron-glial antigen 2

A2B5 – A2B5 antigen

PDGFR-platelet-derived growth factor receptor

GABA-gamma-aminobutyric acid

RA-retinoic acid

bFGF – basic fibroblast growth factor

CXCL12 –chemokine (C-X-C motif) ligand 2

MIP2-alpha – macrophage inflammatory protein 2-alpha

MAP-microtubule-associated protein

CSPG4-chondroitin sulfate proteoglycan

MS – multiple sclerosis

SCI – spinal cord injury

O_2 – oxygen

ATP-adenosine triphosphate

EGF – epidermal growth factor

GDNF – glia cell-derived neurotrophic factor

BMP – bone morphogenic protein

CNTF-ciliary neurotrophic factor

TGF – transforming growth factor

EPO – erythropoietin

G-CSF-granulocyte-colony stimulating factor

MMP – matrix metalloproteinase

SDF-1 – stromal cell derived factor-1

CXCR4 – chemokine receptor type 4

CCL2 – chemokine (C-C motif) ligand 2

MCP-1-monocyte chemoattractant protein-1

EAE-experimental autoimmune encephalomyelitis

MRI-magnetic resonance imaging

NGF-nerve growth factor

NT – neurotrophin

HGF – hepatocyte growth factor

TOR – target of rapamycin

PEGF – pigment-epitheliun derived growth factor

PDGF – platelet-derived growth factors

PDGFR – platelet-derived growth factor receptor

Wnt – wingless-related integration site

HipK1-homeodomain interacting protein kinase 1

Ptc – Patched

GSKβ-glycogen synthase kinase β

LIF – leukemia inhibitory factor

CNTFR-ciliary neurotrophic factor receptor

SDNSF – stem cell-derived neural stem/progenitor cell supporting factor

MIF – macrophage migration inhibitory factor

IL – interleukin

TNF-α – tumor necrosis factor α

TNFR-tumor necrosis factor receptor

mGluRs-metabotropic glutamate receptors

ADAM – A-disintegrin and metalloproteinase

NO-nitric oxid

CSPGs – chondroitin sulfate proteoglycans

HSPGs – heparan sulfate proteoglycans

Dll4 – delta like ligand 4

EGFR – epidermal growth factor receptor

NTF – neurotrophic factor

NT – neurotrophin

Trk-tropomyosin-related kinase

ApoE – apolipoprotein E

PEDF – pigment epithelium-derived factor

http://dx.doi.org/10.5772/58599

hrEPO – human recombinant erythropoietin

HSC-hematopoietic stem cells

MSC – mesenchymal stem cells

CCR, CXCR – chemokine receptor

SLIT2-slit homologue 2

IBA-ionized calcium-binding adapter molecule 1

PDGF-A – platelet derived growth factor-A

ICAM-1 – intercellular adhesion molecule 1

LFA-1 – lymphocyte function-associated antigen 1

FasL – fas ligand

TRAIL – tumor necrosis factor related apoptosis inducing ligand

APO3L-Apo-3 ligand

PGE2 – prostaglandin E2

TCR – reduced T-cell receptor

hESCs-human embryonic stem cells

iPSC-induced pluripotent stem cells

PEG – polyethylene glycol

PLL – poly-L-lysine

IKVAV-isoleucine-lusine-valine-alanine-valine

RGD – tripeptide-Arg-Gly-Asp

Acknowledgements

No conflicts of interest to disclose. The authors wish to thank Dr. Lee-Wickner and Mr. Connor Caples for their critical reading of the chapter. The authors would also like to thank Ms. Alicia Livinski for assistance with the literature review.

This work was supported by grants from the National Multiple Sclerosis Society (NMSS; RG-4001-A1 to SP), the Italian Multiple Sclerosis Foundation (FISM; RG 2010/R/31 to SP), the Italian Ministry of Health (GR08/7 to SP) the European Research Council (ERC) 2010-StG (RG 260511-SEM_SEM to SP), the European Community (EC) 7th Framework Program (FP7/2007-2013; RG 280772-iONE to SP), The Evelyn Trust (RG 69865 to SP). Additionally, JH was supported by the Intramural Research Program of NINDS/NIH. JB was support by a NIH-OxCam fellowship. BH was supported by the China Scholarship Council (No.201306320024).

Author details

Joshua Bernstock[1,2], Jeroen Verheyen[1], Bing Huang[1], John Hallenbeck[2] and Stefano Pluchino[1]

*Address all correspondence to: spp24@cam.ac.uk

1 John van Geest Centre for Brain Repair and Wellcome Trust-Medical Research Council Stem Cell Institute, Dept. of Clinical Neurosciences, University of Cambridge, Cambridge, UK

2 Stroke Branch, National Institute of Neurological Disorders and Stroke, National Institutes of Health (NINDS/NIH), USA

All authors contributed equally to this work.

References

[1] Chklovskii DB, Mel BW, Svoboda K. Cortical rewiring and information storage. Nature. 2004;431(7010):782-8.

[2] Brus M, Keller M, Levy F. Temporal features of adult neurogenesis: differences and similarities across mammalian species. Frontiers in neuroscience. 2013;7:135.

[3] A A-B, DG H, H W. The subventricular zone: source of neuronal precursors for brain repair. Prog Brain Res. (0079-6123):1-11.

[4] Gage FH. Isolation, Characterization, and use of Stem Cells from the CNS. Annu Rev Neurosci. 1995:159-92.

[5] Kirschenbaum B, Doetsch F, Lois C, Alvarez-Buylla A. Adult subventricular zone neuronal precursors continue to proliferate and migrate in the absence of the olfactory bulb. The Journal of neuroscience : the official journal of the Society for Neuroscience. 1999;19(6):2171-80.

[6] Luskin MB, Zigova T, Soteres BJ, Stewart RR. Neuronal progenitor cells derived from the anterior subventricular zone of the neonatal rat forebrain continue to proliferate in vitro and express a neuronal phenotype. Molecular and cellular neurosciences. 1997;8(5):351-66.

[7] Altman J, Das GD. Autoradiographic and Histological Evidence of Postnatal Hippocampal Neurogenesis in Rats. The Journal of comparative neurology. 1965;124:319-36.

[8] Altman J, Das GD. Autoradiographic and Histological Studies of Postnatal Neurogenesis. I. A Longitudinal Investigation of the Kinetics, Migration and Transformation of Cells Incorporating Tritiated Thymidine in Neonate Rats, with Special

Reference to Postnatal Neurogenesis in Some Brain Regions. The Journal of comparative neurology. 1966;126:337-90

[9] Bellenchi GC, Volpicelli F, Piscopo V, Perrone-Capano C, di Porzio U. Adult neural stem cells: an endogenous tool to repair brain injury? Journal of neurochemistry. 2013;124(2):159-67.

[10] Decimo I, Bifari F, Krampera M, Fumagalli G. Neural stem cell niches in health and diseases. Current Pharmaceutical Design. 2012;18(13):1755-83.

[11] Morrison SJ, Shah NM, Anderson DJ. Regulatory mechanisms in stem cell biology. Cell. 1997;88(3):287-98.

[12] Weiss S, Dunne C, Hewson J, Wohl C, Wheatley M, Peterson AC, et al. Multipotent CNS stem cells are present in the adult mammalian spinal cord and ventricular neuroaxis. The Journal of neuroscience : the official journal of the Society for Neuroscience. 1996;16(23):7599-609.

[13] Johansson CB, Momma S, Clarke DL, Risling M, Lendahl U, Frisen J. Identification of a neural stem cell in the adult mammalian central nervous system. Cell. 1999;96(1): 25-34.

[14] Yamamoto S, Yamamoto N, Kitamura T, Nakamura K, Nakafuku M. Proliferation of parenchymal neural progenitors in response to injury in the adult rat spinal cord. Experimental neurology. 2001;172(1):115-27.

[15] Shihabuddin LS, Ray J, Gage FH. FGF-2 is sufficient to isolate progenitors found in the adult mammalian spinal cord. Experimental neurology. 1997;148(2):577-86.

[16] Yamamoto S, Nagao M, Sugimori M, Kosako H, Nakatomi H, Yamamoto N, et al. Transcription factor expression and Notch-dependent regulation of neural progenitors in the adult rat spinal cord. The Journal of neuroscience : the official journal of the Society for Neuroscience. 2001;21(24):9814-23.

[17] Gross CG. Neurogenesis in the adult brain: death of a dogma. Nat Rev Neurosci. 2000;1(1):67-73.

[18] Shihabuddin LS, Horner PJ, Ray J, Gage FH. Adult spinal cord stem cells generate neurons after transplantation in the adult dentate gyrus. The Journal of neuroscience : the official journal of the Society for Neuroscience. 2000;20(23):8727-35.

[19] Horner PJ, Power AE, Kempermann G, Kuhn HG, Palmer TD, Winkler J, et al. Proliferation and differentiation of progenitor cells throughout the intact adult rat spinal cord. The Journal of neuroscience : the official journal of the Society for Neuroscience. 2000;20(6):2218-28.

[20] Lendahl U, Zimmerman LB, McKay RD. CNS stem cells express a new class of intermediate filament protein. Cell. 1990;60(4):585-95.

[21] Doetsch F. The glial identity of neural stem cells. Nature neuroscience. 2003;6(11): 1127-34.

[22] Sakakibara S, Imai T, Hamaguchi K, Okabe M, Aruga J, Nakajima K, et al. Mouse-Musashi-1, a neural RNA-binding protein highly enriched in the mammalian CNS stem cell. Developmental biology. 1996;176(2):230-42.

[23] Ellis P, Fagan BM, Magness ST, Hutton S, Taranova O, Hayashi S, et al. SOX2, a persistent marker for multipotential neural stem cells derived from embryonic stem cells, the embryo or the adult. Dev Neurosci. 2004;26(2-4):148-65.

[24] Lothian C, Prakash N, Lendahl U, Wahlstrom GM. Identification of both general and region-specific embryonic CNS enhancer elements in the nestin promoter. Experimental cell research. 1999;248(2):509-19.

[25] Garcia-Verdugo JM, Ferron S, Flames N, Collado L, Desfilis E, Font E. The proliferative ventricular zone in adult vertebrates: a comparative study using reptiles, birds, and mammals. Brain research bulletin. 2002;57(6):765-75.

[26] Tavazoie M, Van der Veken L, Silva-Vargas V, Louissaint M, Colonna L, Zaidi B, et al. A specialized vascular niche for adult neural stem cells. Cell stem cell. 2008;3(3): 279-88.

[27] Shen Q, Goderie SK, Jin L, Karanth N, Sun Y, Abramova N, et al. Endothelial cells stimulate self-renewal and expand neurogenesis of neural stem cells. Science. 2004;304(5675):1338-40.

[28] Chojnacki A, Weiss S. Pigment epithelium-derived growth factor: modulating adult neural stem cell self-renewal. Nature neuroscience. 2009;12(12):1481-3.

[29] Scott IS, Morris LS, Bird K, Davies RJ, Vowler SL, Rushbrook SM, et al. A novel immunohistochemical method to estimate cell-cycle phase distribution in archival tissue: implications for the prediction of outcome in colorectal cancer. The Journal of pathology. 2003;201(2):187-97.

[30] Cai Y, Xiong K, Chu Y, Luo DW, Luo XG, Yuan XY, et al. Doublecortin expression in adult cat and primate cerebral cortex relates to immature neurons that develop into GABAergic subgroups. Experimental neurology. 2009;216(2):342-56.

[31] Nacher J, Lanuza E, McEwen BS. Distribution of PSA-NCAM expression in the amygdala of the adult rat. Neuroscience. 2002;113(3):479-84.

[32] Pluchino S. Typical and Atypical Neural Stem Cell Niches. Electronic Journal of Biology. 2008;4(2):68-78.

[33] Kempermann G. Adult neurogenesis : stem cells and neuronal development in the adult brain. New York: Oxford University Press; 2006. x, 426 p. p.

[34] Hugnot JP. The Spinal Cord Neural Stem Cell Niche. In: Sun T, editor. Neural Stemm Cells and Therapy: InTech; 2012.

[35] Gould E. How widespread is adult neurogenesis in mammals? Nat Rev Neurosci. 2007;8(6):481-8.

[36] Kondo T, Raff M. Oligodendrocyte precursor cells reprogrammed to become multipotential CNS stem cells. Science. 2000;289(5485):1754-7.

[37] Palmer TD, Markakis EA, Willhoite AR, Safar F, Gage FH. Fibroblast growth factor-2 activates a latent neurogenic program in neural stem cells from diverse regions of the adult CNS. The Journal of neuroscience : the official journal of the Society for Neuroscience. 1999;19(19):8487-97.

[38] Laywell ED, Rakic P, Kukekov VG, Holland EC, Steindler DA. Identification of a multipotent astrocytic stem cell in the immature and adult mouse brain. Proceedings of the National Academy of Sciences of the United States of America. 2000;97(25): 13883-8.

[39] Panayiotou E, Malas S. Adult spinal cord ependymal layer: a promising pool of quiescent stem cells to treat spinal cord injury. Frontiers in physiology. 2013;4:340.

[40] Reynolds BA, Weiss S. Generation of neurons and astrocytes from isolated cells of the adult mammalian central nervous system. Science. 1992;255(5052):1707-10.

[41] Miyata T, Kawaguchi A, Okano H, Ogawa M. Asymmetric inheritance of radial glial fibers by cortical neurons. Neuron. 2001;31(5):727-41.

[42] Merkle FT, Mirzadeh Z, Alvarez-Buylla A. Mosaic organization of neural stem cells in the adult brain. Science. 2007;317(5836):381-4.

[43] Bonfanti L, Peretto P. Radial glial origin of the adult neural stem cells in the subventricular zone. Progress in neurobiology. 2007;83(1):24-36.

[44] Peretto P, Giachino C, Aimar P, Fasolo A, Bonfanti L. Chain formation and glial tube assembly in the shift from neonatal to adult subventricular zone of the rodent forebrain. The Journal of comparative neurology. 2005;487(4):407-27.

[45] Eckenhoff MF, Rakic P. Radial organization of the hippocampal dentate gyrus: a Golgi, ultrastructural, and immunocytochemical analysis in the developing rhesus monkey. The Journal of comparative neurology. 1984;223(1):1-21.

[46] Martino G, Pluchino S. The therapeutic potential of neural stem cells. Nat Rev Neurosci. 2006;7(5):395-406.

[47] Baizabal JM, Furlan-Magaril M, Santa-Olalla J, Covarrubias L. Neural stem cells in development and regenerative medicine. Archives of medical research. 2003;34(6): 572-88.

[48] Schofield R. The relationship between the spleen colony-forming cell and the haemopoietic stem cell. Blood cells. 1978;4(1-2):7-25.

[49] Jones DL, Wagers AJ. No place like home: anatomy and function of the stem cell niche. Nature reviews Molecular cell biology. 2008;9(1):11-21.

[50] Decimo I, Bifari F, Krampera M, Fumagalli G. Neural stem cell niches in health and diseases. Curr Pharm Des. 2012;18(13):1755-83.

[51] Morrison SJ, Spradling AC. Stem cells and niches: mechanisms that promote stem cell maintenance throughout life. Cell. 2008;132(4):598-611.

[52] Jordan JD, Ma DK, Ming GL, Song H. Cellular niches for endogenous neural stem cells in the adult brain. CNS & neurological disorders drug targets. 2007;6(5):336-41.

[53] Bonfanti L, Ponti G. Adult mammalian neurogenesis and the New Zealand white rabbit. Veterinary journal. 2008;175(3):310-31.

[54] Alvarez-Buylla A, Garcia-Verdugo JM. Neurogenesis in adult subventricular zone. The Journal of neuroscience : the official journal of the Society for Neuroscience. 2002;22(3):629-34.

[55] Gage FH. Mammalian neural stem cells. Science. 2000;287(5457):1433-8.

[56] Doetsch F, Caille I, Lim DA, Garcia-Verdugo JM, Alvarez-Buylla A. Subventricular zone astrocytes are neural stem cells in the adult mammalian brain. Cell. 1999;97(6): 703-16.

[57] Malatesta P, Hartfuss E, Gotz M. Isolation of radial glial cells by fluorescent-activated cell sorting reveals a neuronal lineage. Development (Cambridge, England). 2000;127(24):5253-63.

[58] Ponti G, Obernier K, Guinto C, Jose L, Bonfanti L, Alvarez-Buylla A. Cell cycle and lineage progression of neural progenitors in the ventricular-subventricular zones of adult mice. Proceedings of the National Academy of Sciences of the United States of America. 2013;110(11):E1045-54.

[59] Kriegstein A, Alvarez-Buylla A. The glial nature of embryonic and adult neural stem cells. Annual review of neuroscience. 2009;32:149-84.

[60] Altman J. Autoradiographic and histological studies of postnatal neurogenesis. IV. Cell proliferation and migration in the anterior forebrain, with special reference to persisting neurogenesis in the olfactory bulb. The Journal of comparative neurology. 1969;137(4):433-57.

[61] Lois C, Alvarez-Buylla A. Proliferating subventricular zone cells in the adult mammalian forebrain can differentiate into neurons and glia. Proceedings of the National Academy of Sciences of the United States of America. 1993;90(5):2074-7.

[62] Luskin MB. Restricted proliferation and migration of postnatally generated neurons derived from the forebrain subventricular zone. Neuron. 1993;11(1):173-89.

[63] Lois C, Garcia-Verdugo JM, Alvarez-Buylla A. Chain migration of neuronal precursors. Science. 1996;271(5251):978-81.

[64] Merkle FT, Tramontin AD, Garcia-Verdugo JM, Alvarez-Buylla A. Radial glia give rise to adult neural stem cells in the subventricular zone. Proceedings of the National Academy of Sciences of the United States of America. 2004;101(50):17528-32.

[65] Shen Q, Wang Y, Kokovay E, Lin G, Chuang SM, Goderie SK, et al. Adult SVZ stem cells lie in a vascular niche: a quantitative analysis of niche cell-cell interactions. Cell stem cell. 2008;3(3):289-300.

[66] Mirzadeh Z, Merkle FT, Soriano-Navarro M, Garcia-Verdugo JM, Alvarez-Buylla A. Neural stem cells confer unique pinwheel architecture to the ventricular surface in neurogenic regions of the adult brain. Cell stem cell. 2008;3(3):265-78.

[67] Wang C, Liu F, Liu YY, Zhao CH, You Y, Wang L, et al. Identification and characterization of neuroblasts in the subventricular zone and rostral migratory stream of the adult human brain. Cell research. 2011;21(11):1534-50.

[68] Sanai N, Tramontin AD, Quinones-Hinojosa A, Barbaro NM, Gupta N, Kunwar S, et al. Unique astrocyte ribbon in adult human brain contains neural stem cells but lacks chain migration. Nature. 2004;427(6976):740-4.

[69] Curtis MA, Kam M, Nannmark U, Anderson MF, Axell MZ, Wikkelso C, et al. Human neuroblasts migrate to the olfactory bulb via a lateral ventricular extension. Science. 2007;315(5816):1243-9.

[70] Sanai N, Nguyen T, Ihrie RA, Mirzadeh Z, Tsai HH, Wong M, et al. Corridors of migrating neurons in the human brain and their decline during infancy. Nature. 2011;478(7369):382-6.

[71] Goritz C, Frisen J. Neural stem cells and neurogenesis in the adult. Cell stem cell. 2012;10(6):657-9.

[72] Knoth R, Singec I, Ditter M, Pantazis G, Capetian P, Meyer RP, et al. Murine features of neurogenesis in the human hippocampus across the lifespan from 0 to 100 years. PloS one. 2010;5(1):e8809.

[73] Spalding KL, Bergmann O, Alkass K, Bernard S, Salehpour M, Huttner HB, et al. Dynamics of hippocampal neurogenesis in adult humans. Cell. 2013;153(6):1219-27.

[74] Ernst A, Alkass K, Bernard S, Salehpour M, Perl S, Tisdale J, et al. Neurogenesis in the Striatum of the Adult Human Brain. Cell. 2014.

[75] Tong CK, Chen J, Cebrian-Silla A, Mirzadeh Z, Obernier K, Guinto CD, et al. Axonal Control of the Adult Neural Stem Cell Niche. Cell stem cell. 2014.

[76] Kempermann G, Jessberger S, Steiner B, Kronenberg G. Milestones of neuronal development in the adult hippocampus. Trends in neurosciences. 2004;27(8):447-52.

[77] Seri B, Garcia-Verdugo JM, Collado-Morente L, McEwen BS, Alvarez-Buylla A. Cell types, lineage, and architecture of the germinal zone in the adult dentate gyrus. The Journal of comparative neurology. 2004;478(4):359-78.

[78] Ehninger D, Kempermann G. Neurogenesis in the adult hippocampus. Cell and tissue research. 2008;331(1):243-50.

[79] Eriksson PS, Perfilieva E, Bjork-Eriksson T, Alborn AM, Nordborg C, Peterson DA, et al. Neurogenesis in the adult human hippocampus. Nature medicine. 1998;4(11): 1313-7.

[80] Gage FH, Kempermann G, Palmer TD, Peterson DA, Ray J. Multipotent progenitor cells in the adult dentate gyrus. Journal of Neurobiology. 1998;36(2):249-66.

[81] Gould E, Cameron HA. Regulation of neuronal birth, migration and death in the rat dentate gyrus. Dev Neurosci. 1996;18(1-2):22-35.

[82] Manganas LN, Zhang X, Li Y, Hazel RD, Smith SD, Wagshul ME, et al. Magnetic resonance spectroscopy identifies neural progenitor cells in the live human brain. Science. 2007;318(5852):980-5.

[83] Gould E, Tanapat P, McEwen BS, Flugge G, Fuchs E. Proliferation of granule cell precursors in the dentate gyrus of adult monkeys is diminished by stress. Proceedings of the National Academy of Sciences of the United States of America. 1998;95(6): 3168-71.

[84] J A, GD D.-Autoradiographic and histological evidence of postnatal hippocampal neurogenesis. D-0406041. (-0021-9967 (Print)):-319-35.

[85] Sierra A, Encinas JM, Maletic-Savatic M. Adult human neurogenesis: from microscopy to magnetic resonance imaging. Frontiers in neuroscience. 2011;5:47.

[86] Seri B, Garcia-Verdugo JM, McEwen BS, Alvarez-Buylla A. Astrocytes give rise to new neurons in the adult mammalian hippocampus. The Journal of neuroscience : the official journal of the Society for Neuroscience. 2001;21(18):7153-60.

[87] Martino G, Pluchino S, Bonfanti L, Schwartz M. Brain regeneration in physiology and pathology: the immune signature driving therapeutic plasticity of neural stem cells. Physiological reviews. 2011;91(4):1281-304.

[88] Zhao C, Deng W, Gage FH. Mechanisms and functional implications of adult neurogenesis. Cell. 2008;132(4):645-60.

[89] Ming GL, Song H. Adult neurogenesis in the mammalian brain: significant answers and significant questions. Neuron. 2011;70(4):687-702.

[90] Palmer TD, Willhoite AR, Gage FH. Vascular niche for adult hippocampal neurogenesis. The Journal of comparative neurology. 2000;425(4):479-94.

[91] Santarelli L, Saxe M, Gross C, Surget A, Battaglia F, Dulawa S, et al. Requirement of hippocampal neurogenesis for the behavioral effects of antidepressants. Science. 2003;301(5634):805-9.

[92] Yu TS, Zhang G, Liebl DJ, Kernie SG. Traumatic brain injury-induced hippocampal neurogenesis requires activation of early nestin-expressing progenitors. The Journal of neuroscience : the official journal of the Society for Neuroscience. 2008;28(48): 12901-12.

[93] Bonaguidi MA, Song J, Ming GL, Song H. A unifying hypothesis on mammalian neural stem cell properties in the adult hippocampus. Current opinion in neurobiology. 2012;22(5):754-61.

[94] Filippov V, Kronenberg G, Pivneva T, Reuter K, Steiner B, Wang LP, et al. Subpopulation of nestin-expressing progenitor cells in the adult murine hippocampus shows electrophysiological and morphological characteristics of astrocytes. Molecular and Cellular Neuroscience. 2003;23(3):373-82.

[95] Lagace DC, Whitman MC, Noonan MA, Ables JL, DeCarolis NA, Arguello AA, et al. Dynamic contribution of nestin-expressing stem cells to adult Neurogenesis. Journal of Neuroscience. 2007;27(46):12623-9.

[96] Ahn S, Joyner AL. In vivo analysis of quiescent adult neural stem cells responding to Sonic hedgehog. Nature. 2005;437(7060):894-7.

[97] Denise A, Garcia R, Doan NB, Imura T, Bush TG, Sofroniew MV. GFAP-expressing progenitors are the principal source of constitutive neurogenesis in adult mouse forebrain. Nature neuroscience. 2004;7(11):1233-41.

[98] Dhaliwal J, Lagace DC. Visualization and genetic manipulation of adult neurogenesis using transgenic mice. The European journal of neuroscience. 2011;33(6):1025-36.

[99] Bonaguidi MA, Wheeler MA, Shapiro JS, Stadel RP, Sun GJ, Ming GL, et al. In vivo clonal analysis reveals self-renewing and multipotent adult neural stem cell characteristics. Cell. 2011;145(7):1142-55.

[100] Fuentealba LC, Obernier K, Alvarez-Buylla A. Adult neural stem cells bridge their niche. Cell stem cell. 2012;10(6):698-708.

[101] Ehm O, Goritz C, Covic M, Schaffner I, Schwarz TJ, Karaca E, et al. RBPJkappa-dependent signaling is essential for long-term maintenance of neural stem cells in the adult hippocampus. The Journal of neuroscience : the official journal of the Society for Neuroscience. 2010;30(41):13794-807.

[102] Lavado A, Lagutin OV, Chow LM, Baker SJ, Oliver G. Prox1 is required for granule cell maturation and intermediate progenitor maintenance during brain neurogenesis. PLoS biology. 2010;8(8).

[103] Song J, Zhong C, Bonaguidi MA, Sun GJ, Hsu D, Gu Y, et al. Neuronal circuitry mechanism regulating adult quiescent neural stem-cell fate decision. Nature. 2012;489(7414):150-4.

[104] Zhao C, Teng EM, Summers RG, Jr., Ming GL, Gage FH. Distinct morphological stages of dentate granule neuron maturation in the adult mouse hippocampus. The Journal of neuroscience : the official journal of the Society for Neuroscience. 2006;26(1):3-11.

[105] Ge S, Sailor KA, Ming GL, Song H. Synaptic integration and plasticity of new neurons in the adult hippocampus. The Journal of physiology. 2008;586(16):3759-65.

[106] Dayer AG, Ford AA, Cleaver KM, Yassaee M, Cameron HA. Short-term and long-term survival of new neurons in the rat dentate gyrus. Journal of Comparative Neurology. 2003;460(4):563-72.

[107] Kirshblum S, Campagnolo DI. Spinal cord medicine. 2nd ed. Philadelphia: Wolters Kluwer Health/Lippincott Williams & Wilkins; 2011. xiv, 688 p. p.

[108] Holtz A, Levi R. Spinal cord injury. New York: Oxford University Press; 2010. vi, 319 p., 4 p. of plates p.

[109] Smart IH. Proliferative characteristics of the ependymal layer during the early development of the spinal cord in the mouse. Journal of anatomy. 1972;111(Pt 3):365-80.

[110] Del Bigio MR. Ependymal cells: biology and pathology. Acta neuropathologica. 2010;119(1):55-73.

[111] Fu H, Qi Y, Tan M, Cai J, Hu X, Liu Z, et al. Molecular mapping of the origin of postnatal spinal cord ependymal cells: evidence that adult ependymal cells are derived from Nkx6.1+ventral neural progenitor cells. The Journal of comparative neurology. 2003;456(3):237-44.

[112] Sabourin JC, Ackema KB, Ohayon D, Guichet PO, Perrin FE, Garces A, et al. A mesenchymal-like ZEB1(+) niche harbors dorsal radial glial fibrillary acidic protein-positive stem cells in the spinal cord. Stem cells. 2009;27(11):2722-33.

[113] Kehl LJ, Fairbanks CA, Laughlin TM, Wilcox GL. Neurogenesis in postnatal rat spinal cord: a study in primary culture. Science. 1997;276(5312):586-9.

[114] Barnabe-Heider F, Goritz C, Sabelstrom H, Takebayashi H, Pfrieger FW, Meletis K, et al. Origin of new glial cells in intact and injured adult spinal cord. Cell stem cell. 2010;7(4):470-82.

[115] Alfaro-Cervello C, Soriano-Navarro M, Mirzadeh Z, Alvarez-Buylla A, Garcia-Verdugo JM. Biciliated ependymal cell proliferation contributes to spinal cord growth. The Journal of comparative neurology. 2012;520(15):3528-52.

[116] Nolte C, Matyash M, Pivneva T, Schipke CG, Ohlemeyer C, Hanisch UK, et al. GFAP promoter-controlled EGFP-expressing transgenic mice: a tool to visualize astrocytes and astrogliosis in living brain tissue. Glia. 2001;33(1):72-86.

[117] Hamilton LK, Truong MK, Bednarczyk MR, Aumont A, Fernandes KJ. Cellular organization of the central canal ependymal zone, a niche of latent neural stem cells in the adult mammalian spinal cord. Neuroscience. 2009;164(3):1044-56.

[118] Seitz R, Lohler J, Schwendemann G. Ependyma and meninges of the spinal cord of the mouse. A light-and electron-microscopic study. Cell and tissue research. 1981;220(1):61-72.

[119] Meletis K, Barnabe-Heider F, Carlen M, Evergren E, Tomilin N, Shupliakov O, et al. Spinal cord injury reveals multilineage differentiation of ependymal cells. PLoS biology. 2008;6(7):e182.

[120] Bodega G, Suarez I, Rubio M, Fernandez B. Ependyma: phylogenetic evolution of glial fibrillary acidic protein (GFAP) and vimentin expression in vertebrate spinal cord. Histochemistry. 1994;102(2):113-22.

[121] Sabelstrom H, Stenudd M, Reu P, Dias DO, Elfineh M, Zdunek S, et al. Resident neural stem cells restrict tissue damage and neuronal loss after spinal cord injury in mice. Science. 2013;342(6158):637-40.

[122] Rafols JA, Goshgarian HG. Spinal tanycytes in the adult rat: a correlative Golgi gold-toning study. The Anatomical record. 1985;211(1):75-86.

[123] Huang AL, Chen X, Hoon MA, Chandrashekar J, Guo W, Trankner D, et al. The cells and logic for mammalian sour taste detection. Nature. 2006;442(7105):934-8.

[124] Marichal N, Garcia G, Radmilovich M, Trujillo-Cenoz O, Russo RE. Enigmatic central canal contacting cells: immature neurons in "standby mode"? The Journal of neuroscience : the official journal of the Society for Neuroscience. 2009;29(32):10010-24.

[125] Goritz C, Dias DO, Tomilin N, Barbacid M, Shupliakov O, Frisen J. A pericyte origin of spinal cord scar tissue. Science. 2011;333(6039):238-42.

[126] Moreno-Manzano V, Rodriguez-Jimenez FJ, Garcia-Rosello M, Lainez S, Erceg S, Calvo MT, et al. Activated spinal cord ependymal stem cells rescue neurological function. Stem cells. 2009;27(3):733-43.

[127] Kulbatski I, Tator CH. Region-specific differentiation potential of adult rat spinal cord neural stem/precursors and their plasticity in response to in vitro manipulation. The journal of histochemistry and cytochemistry : official journal of the Histochemistry Society. 2009;57(5):405-23.

[128] Zhao M, Momma S, Delfani K, Carlen M, Cassidy RM, Johansson CB, et al. Evidence for neurogenesis in the adult mammalian substantia nigra. Proceedings of the National Academy of Sciences of the United States of America. 2003;100(13):7925-30.

[129] Bernier PJ, Bedard A, Vinet J, Levesque M, Parent A. Newly generated neurons in the amygdala and adjoining cortex of adult primates. Proceedings of the National Academy of Sciences of the United States of America. 2002;99(17):11464-9.

[130] Gould E, Reeves AJ, Graziano MS, Gross CG. Neurogenesis in the neocortex of adult primates. Science. 1999;286(5439):548-52.

[131] Migaud M, Batailler M, Segura S, Duittoz A, Franceschini I, Pillon D. Emerging new sites for adult neurogenesis in the mammalian brain: a comparative study between the hypothalamus and the classical neurogenic zones. The European journal of neuroscience. 2010;32(12):2042-52.

[132] Kempermann G, Wiskott L, Gage FH. Functional significance of adult neurogenesis. Current opinion in neurobiology. 2004;14(2):186-91.

[133] Lledo PM, Alonso M, Grubb MS. Adult neurogenesis and functional plasticity in neuronal circuits. Nature reviews Neuroscience. 2006;7(3):179-93.

[134] Kokoeva MV, Yin H, Flier JS. Evidence for constitutive neural cell proliferation in the adult murine hypothalamus. J Comp Neurol. 2007;505(2):209-20.

[135] Pencea V, Bingaman KD, Freedman LJ, Luskin MB. Neurogenesis in the subventricular zone and rostral migratory stream of the neonatal and adult primate forebrain. Experimental neurology. 2001;172(1):1-16.

[136] Kokoeva MV, Yin H, Flier JS. Neurogenesis in the hypothalamus of adult mice: potential role in energy balance. Science. 2005;310(5748):679-83.

[137] Bennett L, Yang M, Enikolopov G, Iacovitti L. Circumventricular organs: a novel site of neural stem cells in the adult brain. Molecular and cellular neurosciences. 2009;41(3):337-47.

[138] Mercier F, Kitasako JT, Hatton GI. Anatomy of the brain neurogenic zones revisited: fractones and the fibroblast/macrophage network. The Journal of comparative neurology. 2002;451(2):170-88.

[139] Bifari F, Decimo I, Chiamulera C, Bersan E, Malpeli G, Johansson J, et al. Novel stem/progenitor cells with neuronal differentiation potential reside in the leptomeningeal niche. Journal of cellular and molecular medicine. 2009;13(9B):3195-208.

[140] Itokazu Y, Kitada M, Dezawa M, Mizoguchi A, Matsumoto N, Shimizu A, et al. Choroid plexus ependymal cells host neural progenitor cells in the rat. Glia. 2006;53(1):32-42.

[141] DeGiorgio LA, Sheu KF, Blass JP. Culture from human leptomeninges of cells containing neurofilament protein and neuron-specific enolase. J Neurol Sci. 1994;124(2):141-8.

[142] Petricevic J, Forempoher G, Ostojic L, Mardesic-Brakus S, Andjelinovic S, Vukojevic K, et al. Expression of nestin, mesothelin and epithelial membrane antigen (EMA) in

developing and adult human meninges and meningiomas. Acta histochemica. 2011;113(7):703-11.

[143] Decimo I, Bifari F, Rodriguez FJ, Malpeli G, Dolci S, Lavarini V, et al. Nestin-and doublecortin-positive cells reside in adult spinal cord meninges and participate in injury-induced parenchymal reaction. Stem cells. 2011;29(12):2062-76.

[144] Halfter W, Dong S, Yip YP, Willem M, Mayer U. A critical function of the pial basement membrane in cortical histogenesis. The Journal of neuroscience : the official journal of the Society for Neuroscience. 2002;22(14):6029-40.

[145] Mercier F, Hatton GI. Connexin 26 and basic fibroblast growth factor are expressed primarily in the subpial and subependymal layers in adult brain parenchyma: roles in stem cell proliferation and morphological plasticity? The Journal of comparative neurology. 2001;431(1):88-104.

[146] Stumm R, Kolodziej A, Schulz S, Kohtz JD, Hollt V. Patterns of SDF-1alpha and SDF-1gamma mRNAs, migration pathways, and phenotypes of CXCR4-expressing neurons in the developing rat telencephalon. The Journal of comparative neurology. 2007;502(3):382-99.

[147] Siegenthaler JA, Ashique AM, Zarbalis K, Patterson KP, Hecht JH, Kane MA, et al. Retinoic acid from the meninges regulates cortical neuron generation. Cell. 2009;139(3):597-609.

[148] Lee A, Kessler JD, Read TA, Kaiser C, Corbeil D, Huttner WB, et al. Isolation of neural stem cells from the postnatal cerebellum. Nature neuroscience. 2005;8(6):723-9.

[149] Ponti G, Crociara P, Armentano M, Bonfanti L. Adult neurogenesis without germinal layers: the "atypical" cerebellum of rabbits. Arch Ital Biol. 2010;148(2):147-58.

[150] Ponti G, Peretto P, Bonfanti L. Genesis of neuronal and glial progenitors in the cerebellar cortex of peripuberal and adult rabbits. PLoS One. 2008;3(6):e2366.

[151] Ponti G, Peretto P, Bonfanti L. A subpial, transitory germinal zone forms chains of neuronal precursors in the rabbit cerebellum. Developmental biology. 2006;294(1): 168-80.

[152] McDonough A, Martinez-Cerdeno V. Endogenous proliferation after spinal cord injury in animal models. Stem Cells Int. 2012;2012:387513.

[153] Bonfanti L, Peretto P. Adult neurogenesis in mammals--a theme with many variations. The European journal of neuroscience. 2011;34(6):930-50.

[154] Nunes MC, Roy NS, Keyoung HM, Goodman RR, McKhann G, 2nd, Jiang L, et al. Identification and isolation of multipotential neural progenitor cells from the subcortical white matter of the adult human brain. Nature medicine. 2003;9(4):439-47.

[155] Zhu X, Bergles DE, Nishiyama A. NG2 cells generate both oligodendrocytes and gray matter astrocytes. Development (Cambridge, England). 2008;135(1):145-57.

[156] Horner PJ, Thallmair M, Gage FH. Defining the NG2-expressing cell of the adult CNS. J Neurocytol. 2002;31(6-7):469-80.

[157] Shechter R, Ziv Y, Schwartz M. New GABAergic interneurons supported by myelin-specific T cells are formed in intact adult spinal cord. Stem Cells. 2007;25(9):2277-82.

[158] Kokaia Z, Thored P, Arvidsson A, Lindvall O. Regulation of stroke-induced neurogenesis in adult brain--recent scientific progress. Cerebral cortex. 2006;16 Suppl 1:i162-7.

[159] Jin K, Wang X, Xie L, Mao XO, Zhu W, Wang Y, et al. Evidence for stroke-induced neurogenesis in the human brain. Proceedings of the National Academy of Sciences of the United States of America. 2006;103(35):13198-202.

[160] Jin K, Minami M, Lan JQ, Mao XO, Batteur S, Simon RP, et al. Neurogenesis in dentate subgranular zone and rostral subventricular zone after focal cerebral ischemia in the rat. Proceedings of the National Academy of Sciences of the United States of America. 2001;98(8):4710-5.

[161] Monje ML, Toda H, Palmer TD. Inflammatory blockade restores adult hippocampal neurogenesis. Science. 2003;302(5651):1760-5.

[162] Gould E, Tanapat P. Lesion-induced proliferation of neuronal progenitors in the dentate gyrus of the adult. Neuroscience. 1997;80(2):427-36.

[163] Pluchino S, Muzio L, Imitola J, Deleidi M, Alfaro-Cervello C, Salani G, et al. Persistent inflammation alters the function of the endogenous brain stem cell compartment. Brain : a journal of neurology. 2008;131:2564-78.

[164] Zheng WM, Zhuge QC, Zhong M, Chen GR, Shao B, Wang H, et al. Neurogenesis in Adult Human Brain after Traumatic Brain Injury. J Neurotraum. 2013;30(22):1872-80.

[165] Urrea C, Castellanos DA, Sagen J, Tsoulfas P, Bramlett HM, Dietrich WD. Widespread cellular proliferation and focal neurogenesis after traumatic brain injury in the rat. Restor Neurol Neuros. 2007;25(1):65-76.

[166] Brundin L, Brismar H, Danilov AI, Olsson T, Johansson CB. Neural stem cells: A potential source for remyelination in neuroinflammatory disease. Brain Pathol. 2003;13(3):322-8.

[167] Yagita Y, Kitagawa K, Ohtsuki T, Takasawa K, Miyata T, Okano H, et al. Neurogenesis by progenitor cells in the ischemic adult rat hippocampus. Stroke; a journal of cerebral circulation. 2001;32(8):1890-6.

[168] Donnan GA, Fisher M, Macleod M, Davis SM. Stroke. Lancet. 2008;371(9624):1612-23.

[169] Go AS, Mozaffarian D, Roger VL, Benjamin EJ, Berry JD, Borden WB, et al. Heart Disease and Stroke Statistics-2013 Update A Report From the American Heart Association. Circulation. 2013;127(1):E6-E245.

[170] Endres M, Dirnagl U, Moskowitz MA. The ischemic cascade and mediators of ischemic injury. Handbook of clinical neurology. 2009;92:31-41.

[171] Escuret E. [Cerebral ischemic cascade]. Annales francaises d'anesthesie et de reanimation. 1995;14(1):103-13.

[172] Felberg RA, Burgin WS, Grotta JC. Neuroprotection and the ischemic cascade. CNS spectrums. 2000;5(3):52-8.

[173] Dirnagl U, Iadecola C, Moskowitz MA. Pathobiology of ischaemic stroke: an integrated view. Trends in neurosciences. 1999;22(9):391-7.

[174] Zhang RL, Zhang ZG, Lu M, Wang Y, Yang JJ, Chopp M. Reduction of the cell cycle length by decreasing G1 phase and cell cycle reentry expand neuronal progenitor cells in the subventricular zone of adult rat after stroke. Journal of cerebral blood flow and metabolism : official journal of the International Society of Cerebral Blood Flow and Metabolism. 2006;26(6):857-63.

[175] Arvidsson A, Collin T, Kirik D, Kokaia Z, Lindvall O. Neuronal replacement from endogenous precursors in the adult brain after stroke. Nature medicine. 2002;8(9): 963-70.

[176] Thored P, Arvidsson A, Cacci E, Ahlenius H, Kallur T, Darsalia V, et al. Persistent production of neurons from adult brain stem cells during recovery after stroke. Stem cells. 2006;24(3):739-47.

[177] Nakatomi H, Kuriu T, Okabe S, Yamamoto S, Hatano O, Kawahara N, et al. Regeneration of hippocampal pyramidal neurons after ischemic brain injury by recruitment of endogenous neural progenitors. Cell. 2002;110(4):429-41.

[178] Darsalia V, Heldmann U, Lindvall O, Kokaia Z. Stroke-induced neurogenesis in aged brain. Stroke; a journal of cerebral circulation. 2005;36(8):1790-5.

[179] Tanaka R, Yamashiro K, Mochizuki H, Cho N, Onodera M, Mizuno Y, et al. Neurogenesis after transient global ischemia in the adult hippocampus visualized by improved retroviral vector. Stroke; a journal of cerebral circulation. 2004;35(6):1454-9.

[180] Nakagomi T, Taguchi A, Fujimori Y, Saino O, Nakano-Doi A, Kubo S, et al. Isolation and characterization of neural stem/progenitor cells from post-stroke cerebral cortex in mice. European Journal of Neuroscience. 2009;29(9):1842-52.

[181] Zhang RL, Zhang ZG, Chopp M. Neurogenesis in the adult ischemic brain: Generation, migration, survival, and restorative therapy. The Neuroscientist : a review journal bringing neurobiology, neurology and psychiatry. 2005;11(5):408-16.

[182] Kawai T, Takagi N, Miyake-Takagi K, Okuyama N, Mochizuki N, Takeo S. Characterization of BrdU-positive neurons induced by transient global ischemia in adult hippocampus. J Cerebr Blood F Met. 2004;24(5):548-55.

[183] Andres RH, Horie N, Slikker W, Keren-Gill H, Zhan K, Sun G, et al. Human neural stem cells enhance structural plasticity and axonal transport in the ischaemic brain. Brain : a journal of neurology. 2011;134(Pt 6):1777-89.

[184] Lichtenwalner RJ, Parent JM. Adult neurogenesis and the ischemic forebrain. J Cerebr Blood F Met. 2006;26(1):1-20.

[185] Kunze A, Grass S, Witte OW, Yamaguchi M, Kempermann G, Redecker C. Proliferative response of distinct hippocampal progenitor cell populations after cortical infarcts in the adult brain. Neurobiol Dis. 2006;21(2):324-32.

[186] Vandenbosch R, Borgs L, Beukelaers P, Belachew S, Moonen G, Nguyen L, et al. Adult neurogenesis and the diseased brain. Curr Med Chem. 2009;16(6):652-66.

[187] Zhang RL, Zhang ZG, Zhang CL, Zhang L, Robin A, Wang Y, et al. Stroke transiently increases subventricular zone cell division from asymmetric to symmetric and increases neuronal differentiation in the adult rat. Journal of Neuroscience. 2004;24(25): 5810-5.

[188] Zhang RL, Zhang ZG, Wang Y, LeTourneau Y, Liu XS, Zhang XG, et al. Stroke induces ependymal cell transformation into radial glia in the subventricular zone of the adult rodent brain. J Cerebr Blood F Met. 2007;27(6):1201-12.

[189] Vandenbosch R. Adult Neurogenesis and the Diseased Brain. Current Medicinal Chemistry. 2009;16:652-66.

[190] Parent JM, Vexler ZS, Gong C, Derugin N, Ferriero DM. Rat forebrain neurogenesis and striatal neuron replacement after focal stroke. Annals of neurology. 2002;52(6): 802-13.

[191] Zhang R, Zhang Z, Wang L, Wang Y, Gousev A, Zhang L, et al. Activated neural stem cells contribute to stroke-induced neurogenesis and neuroblast migration toward the infarct boundary in adult rats. Journal of cerebral blood flow and metabolism : official journal of the International Society of Cerebral Blood Flow and Metabolism. 2004;24(4):441-8.

[192] Decimo I, Bifari F, Krampera M, Fumagalli G. Neural Stem Cell Niches in Health and Diseases. Current Pharmaceutical Design. 2012;18:1755-83.

[193] Planas AM, Justicia C, Soriano MA, Ferrer I. Epidermal growth factor receptor in proliferating reactive glia following transient focal ischemia in the rat brain. Glia. 1998;23(2):120-9.

[194] Tureyen K, Vemuganti R, Bowen KK, Sailor KA, Dempsey RJ. EGF and FGF-2 infusion increases post-ischemic neural progenitor cell proliferation in the adult rat brain. Neurosurgery. 2005;57(6):1254-62.

[195] Tokumine J, Kakinohana O, Cizkova D, Smith DW, Marsala M. Changes in spinal GDNF, BDNF, and NT-3 expression after transient ischemia in the rat. Journal of neuroscience research. 2003;74(4):552-61.

[196] Kitagawa H, Sasaki C, Zhang WR, Sakai K, Shiro Y, Warita H, et al. Induction of glial cell line-derived neurotrophic factor receptor proteins in cerebral cortex and striatum after permanent middle cerebral artery occlusion in rats. Brain research. 1999;834(1-2):190-5.

[197] Takami K, Kiyota Y, Iwane M, Miyamoto M, Tsukuda R, Igarashi K, et al. Upregulation of fibroblast growth factor-receptor messenger RNA expression in rat brain following transient forebrain ischemia. Experimental brain research. 1993;97(2):185-94.

[198] Kokaia Z, Zhao Q, Kokaia M, Elmer E, Metsis M, Smith ML, et al. Regulation of Brain-Derived Neurotrophic Factor Gene-Expression after Transient Middle Cerebral-Artery Occlusion with and without Brain-Damage. Experimental neurology. 1995;136(1):73-88.

[199] Lin TN, Te J, Lee M, Sun GY, Hsu CY. Induction of basic fibroblast growth factor (bFGF) expression following focal cerebral ischemia. Brain research Molecular brain research. 1997;49(1-2):255-65.

[200] Lin TN, Wong YP, Chen JJ, Cheng JT, Yu SF, Sun SH, et al. Elevated basic fibroblast growth factor levels in stroke-prone spontaneously hypertensive rats. Neuroscience. 1997;76(2):557-70.

[201] Heldmann U, Thored P, Claasen JH, Arvidsson A, Kokaia Z, Lindvall O. TNF-alpha antibody infusion impairs survival of stroke-generated neuroblasts in adult rat brain. Experimental neurology. 2005;196(1):204-8.

[202] Tsai PT, Ohab JJ, Kertesz N, Groszer M, Matter C, Gao J, et al. A critical role of erythropoietin receptor in neurogenesis and post-stroke recovery. Journal of Neuroscience. 2006;26(4):1269-74.

[203] Chou J, Harvey BK, Chang CF, Shen H, Morales M, Wang Y. Neuroregenerative effects of BMP7 after stroke in rats. J Neurol Sci. 2006;240(1-2):21-9.

[204] Kang SS, Keasey MP, Arnold SA, Reid R, Geralds J, Hagg T. Endogenous CNTF mediates stroke-induced adult CNS neurogenesis in mice. Neurobiol Dis. 2012;49C: 68-78.

[205] Guerra-Crespo M, Gleason D, Sistos A, Toosky T, Solaroglu I, Zhang JH, et al. Transforming growth factor-alpha induces neurogenesis and behavioral improvement in a chronic stroke model. Neuroscience. 2009;160(2):470-83.

[206] Leker RR, Toth ZE, Shahar T, Cassiani-Ingoni R, Szalayova I, Key S, et al. Transforming growth factor alpha induces angiogenesis and neurogenesis following stroke. Neuroscience. 2009;163(1):233-43.

[207] Sun Y, Jin K, Xie L, Childs J, Mao XO, Logvinova A, et al. VEGF-induced neuroprotection, neurogenesis, and angiogenesis after focal cerebral ischemia. The Journal of clinical investigation. 2003;111(12):1843-51.

[208] Jin K, Mao XO, Sun Y, Xie L, Greenberg DA. Stem cell factor stimulates neurogenesis in vitro and in vivo. The Journal of clinical investigation. 2002;110(3):311-9.

[209] Woodbury ME, Ikezu T. Fibroblast Growth Factor-2 Signaling in Neurogenesis and Neurodegeneration. Journal of neuroimmune pharmacology : the official journal of the Society on NeuroImmune Pharmacology. 2013.

[210] Shingo T, Sorokan ST, Shimazaki T, Weiss S. Erythropoietin regulates the in vitro and in vivo production of neuronal progenitors by mammalian forebrain neural stem cells. The Journal of neuroscience : the official journal of the Society for Neuroscience. 2001;21(24):9733-43.

[211] Zhu W, Fan Y, Frenzel T, Gasmi M, Bartus RT, Young WL, et al. Insulin growth factor-1 gene transfer enhances neurovascular remodeling and improves long-term stroke outcome in mice. Stroke; a journal of cerebral circulation. 2008;39(4):1254-61.

[212] Schneider A, Kruger C, Steigleder T, Weber D, Pitzer C, Laage R, et al. The hematopoietic factor G-CSF is a neuronal ligand that counteracts programmed cell death and drives neurogenesis. Journal of Clinical Investigation. 2005;115(8):2083-98.

[213] Plane JM, Andjelkovic AV, Keep RF, Parent JM. Intact and injured endothelial cells differentially modulate postnatal murine forebrain neural stem cells. Neurobiol Dis. 2010;37(1):218-27.

[214] Chen JL, Zacharek A, Zhang CL, Jiang H, Li Y, Roberts C, et al. Endothelial nitric oxide synthase regulates brain-derived neurotrophic factor expression and neurogenesis after stroke in mice. Journal of Neuroscience. 2005;25(9):2366-75.

[215] Wang Q, Tang XN, Yenari MA. The inflammatory response in stroke. J Neuroimmunol. 2007;184(1-2):53-68.

[216] Hoehn BD, Palmer TD, Steinberg GK. Neurogenesis in rats after focal cerebral ischemia is enhanced by indomethacin. Stroke; a journal of cerebral circulation. 2005;36(12):2718-24.

[217] Kluska MM, Witte OW, Bolz J, Redecker C. Neurogenesis in the adult dentate gyrus after cortical infarcts: Effects of infarct location, N-methyl-D-aspartate receptor blockade and anti-inflammatory treatment. Neuroscience. 2005;135(3):723-35.

[218] Rahpeymai Y, Hietala MA, Wilhelmsson U, Fotheringham A, Davies I, Nilsson AK, et al. Complement: a novel factor in basal and ischemia-induced neurogenesis. Embo Journal. 2006;25(6):1364-74.

[219] Kohman RA, Rhodes JS. Neurogenesis, inflammation and behavior. Brain, behavior, and immunity. 2013;27(1):22-32.

[220] Dooley D, Vidal P, Hendrix S. Immunopharmacological intervention for successful neural stem cell therapy: New perspectives in CNS neurogenesis and repair. Pharmacology & therapeutics. 2014;141(1):21-31.

[221] An C, Shi Y, Li P, Hu X, Gan Y, Stetler RA, et al. Molecular dialogs between the ischemic brain and the peripheral immune system: Dualistic roles in injury and repair. Progress in neurobiology. 2013.

[222] Ekdahl CT, Kokaia Z, Lindvall O. Brain inflammation and adult neurogenesis: the dual role of microglia. Neuroscience. 2009;158(3):1021-9.

[223] Bernabeu R, Sharp FR. NMDA and AMPA/kainate glutamate receptors modulate dentate neurogenesis and CA3 synapsin-I in normal and ischemic hippocampus. Journal of cerebral blood flow and metabolism : official journal of the International Society of Cerebral Blood Flow and Metabolism. 2000;20(12):1669-80.

[224] Arvidsson A, Kokaia Z, Lindvall O. N-methyl-D-aspartate receptor-mediated increase of neurogenesis in adult rat dentate gyrus following stroke. The European journal of neuroscience. 2001;14(1):10-8.

[225] Kojima T, Hirota Y, Ema M, Takahashi S, Miyoshi I, Okano H, et al. Subventricular Zone-Derived Neural Progenitor Cells Migrate Along a Blood Vessel Scaffold Toward the Post-stroke Striatum. Stem cells. 2010;28(3):545-54.

[226] Zhang RL, Zhang ZG, Zhang L, Chopp M. Proliferation and differentiation of progenitor cells in the cortex and the subventricular zone in the adult rat after focal cerebral ischemia. Neuroscience. 2001;105(1):33-41.

[227] Hou SW, Wang YQ, Xu M, Shen DH, Wang JJ, Huang F, et al. Functional integration of newly generated neurons into striatum after cerebral ischemia in the adult rat brain. Stroke; a journal of cerebral circulation. 2008;39(10):2837-44.

[228] Kadam SD, Smith-Hicks CL, Smith DR, Worley PF, Comi AM. Functional integration of new neurons into hippocampal networks and poststroke comorbidities following neonatal stroke in mice. Epilepsy Behav. 2010;18(4):344-57.

[229] Zhang RL, Chopp M, Gregg SR, Toh Y, Roberts C, Letourneau Y, et al. Patterns and dynamics of subventricular zone neuroblast migration in the ischemic striatum of the adult mouse. Journal of cerebral blood flow and metabolism : official journal of the International Society of Cerebral Blood Flow and Metabolism. 2009;29(7):1240-50.

[230] Jin K, Sun Y, Xie L, Peel A, Mao XO, Batteur S, et al. Directed migration of neuronal precursors into the ischemic cerebral cortex and striatum. Molecular and Cellular Neuroscience. 2003;24(1):171-89.

[231] Wang L, Zhang ZG, Zhang RL, Gregg SR, Hozeska-Solgot A, LeTourneau Y, et al. Matrix metalloproteinase 2 (MMP2) and MMP9 secreted by erythropoietin-activated endothelial cells promote neural progenitor cell migration. Journal of Neuroscience. 2006;26(22):5996-6003.

[232] Hill WD, Hess DC, Martin-Studdard A, Carothers JJ, Zheng JQ, Hale D, et al. SDF-1 (CXCL12) is upregulated in the ischemic penumbra following stroke: Association with bone marrow cell homing to injury. Journal of neuropathology and experimental neurology. 2004;63(1):84-96.

[233] Miller JT, Bartley JH, Wimborne HJC, Walker AL, Hess DC, Hill WD, et al. The neuroblast and angioblast chemotaxic factor SDF-1 (CXCL12) expression is briefly up regulated by reactive astrocytes in brain following neonatal hypoxic-ischemic injury. Bmc Neurosci. 2005;6.

[234] Robin AM, Zhang ZG, Wang L, Zhang RL, Katakowski M, Zhang L, et al. Stromal cell-derived factor 1alpha mediates neural progenitor cell motility after focal cerebral ischemia. Journal of cerebral blood flow and metabolism : official journal of the International Society of Cerebral Blood Flow and Metabolism. 2006;26(1):125-34.

[235] Wang YT, Huang J, Li YN, Yang GY. Roles of Chemokine CXCL12 and its Receptors in Ischemic Stroke. Current drug targets. 2012;13(2):166-72.

[236] Li M, Chang CJ, Lathia JD, Wang L, Pacenta HL, Cotleur A, et al. Chemokine receptor CXCR4 signaling modulates the growth factor-induced cell cycle of self-renewing and multipotent neural progenitor cells. Glia. 2011;59(1):108-18.

[237] Yan YP, Sailor KA, Lang BT, Park SW, Vemuganti R, Dempsey RJ. Monocyte chemoattractant protein-1 plays a critical role in neuroblast migration after focal cerebral ischemia. Journal of cerebral blood flow and metabolism : official journal of the International Society of Cerebral Blood Flow and Metabolism. 2007;27(6):1213-24.

[238] Young CC, Brooks KJ, Buchan AM, Szele FG. Cellular and molecular determinants of stroke-induced changes in subventricular zone cell migration. Antioxidants & redox signaling. 2011;14(10):1877-88.

[239] Jin KL, Wang XM, Xie L, Mao XO, Greenberg DA. Transgenic ablation of doublecortin-expressing cells suppresses adult neurogenesis and worsens stroke outcome in mice. Proceedings of the National Academy of Sciences of the United States of America. 2010;107(17):7993-8.

[240] Wang XM, Mao XO, Xie L, Sun F, Greenberg DA, Jin KL. Conditional Depletion of Neurogenesis Inhibits Long-Term Recovery after Experimental Stroke in Mice. Plos One. 2012;7(6).

[241] Kumar S, Selim MH, Caplan LR. Medical complications after stroke. Lancet neurology. 2010;9(1):105-18.

[242] Schaapsmeerders P, Maaijwee NAM, van Dijk EJ, Rutten-Jacobs LCA, Arntz RM, Schoonderwaldt HC, et al. Long-Term Cognitive Impairment After First-Ever Ischemic Stroke in Young Adults. Stroke; a journal of cerebral circulation. 2013;44(6): 1621-8.

[243] Wolfe CDA, Crichton SL, Heuschmann PU, McKevitt CJ, Toschke AM, Grieve AP, et al. Estimates of Outcomes Up to Ten Years after Stroke: Analysis from the Prospective South London Stroke Register. PLoS medicine. 2011;8(5).

[244] Gemma C, Bachstetter AD. The role of microglia in adult hippocampal neurogenesis. Frontiers in cellular neuroscience. 2013;7:229.

[245] Nadareishvili Z, Hallenbeck J. Neuronal regeneration after stroke. New Engl J Med. 2003;348(23):2355-6.

[246] Kokaia Z, Lindvall O. Stem cell repair of striatal ischemia. Progress in brain research. 2012;201:35-53.

[247] Hauser SL, Oksenberg JR. The neurobiology of multiple sclerosis: genes, inflammation, and neurodegeneration. Neuron. 2006;52(1):61-76.

[248] Wingerchuk DM, Carter JL. Multiple sclerosis: current and emerging disease-modifying therapies and treatment strategies. Mayo Clinic proceedings. 2014;89(2):225-40.

[249] Noseworthy JH, Lucchinetti C, Rodriguez M, Weinshenker BG. Multiple sclerosis. The New England journal of medicine. 2000;343(13):938-52.

[250] Handel AE, Giovannoni G, Ebers GC, Ramagopalan SV. Environmental factors and their timing in adult-onset multiple sclerosis. Nat Rev Neurol. 2010;6(3):156-66.

[251] Tullman MJ. Overview of the epidemiology, diagnosis, and disease progression associated with multiple sclerosis. The American journal of managed care. 2013;19(2 Suppl):S15-20.

[252] Popescu BF, Lucchinetti CF. Pathology of demyelinating diseases. Annual review of pathology. 2012;7:185-217.

[253] Lucchinetti CF, Popescu BF, Bunyan RF, Moll NM, Roemer SF, Lassmann H, et al. Inflammatory cortical demyelination in early multiple sclerosis. The New England journal of medicine. 2011;365(23):2188-97.

[254] Geurts JJ, Barkhof F. Grey matter pathology in multiple sclerosis. Lancet neurology. 2008;7(9):841-51.

[255] Nait-Oumesmar B, Picard-Riera N, Kerninon C, Baron-Van Evercooren A. The role of SVZ-derived neural precursors in demyelinating diseases: from animal models to multiple sclerosis. J Neurol Sci. 2008;265(1-2):26-31.

[256] Rasmussen S, Imitola J, Ayuso-Sacido A, Wang Y, Starossom SC, Kivisakk P, et al. Reversible neural stem cell niche dysfunction in a model of multiple sclerosis. Annals of neurology. 2011;69(5):878-91.

[257] Franklin RJ. Why does remyelination fail in multiple sclerosis? Nat Rev Neurosci. 2002;3(9):705-14.

[258] Maki T, Liang AC, Miyamoto N, Lo EH, Arai K. Mechanisms of oligodendrocyte regeneration from ventricular-subventricular zone-derived progenitor cells in white matter diseases. Frontiers in cellular neuroscience. 2013;7:275.

[259] Whitney NP, Eidem TM, Peng H, Huang Y, Zheng JC. Inflammation mediates varying effects in neurogenesis: relevance to the pathogenesis of brain injury and neurodegenerative disorders. Journal of neurochemistry. 2009;108(6):1343-59.

[260] Picard-Riera N, Decker L, Delarasse C, Goude K, Nait-Oumesmar B, Liblau R, et al. Experimental autoimmune encephalomyelitis mobilizes neural progenitors from the subventricular zone to undergo oligodendrogenesis in adult mice. Proceedings of the National Academy of Sciences of the United States of America. 2002;99(20):13211-6.

[261] Nait-Oumesmar B, Picard-Riera N, Kerninon C, Decker L, Seilhean D, Hoglinger GU, et al. Activation of the subventricular zone in multiple sclerosis: evidence for early glial progenitors. Proceedings of the National Academy of Sciences of the United States of America. 2007;104(11):4694-9.

[262] Huehnchen P, Prozorovski T, Klaissle P, Lesemann A, Ingwersen J, Wolf SA, et al. Modulation of adult hippocampal neurogenesis during myelin-directed autoimmune neuroinflammation. Glia. 2011;59(1):132-42.

[263] Giannakopoulou A, Grigoriadis N, Bekiari C, Lourbopoulos A, Dori I, Tsingotjidou AS, et al. Acute inflammation alters adult hippocampal neurogenesis in a multiple sclerosis mouse model. Journal of neuroscience research. 2013;91(7):890-900.

[264] Geurts JJG, Bo L, Roosendaal SD, Hazes T, Daniels R, Barkhof F, et al. Extensive hippocampal demyelination in multiple sclerosis. Journal of neuropathology and experimental neurology. 2007;66(9):819-27.

[265] Papadopoulos D, Dukes S, Patel R, Nicholas R, Vora A, Reynolds R. Substantial Archaeocortical Atrophy and Neuronal Loss in Multiple Sclerosis. Brain Pathol. 2009;19(2):238-53.

[266] Roosendaal SD, Moraal B, Vrenken H, Castelijns JA, Pouwels PJW, Barkhof F, et al. In vivo MR imaging of hippocampal lesions in multiple sclerosis. J Magn Reson Imaging. 2008;27(4):726-31.

[267] Sicotte NL, Kern KC, Giesser BS, Arshanapalli A, Schultz A, Montag M, et al. Regional hippocampal atrophy in multiple sclerosis. Brain : a journal of neurology. 2008;131:1134-41.

[268] Danilov AI, Covacu R, Moe MC, Langmoen IA, Johansson CB, Olsson T, et al. Neurogenesis in the adult spinal cord in an experimental model of multiple sclerosis. The European journal of neuroscience. 2006;23(2):394-400.

[269] Wolswijk G. Oligodendrocyte precursor cells in the demyelinated multiple sclerosis spinal cord. Brain : a journal of neurology. 2002;125:338-49.

[270] Chang A, Nishiyama A, Peterson J, Prineas J, Trapp BD. NG2-positive oligodendrocyte progenitor cells in adult human brain and multiple sclerosis lesions. The Journal of neuroscience : the official journal of the Society for Neuroscience. 2000;20(17): 6404-12.

[271] Scolding N, Franklin R, Stevens S, Heldin CH, Compston A, Newcombe J. Oligodendrocyte progenitors are present in the normal adult human CNS and in the lesions of multiple sclerosis. Brain : a journal of neurology. 1998;121 (Pt 12):2221-8.

[272] Snethen H, Love S, Scolding N. Disease-responsive neural precursor cells are present in multiple sclerosis lesions. Regenerative medicine. 2008;3(6):835-47.

[273] Goldman S. Stem and progenitor cell-based therapy of the human central nervous system. Nature biotechnology. 2005;23(7):862-71.

[274] Cripps RA, Lee BB, Wing P, Weerts E, Mackay J, Brown D. A global map for traumatic spinal cord injury epidemiology: towards a living data repository for injury prevention. Spinal cord. 2011;49(4):493-501.

[275] Karlsson AK. Autonomic dysfunction in spinal cord injury: clinical presentation of symptoms and signs. Autonomic Dysfunction after Spinal Cord Injury. 2006;152:1-8.

[276] Oyinbo CA. Secondary injury mechanisms in traumatic spinal cord injury: a nugget of this multiply cascade. Acta neurobiologiae experimentalis. 2011;71(2):281-99.

[277] Tator CH. Biology of neurological recovery and functional restoration after spinal cord injury. Neurosurgery. 1998;42(4):696-707; discussion-8.

[278] Bareyre FM, Schwab ME. Inflammation, degeneration and regeneration in the injured spinal cord: insights from DNA microarrays. Trends in neurosciences. 2003;26(10):555-63.

[279] Schnell L, Fearn S, Klassen H, Schwab ME, Perry VH. Acute inflammatory responses to mechanical lesions in the CNS: differences between brain and spinal cord. The European journal of neuroscience. 1999;11(10):3648-58.

[280] Cao HQ, Dong ED. An update on spinal cord injury research. Neuroscience bulletin. 2013;29(1):94-102.

[281] Beattie MS, Hermann GE, Rogers RC, Bresnahan JC. Cell death in models of spinal cord injury. Progress in brain research. 2002;137:37-47.

[282] Fitch MT, Silver J. Activated macrophages and the blood-brain barrier: inflammation after CNS injury leads to increases in putative inhibitory molecules. Exp Neurol. 1997;148:587-603.

[283] Reier PJ, Houle JD. The glial scar: its bearing on axonal elongation and transplantation approaches to CNS repair. Adv Neurol. 1988;47:87-138.

[284] Rolls A, Shechter R, Schwartz M. The bright side of the glial scar in CNS repair. Nature reviews Neuroscience. 2009;10(3):235-41.

[285] Pekny M, Wilhelmsson U, Pekna M. The dual role of astrocyte activation and reactive gliosis. Neuroscience letters. 2014.

[286] Yuan YM, He C. The glial scar in spinal cord injury and repair. Neuroscience bulletin. 2013;29(4):421-35.

[287] Yiu G, He Z. Glial inhibition of CNS axon regeneration. Nature reviews Neuroscience. 2006;7(8):617-27.

[288] Silver J, Miller JH. Regeneration beyond the glial scar. Nature reviews Neuroscience. 2004;5(2):146-56.

[289] Faulkner JR. Reactive astrocytes protect tissue and preserve function after spinal cord injury. J Neurosci. 2004;24:2143-55.

[290] Sofroniew MV. Reactive astrocytes in neural repair and protection. Neuroscientist. 2005;11:400-7.

[291] Fawcett JW. Overcoming inhibition in the damaged spinal cord. Journal of neurotrauma. 2006;23(3-4):371-83.

[292] do Carmo Cunha J. Responses of reactive astrocytes containing S100[beta] protein and fibroblast growth factor-2 in the border and in the adjacent preserved tissue after a contusion injury of the spinal cord in rats: implications for wound repair and neuroregeneration. Wound Repair Regen. 2007;15:134-46.

[293] White RE, Yin FQ, Jakeman LB. TGF-a increases astrocyte invasion and promotes axonal growth into the lesion following spinal cord injury in mice. Exp Neurol. 2008;214:10-24.

[294] Wu VW, Nishiyama N, Schwartz JP. A culture model of reactive astrocytes: increased nerve growth factor synthesis and reexpression of cytokine responsiveness. J Neurochem. 1998;71:749-56.

[295] Schwartz JP, Nishiyama N. Neurotrophic factor gene expression in astrocytes during development and following injury. Brain Res Bull. 1994;35:403-7.

[296] Hagg T, Oudega M. Degenerative and spontaneous regenerative processes after spinal cord injury. J Neurotrauma. 2006;23(3-4):264-80.

[297] Horky LL, Galimi F, Gage FH, Horner PJ. Fate of endogenous stem/progenitor cells following spinal cord injury. The Journal of comparative neurology. 2006;498(4):525-38.

[298] Takahashi M, Arai Y, Kurosawa H, Sueyoshi N, Shirai S. Ependymal cell reactions in spinal cord segments after compression injury in adult rat. Journal of neuropathology and experimental neurology. 2003;62(2):185-94.

[299] Hofstetter CP, Holmstrom NA, Lilja JA, Schweinhardt P, Hao J, Spenger C, et al. Allodynia limits the usefulness of intraspinal neural stem cell grafts; directed differentiation improves outcome. Nature neuroscience. 2005;8(3):346-53.

[300] Ramasamy S, Narayanan G, Sankaran S, Yu YH, Ahmed S. Neural stem cell survival factors. Archives of biochemistry and biophysics. 2013;534(1-2):71-87.

[301] Pluchino S, Cossetti C. How stem cells speak with host immune cells in inflammatory brain diseases. Glia. 2013;61(9):1379-401.

[302] Gritti A, Frolichsthal-Schoeller P, Galli R, Parati EA, Cova L, Pagano SF, et al. Epidermal and fibroblast growth factors behave as mitogenic regulators for a single multipotent stem cell-like population from the subventricular region of the adult mouse forebrain. The Journal of neuroscience : the official journal of the Society for Neuroscience. 1999;19(9):3287-97.

[303] Morshead CM, Reynolds BA, Craig CG, McBurney MW, Staines WA, Morassutti D, et al. Neural stem cells in the adult mammalian forebrain: a relatively quiescent subpopulation of subependymal cells. Neuron. 1994;13(5):1071-82.

[304] Maina F, Klein R. Hepatocyte growth factor, a versatile signal for developing neurons. Nature neuroscience. 1999;2(3):213-7.

[305] Wang TW, Zhang H, Gyetko MR, Parent JM. Hepatocyte growth factor acts as a mitogen and chemoattractant for postnatal subventricular zone-olfactory bulb neurogenesis. Molecular and cellular neurosciences. 2011;48(1):38-50.

[306] Hu ZX, Geng JM, Liang DM, Luo M, Li ML. Hepatocyte growth factor protects human embryonic stem cell derived-neural progenitors from hydrogen peroxide-induced apoptosis. European journal of pharmacology. 2010;645(1-3):23-31.

[307] Doeppner TR, Kaltwasser B, ElAli A, Zechariah A, Hermann DM, Bahr M. Acute hepatocyte growth factor treatment induces long-term neuroprotection and stroke recovery via mechanisms involving neural precursor cell proliferation and differentiation. Journal of cerebral blood flow and metabolism : official journal of the International Society of Cerebral Blood Flow and Metabolism. 2011;31(5):1251-62.

[308] Shalaby F, Rossant J, Yamaguchi TP, Gertsenstein M, Wu XF, Breitman ML, et al. Failure of blood-island formation and vasculogenesis in Flk-1-deficient mice. Nature. 1995;376(6535):62-6.

[309] Shalaby F, Ho J, Stanford WL, Fischer KD, Schuh AC, Schwartz L, et al. A requirement for Flk1 in primitive and definitive hematopoiesis and vasculogenesis. Cell. 1997;89(6):981-90.

[310] Ema M, Faloon P, Zhang WJ, Hirashima M, Reid T, Stanford WL, et al. Combinatorial effects of Flk1 and Tal1 on vascular and hematopoietic development in the mouse. Genes & development. 2003;17(3):380-93.

[311] Jin K, Zhu Y, Sun Y, Mao XO, Xie L, Greenberg DA. Vascular endothelial growth factor (VEGF) stimulates neurogenesis in vitro and in vivo. Proceedings of the National Academy of Sciences of the United States of America. 2002;99(18):11946-50.

[312] Calvo CF, Fontaine RH, Soueid J, Tammela T, Makinen T, Alfaro-Cervello C, et al. Vascular endothelial growth factor receptor 3 directly regulates murine neurogenesis. Genes & development. 2011;25(8):831-44.

[313] Mackenzie F, Ruhrberg C. Diverse roles for VEGF-A in the nervous system. Development. 2012;139(8):1371-80.

[314] Wada T, Haigh JJ, Ema M, Hitoshi S, Chaddah R, Rossant J, et al. Vascular endothelial growth factor directly inhibits primitive neural stem cell survival but promotes definitive neural stem cell survival. The Journal of neuroscience : the official journal of the Society for Neuroscience. 2006;26(25):6803-12.

[315] Darland DC, Cain JT, Berosik MA, Saint-Geniez M, Odens PW, Schaubhut GJ, et al. Vascular endothelial growth factor (VEGF) isoform regulation of early forebrain development. Developmental biology. 2011;358(1):9-22.

[316] Zheng XR, Zhang SS, Yin F, Tang JL, Yang YJ, Wang X, et al. Neuroprotection of VEGF-expression neural stem cells in neonatal cerebral palsy rats. Behavioural brain research. 2012;230(1):108-15.

[317] Lian Jin H, Pennant WA, Hyung Lee M, Su S, Ah Kim H, Lu Liu M, et al. Neural stem cells modified by a hypoxia-inducible VEGF gene expression system improve cell viability under hypoxic conditions and spinal cord injury. Spine. 2011;36(11):857-64.

[318] Goberdhan DC, Wilson C. The functions of insulin signaling: size isn't everything, even in Drosophila. Differentiation; research in biological diversity. 2003;71(7):375-97.

[319] Rotwein P, Burgess SK, Milbrandt JD, Krause JE. Differential expression of insulin-like growth factor genes in rat central nervous system. Proceedings of the National Academy of Sciences of the United States of America. 1988;85(1):265-9.

[320] Werther GA, Abate M, Hogg A, Cheesman H, Oldfield B, Hards D, et al. Localization of insulin-like growth factor-I mRNA in rat brain by in situ hybridization--relationship to IGF-I receptors. Molecular endocrinology. 1990;4(5):773-8.

[321] D'Ercole AJ, Ye P, O'Kusky JR. Mutant mouse models of insulin-like growth factor actions in the central nervous system. Neuropeptides. 2002;36(2-3):209-20.

[322] Renault VM, Rafalski VA, Morgan AA, Salih DA, Brett JO, Webb AE, et al. FoxO3 regulates neural stem cell homeostasis. Cell stem cell. 2009;5(5):527-39.

[323] Tombran-Tink J, Chader GG, Johnson LV. PEDF: a pigment epithelium-derived factor with potent neuronal differentiative activity. Experimental eye research. 1991;53(3):411-4.

[324] Tombran-Tink J, Johnson LV. Neuronal differentiation of retinoblastoma cells induced by medium conditioned by human RPE cells. Investigative ophthalmology & visual science. 1989;30(8):1700-7.

[325] Ramirez-Castillejo C, Sanchez-Sanchez F, Andreu-Agullo C, Ferron SR, Aroca-Aguilar JD, Sanchez P, et al. Pigment epithelium-derived factor is a niche signal for neural stem cell renewal. Nature neuroscience. 2006;9(3):331-9.

[326] Andreu-Agullo C, Morante-Redolat JM, Delgado AC, Farinas I. Vascular niche factor PEDF modulates Notch-dependent stemness in the adult subependymal zone. Nature neuroscience. 2009;12(12):1514-23.

[327] Jackson EL, Garcia-Verdugo JM, Gil-Perotin S, Roy M, Quinones-Hinojosa A, VandenBerg S, et al. PDGFR alpha-positive B cells are neural stem cells in the adult SVZ that form glioma-like growths in response to increased PDGF signaling. Neuron. 2006;51(2):187-99.

[328] Arimura K, Ago T, Kamouchi M, Nakamura K, Ishitsuka K, Kuroda J, et al. PDGF receptor beta signaling in pericytes following ischemic brain injury. Current neurovascular research. 2012;9(1):1-9.

[329] McMahon AP, Joyner AL, Bradley A, McMahon JA. The midbrain-hindbrain phenotype of Wnt-1-/Wnt-1-mice results from stepwise deletion of engrailed-expressing cells by 9.5 days postcoitum. Cell. 1992;69(4):581-95.

[330] Lee SM, Tole S, Grove E, McMahon AP. A local Wnt-3a signal is required for development of the mammalian hippocampus. Development. 2000;127(3):457-67.

[331] Brault V, Moore R, Kutsch S, Ishibashi M, Rowitch DH, McMahon AP, et al. Inactivation of the beta-catenin gene by Wnt1-Cre-mediated deletion results in dramatic brain malformation and failure of craniofacial development. Development (Cambridge, England). 2001;128(8):1253-64.

[332] Chenn A, Walsh CA. Regulation of cerebral cortical size by control of cell cycle exit in neural precursors. Science. 2002;297(5580):365-9.

[333] Lie DC, Colamarino SA, Song HJ, Desire L, Mira H, Consiglio A, et al. Wnt signalling regulates adult hippocampal neurogenesis. Nature. 2005;437(7063):1370-5.

[334] Nusse R. Wnt signaling and stem cell control. Cell research. 2008;18(5):523-7.

[335] Andersson ER, Bryjova L, Biris K, Yamaguchi TP, Arenas E, Bryja V. Genetic interaction between Lrp6 and Wnt5a during mouse development. Developmental dynamics : an official publication of the American Association of Anatomists. 2010;239(1):237-45.

[336] Wang TW, Zhang H, Parent JM. Retinoic acid regulates postnatal neurogenesis in the murine subventricular zone-olfactory bulb pathway. Development (Cambridge, England). 2005;132(12):2721-32.

[337] Jacobs S, Lie DC, DeCicco KL, Shi Y, DeLuca LM, Gage FH, et al. Retinoic acid is required early during adult neurogenesis in the dentate gyrus. Proceedings of the National Academy of Sciences of the United States of America. 2006;103(10):3902-7.

[338] Marinaro C, Pannese M, Weinandy F, Sessa A, Bergamaschi A, Taketo MM, et al. Wnt signaling has opposing roles in the developing and the adult brain that are modulated by Hipk1. Cerebral cortex. 2012;22(10):2415-27.

[339] Mazumdar J, O'Brien WT, Johnson RS, LaManna JC, Chavez JC, Klein PS, et al. O2 regulates stem cells through Wnt/beta-catenin signalling. Nature cell biology. 2010;12(10):1007-13.

[340] Lim DA, Tramontin AD, Trevejo JM, Herrera DG, Garcia-Verdugo JM, Alvarez-Buylla A. Noggin antagonizes BMP signaling to create a niche for adult neurogenesis. Neuron. 2000;28(3):713-26.

[341] Zimmerman LB, De Jesus-Escobar JM, Harland RM. The Spemann organizer signal noggin binds and inactivates bone morphogenetic protein 4. Cell. 1996;86(4):599-606.

[342] Bonaguidi MA, Peng CY, McGuire T, Falciglia G, Gobeske KT, Czeisler C, et al. Noggin expands neural stem cells in the adult hippocampus. The Journal of neuroscience : the official journal of the Society for Neuroscience. 2008;28(37):9194-204.

[343] Lai K, Kaspar BK, Gage FH, Schaffer DV. Sonic hedgehog regulates adult neural progenitor proliferation in vitro and in vivo. Nature neuroscience. 2003;6(1):21-7.

[344] Machold R, Hayashi S, Rutlin M, Muzumdar MD, Nery S, Corbin JG, et al. Sonic hedgehog is required for progenitor cell maintenance in telencephalic stem cell niches. Neuron. 2003;39(6):937-50.

[345] Diederich K, Schabitz WR, Minnerup J. Seeing old friends from a different angle: novel properties of hematopoietic growth factors in the healthy and diseased brain. Hippocampus. 2012;22(5):1051-7.

[346] Tan CC, Eckardt KU, Firth JD, Ratcliffe PJ. Feedback modulation of renal and hepatic erythropoietin mRNA in response to graded anemia and hypoxia. The American journal of physiology. 1992;263(3 Pt 2):F474-81.

[347] Wang Y, Yao M, Zhou C, Dong D, Jiang Y, Wei G, et al. Erythropoietin promotes spinal cord-derived neural progenitor cell proliferation by regulating cell cycle. Neuroscience. 2010;167(3):750-7.

[348] Ransome MI, Turnley AM. Systemically delivered Erythropoietin transiently enhances adult hippocampal neurogenesis. Journal of neurochemistry. 2007;102(6):1953-65.

[349] Pavlica S, Milosevic J, Keller M, Schulze M, Peinemann F, Piscioneri A, et al. Erythropoietin enhances cell proliferation and survival of human fetal neuronal progenitors in normoxia. Brain research. 2012;1452:18-28.

[350] Giese AK, Frahm J, Hubner R, Luo J, Wree A, Frech MJ, et al. Erythropoietin and the effect of oxygen during proliferation and differentiation of human neural progenitor cells. BMC cell biology. 2010;11:94.

[351] Wang L, Zhang ZG, Gregg SR, Zhang RL, Jiao Z, LeTourneau Y, et al. The Sonic hedgehog pathway mediates carbamylated erythropoietin-enhanced proliferation and differentiation of adult neural progenitor cells. The Journal of biological chemistry. 2007;282(44):32462-70.

[352] Yu X, Shacka JJ, Eells JB, Suarez-Quian C, Przygodzki RM, Beleslin-Cokic B, et al. Erythropoietin receptor signalling is required for normal brain development. Development (Cambridge, England). 2002;129(2):505-16.

[353] Kuwabara T, Asashima M. Regenerative medicine using adult neural stem cells: the potential for diabetes therapy and other pharmaceutical applications. Journal of molecular cell biology. 2012;4(3):133-9.

[354] Erickson RI, Paucar AA, Jackson RL, Visnyei K, Kornblum H. Roles of insulin and transferrin in neural progenitor survival and proliferation. Journal of neuroscience research. 2008;86(8):1884-94.

[355] Zhao Y, Xiao Z, Gao Y, Chen B, Zhao Y, Zhang J, et al. Insulin rescues ES cell-derived neural progenitor cells from apoptosis by differential regulation of Akt and ERK pathways. Neuroscience letters. 2007;429(1):49-54.

[356] Yu SW, Baek SH, Brennan RT, Bradley CJ, Park SK, Lee YS, et al. Autophagic death of adult hippocampal neural stem cells following insulin withdrawal. Stem cells. 2008;26(10):2602-10.

[357] Shiraishi H, Okamoto H, Hara H, Yoshida H. Alternative cell death of Apaf1-deficient neural progenitor cells induced by withdrawal of EGF or insulin. Biochimica et biophysica acta. 2010;1800(3):405-15.

[358] Zhang D, Guo M, Zhang W, Lu XY. Adiponectin stimulates proliferation of adult hippocampal neural stem/progenitor cells through activation of p38 mitogen-activated protein kinase (p38MAPK)/glycogen synthase kinase 3beta (GSK-3beta)/beta-catenin signaling cascade. The Journal of biological chemistry. 2011;286(52):44913-20.

[359] Garza JC, Guo M, Zhang W, Lu XY. Leptin restores adult hippocampal neurogenesis in a chronic unpredictable stress model of depression and reverses glucocorticoid-induced inhibition of GSK-3beta/beta-catenin signaling. Molecular psychiatry. 2012;17(8):790-808.

[360] Avraham Y, Davidi N, Lassri V, Vorobiev L, Kabesa M, Dayan M, et al. Leptin induces neuroprotection neurogenesis and angiogenesis after stroke. Current neurovascular research. 2011;8(4):313-22.

[361] Udagawa J, Ono A, Kawamoto M, Otani H. Leptin and its intracellular signaling pathway maintains the neurosphere. Neuroreport. 2010;21(18):1140-5.

[362] Udagawa J, Hashimoto R, Suzuki H, Hatta T, Sotomaru Y, Hioki K, et al. The role of leptin in the development of the cerebral cortex in mouse embryos. Endocrinology. 2006;147(2):647-58.

[363] Bauer S, Rasika S, Han J, Mauduit C, Raccurt M, Morel G, et al. Leukemia inhibitory factor is a key signal for injury-induced neurogenesis in the adult mouse olfactory epithelium. The Journal of neuroscience : the official journal of the Society for Neuroscience. 2003;23(5):1792-803.

[364] Bauer S, Patterson PH. Leukemia inhibitory factor promotes neural stem cell self-renewal in the adult brain. The Journal of neuroscience : the official journal of the Society for Neuroscience. 2006;26(46):12089-99.

[365] Bauer S. Cytokine control of adult neural stem cells. Annals of the New York Academy of Sciences. 2009;1153:48-56.

[366] Shimazaki T, Shingo T, Weiss S. The ciliary neurotrophic factor/leukemia inhibitory factor/gp130 receptor complex operates in the maintenance of mammalian forebrain neural stem cells. The Journal of neuroscience : the official journal of the Society for Neuroscience. 2001;21(19):7642-53.

[367] Muller S, Chakrapani BP, Schwegler H, Hofmann HD, Kirsch M. Neurogenesis in the dentate gyrus depends on ciliary neurotrophic factor and signal transducer and activator of transcription 3 signaling. Stem cells. 2009;27(2):431-41.

[368] Toda H, Tsuji M, Nakano I, Kobuke K, Hayashi T, Kasahara H, et al. Stem cell-derived neural stem/progenitor cell supporting factor is an autocrine/paracrine survival factor for adult neural stem/progenitor cells. The Journal of biological chemistry. 2003;278(37):35491-500.

[369] Dziembowska M, Tham TN, Lau P, Vitry S, Lazarini F, Dubois-Dalcq M. A role for CXCR4 signaling in survival and migration of neural and oligodendrocyte precursors. Glia. 2005;50(3):258-69.

[370] Kokovay E, Goderie S, Wang Y, Lotz S, Lin G, Sun Y, et al. Adult SVZ lineage cells home to and leave the vascular niche via differential responses to SDF1/CXCR4 signaling. Cell stem cell. 2010;7(2):163-73.

[371] Cui X, Chen J, Zacharek A, Roberts C, Yang Y, Chopp M. Nitric oxide donor up-regulation of SDF1/CXCR4 and Ang1/Tie2 promotes neuroblast cell migration after stroke. Journal of neuroscience research. 2009;87(1):86-95.

[372] Ohta S, Misawa A, Fukaya R, Inoue S, Kanemura Y, Okano H, et al. Macrophage migration inhibitory factor (MIF) promotes cell survival and proliferation of neural stem/progenitor cells. Journal of cell science. 2012;125(Pt 13):3210-20.

[373] Mikami Y, Okano H, Sakaguchi M, Nakamura M, Shimazaki T, Okano HJ, et al. Implantation of dendritic cells in injured adult spinal cord results in activation of endogenous neural stem/progenitor cells leading to de novo neurogenesis and functional recovery. Journal of neuroscience research. 2004;76(4):453-65.

[374] de la Mano A, Gato A, Alonso MI, Carnicero E, Martin C, Moro JA. Role of interleukin-1beta in the control of neuroepithelial proliferation and differentiation of the spinal cord during development. Cytokine. 2007;37(2):128-37.

[375] Ben Menachem-Zidon O, Goshen I, Kreisel T, Ben Menahem Y, Reinhartz E, Ben Hur T, et al. Intrahippocampal transplantation of transgenic neural precursor cells overexpressing interleukin-1 receptor antagonist blocks chronic isolation-induced impairment in memory and neurogenesis. Neuropsychopharmacology : official publication of the American College of Neuropsychopharmacology. 2008;33(9):2251-62.

[376] Vawter MP, Basaric-Keys J, Li Y, Lester DS, Lebovics RS, Lesch KP, et al. Human olfactory neuroepithelial cells: tyrosine phosphorylation and process extension are increased by the combination of IL-1beta, IL-6, NGF, and bFGF. Experimental neurology. 1996;142(1):179-94.

[377] Koo JW, Duman RS. IL-1beta is an essential mediator of the antineurogenic and anhedonic effects of stress. Proceedings of the National Academy of Sciences of the United States of America. 2008;105(2):751-6.

[378] Kim TS, Misumi S, Jung CG, Masuda T, Isobe Y, Furuyama F, et al. Increase in dopaminergic neurons from mouse embryonic stem cell-derived neural progenitor/stem cells is mediated by hypoxia inducible factor-1alpha. Journal of neuroscience research. 2008;86(11):2353-62.

[379] Barkho BZ, Song H, Aimone JB, Smrt RD, Kuwabara T, Nakashima K, et al. Identification of astrocyte-expressed factors that modulate neural stem/progenitor cell differentiation. Stem Cells Dev. 2006;15(3):407-21.

[380] Kahn MA, De Vellis J. Regulation of an oligodendrocyte progenitor cell line by the interleukin-6 family of cytokines. Glia. 1994;12(2):87-98.

[381] Murphy M, Dutton R, Koblar S, Cheema S, Bartlett P. Cytokines which signal through the LIF receptor and their actions in the nervous system. Progress in neurobiology. 1997;52(5):355-78.

[382] Martino G, Pluchino S. The therapeutic potential of neural stem cells. Nat Rev Neurosci. 2006;7(5):395-406.

[383] Kokaia Z, Martino G, Schwartz M, Lindvall O. Cross-talk between neural stem cells and immune cells: the key to better brain repair? Nature neuroscience. 2012;15(8): 1078-87.

[384] Brazel CY, Nunez JL, Yang Z, Levison SW. Glutamate enhances survival and proliferation of neural progenitors derived from the subventricular zone. Neuroscience. 2005;131(1):55-65.

[385] Castiglione M, Calafiore M, Costa L, Sortino MA, Nicoletti F, Copani A. Group I metabotropic glutamate receptors control proliferation, survival and differentiation of cultured neural progenitor cells isolated from the subventricular zone of adult mice. Neuropharmacology. 2008;55(4):560-7.

[386] Tian Y, Liu Y, Chen X, Kang Q, Zhang J, Shi Q, et al. AMN082 promotes the proliferation and differentiation of neural progenitor cells with influence on phosphorylation of MAPK signaling pathways. Neurochemistry international. 2010;57(1):8-15.

[387] Zhao L, Jiao Q, Yang P, Chen X, Zhang J, Zhao B, et al. Metabotropic glutamate receptor 5 promotes proliferation of human neural stem/progenitor cells with activation of mitogen-activated protein kinases signaling pathway in vitro. Neuroscience. 2011;192:185-94.

[388] Di Giorgi-Gerevini V, Melchiorri D, Battaglia G, Ricci-Vitiani L, Ciceroni C, Busceti CL, et al. Endogenous activation of metabotropic glutamate receptors supports the proliferation and survival of neural progenitor cells. Cell death and differentiation. 2005;12(8):1124-33.

[389] Nochi R, Kato T, Kaneko J, Itou Y, Kuribayashi H, Fukuda S, et al. Involvement of metabotropic glutamate receptor 5 signaling in activity-related proliferation of adult hippocampal neural stem cells. The European journal of neuroscience. 2012;36(3): 2273-83.

[390] Liu XX, Wang Q, Haydar TF, Bordey A. Nonsynaptic GABA signaling in postnatal subventricular zone controls proliferation of GFAP-expressing progenitors. Nature neuroscience. 2005;8(9):1179-87.

[391] Fernando RN, Eleuteri B, Abdelhady S, Nussenzweig A, Andang M, Ernfors P. Cell cycle restriction by histone H2AX limits proliferation of adult neural stem cells. Proceedings of the National Academy of Sciences of the United States of America. 2011;108(14):5837-42.

[392] Banasr M, Hery M, Printemps R, Daszuta A. Serotonin-induced increases in adult cell proliferation and neurogenesis are mediated through different and common 5-HT receptor subtypes in the dentate gyrus and the subventricular zone. Neuropsychopharmacology : official publication of the American College of Neuropsychopharmacology. 2004;29(3):450-60.

[393] Brezun JM, Daszuta A. Depletion in serotonin decreases neurogenesis in the dentate gyrus and the subventricular zone of adult rats. Neuroscience. 1999;89(4):999-1002.

[394] Brezun JM, Daszuta A. Serotonin depletion in the adult rat produces differential changes in highly polysialylated form of neural cell adhesion molecule and tenascin-C immunoreactivity. Journal of neuroscience research. 1999;55(1):54-70.

[395] Encinas JM, Vaahtokari A, Enikolopov G. Fluoxetine targets early progenitor cells in the adult brain. Proceedings of the National Academy of Sciences of the United States of America. 2006;103(21):8233-8.

[396] Benninghoff J, Gritti A, Rizzi M, Lamorte G, Schloesser RJ, Schmitt A, et al. Serotonin depletion hampers survival and proliferation in neurospheres derived from adult neural stem cells. Neuropsychopharmacology : official publication of the American College of Neuropsychopharmacology. 2010;35(4):893-903.

[397] O'Keeffe GC, Barker RA. Dopamine stimulates epidermal growth factor release from adult neural precursor cells derived from the subventricular zone by a disintegrin and metalloprotease. Neuroreport. 2011;22(18):956-8.

[398] Blobel CP. ADAMs: key components in EGFR signalling and development. Nature reviews Molecular cell biology. 2005;6(1):32-43.

[399] Kippin TE, Kapur S, van der Kooy D. Dopamine specifically inhibits forebrain neural stem cell proliferation, suggesting a novel effect of antipsychotic drugs. The Journal of neuroscience : the official journal of the Society for Neuroscience. 2005;25(24): 5815-23.

[400] Yang P, Arnold SA, Habas A, Hetman M, Hagg T. Ciliary neurotrophic factor mediates dopamine D2 receptor-induced CNS neurogenesis in adult mice. The Journal of neuroscience : the official journal of the Society for Neuroscience. 2008;28(9):2231-41.

[401] Wolosker H, Panizzutti R, De Miranda J. Neurobiology through the looking-glass: D-serine as a new glial-derived transmitter. Neurochemistry international. 2002;41(5): 327-32.

[402] Schell MJ, Molliver ME, Snyder SH. D-serine, an endogenous synaptic modulator: localization to astrocytes and glutamate-stimulated release. Proceedings of the National Academy of Sciences of the United States of America. 1995;92(9):3948-52.

[403] Schell MJ, Brady RO, Jr., Molliver ME, Snyder SH. D-serine as a neuromodulator: regional and developmental localizations in rat brain glia resemble NMDA receptors. The Journal of neuroscience : the official journal of the Society for Neuroscience. 1997;17(5):1604-15.

[404] Mothet JP, Pollegioni L, Ouanounou G, Martineau M, Fossier P, Baux G. Glutamate receptor activation triggers a calcium-dependent and SNARE protein-dependent release of the gliotransmitter D-serine. Proceedings of the National Academy of Sciences of the United States of America. 2005;102(15):5606-11.

[405] Kartvelishvily E, Shleper M, Balan L, Dumin E, Wolosker H. Neuron-derived D-serine release provides a novel means to activate N-methyl-D-aspartate receptors. The Journal of biological chemistry. 2006;281(20):14151-62.

[406] Yasuda E, Ma N, Semba R. Immunohistochemical evidences for localization and production of D-serine in some neurons in the rat brain. Neuroscience letters. 2001;299(1-2):162-4.

[407] Yoshikawa M, Takayasu N, Hashimoto A, Sato Y, Tamaki R, Tsukamoto H, et al. The serine racemase mRNA is predominantly expressed in rat brain neurons. Archives of histology and cytology. 2007;70(2):127-34.

[408] Yoshikawa M, Nakajima K, Takayasu N, Noda S, Sato Y, Kawaguchi M, et al. Expression of the mRNA and protein of serine racemase in primary cultures of rat neurons. European journal of pharmacology. 2006;548(1-3):74-6.

[409] Maeda K, Sugino H, Hirose T, Kitagawa H, Nagai T, Mizoguchi H, et al. Clozapine prevents a decrease in neurogenesis in mice repeatedly treated with phencyclidine. Journal of pharmacological sciences. 2007;103(3):299-308.

[410] Huang X, Kong H, Tang M, Lu M, Ding JH, Hu G. D-Serine regulates proliferation and neuronal differentiation of neural stem cells from postnatal mouse forebrain. CNS neuroscience & therapeutics. 2012;18(1):4-13.

[411] Packer MA, Stasiv Y, Benraiss A, Chmielnicki E, Grinberg A, Westphal H, et al. Nitric oxide negatively regulates mammalian adult neurogenesis. Proceedings of the National Academy of Sciences of the United States of America. 2003;100(16):9566-71.

[412] Moreno-Lopez B, Romero-Grimaldi C, Noval JA, Murillo-Carretero M, Matarredona ER, Estrada C. Nitric oxide is a physiological inhibitor of neurogenesis in the adult mouse subventricular zone and olfactory bulb. The Journal of neuroscience : the official journal of the Society for Neuroscience. 2004;24(1):85-95.

[413] Carreira BP, Morte MI, Inacio A, Costa G, Rosmaninho-Salgado J, Agasse F, et al. Nitric oxide stimulates the proliferation of neural stem cells bypassing the epidermal growth factor receptor. Stem cells. 2010;28(7):1219-30.

[414] Maeda N. Structural variation of chondroitin sulfate and its roles in the central nervous system. Central nervous system agents in medicinal chemistry. 2010;10(1):22-31.

[415] von Holst A, Sirko S, Faissner A. The unique 473HD-Chondroitinsulfate epitope is expressed by radial glia and involved in neural precursor cell proliferation. The Journal of neuroscience : the official journal of the Society for Neuroscience. 2006;26(15): 4082-94.

[416] Sirko S, Akita K, Von Holst A, Faissner A. Structural and functional analysis of chondroitin sulfate proteoglycans in the neural stem cell niche. Methods in enzymology. 2010;479:37-71.

[417] Sirko S, von Holst A, Wizenmann A, Gotz M, Faissner A. Chondroitin sulfate glycosaminoglycans control proliferation, radial glia cell differentiation and neurogenesis in neural stem/progenitor cells. Development. 2007;134(15):2727-38.

[418] Gu WL, Fu SL, Wang YX, Li Y, Lu HZ, Xu XM, et al. Chondroitin sulfate proteoglycans regulate the growth, differentiation and migration of multipotent neural precursor cells through the integrin signaling pathway. BMC neuroscience. 2009;10:128.

[419] Joseph SJ, Ford MD, Barth C, Portbury S, Bartlett PF, Nurcombe V, et al. A proteoglycan that activates fibroblast growth factors during early neuronal development is a perlecan variant. Development. 1996;122(11):3443-52.

[420] Inatani M, Irie F, Plump AS, Tessier-Lavigne M, Yamaguchi Y. Mammalian brain morphogenesis and midline axon guidance require heparan sulfate. Science. 2003;302(5647):1044-6.

[421] Akita K, von Holst A, Furukawa Y, Mikami T, Sugahara K, Faissner A. Expression of multiple chondroitin/dermatan sulfotransferases in the neurogenic regions of the embryonic and adult central nervous system implies that complex chondroitin sulfates have a role in neural stem cell maintenance. Stem cells. 2008;26(3):798-809.

[422] Kazanis I, ffrench-Constant C. Extracellular matrix and the neural stem cell niche. Developmental neurobiology. 2011;71(11):1006-17.

[423] Lathia JD, Rao MS, Mattson MP, Ffrench-Constant C. The microenvironment of the embryonic neural stem cell: lessons from adult niches? Developmental dynamics : an official publication of the American Association of Anatomists. 2007;236(12):3267-82.

[424] Staquicini FI, Dias-Neto E, Li J, Snyder EY, Sidman RL, Pasqualini R, et al. Discovery of a functional protein complex of netrin-4, laminin gamma1 chain, and integrin alpha6beta1 in mouse neural stem cells. Proceedings of the National Academy of Sciences of the United States of America. 2009;106(8):2903-8.

[425] Kazanis I, Lathia JD, Vadakkan TJ, Raborn E, Wan R, Mughal MR, et al. Quiescence and activation of stem and precursor cell populations in the subependymal zone of the mammalian brain are associated with distinct cellular and extracellular matrix signals. The Journal of neuroscience : the official journal of the Society for Neuroscience. 2010;30(29):9771-81.

[426] Hitoshi S, Alexson T, Tropepe V, Donoviel D, Elia AJ, Nye JS, et al. Notch pathway molecules are essential for the maintenance, but not the generation, of mammalian neural stem cells. Genes & development. 2002;16(7):846-58.

[427] Nakamura Y, Sakakibara S, Miyata T, Ogawa M, Shimazaki T, Weiss S, et al. The bHLH gene hes1 as a repressor of the neuronal commitment of CNS stem cells. The Journal of neuroscience : the official journal of the Society for Neuroscience. 2000;20(1):283-93.

[428] Nyfeler Y, Kirch RD, Mantei N, Leone DP, Radtke F, Suter U, et al. Jagged1 signals in the postnatal subventricular zone are required for neural stem cell self-renewal. The EMBO journal. 2005;24(19):3504-15.

[429] Androutsellis-Theotokis A, Leker RR, Soldner F, Hoeppner DJ, Ravin R, Poser SW, et al. Notch signalling regulates stem cell numbers in vitro and in vivo. Nature. 2006;442(7104):823-6.

[430] Mizutani K, Yoon K, Dang L, Tokunaga A, Gaiano N. Differential Notch signalling distinguishes neural stem cells from intermediate progenitors. Nature. 2007;449(7160):351-5.

[431] Aguirre A, Rubio ME, Gallo V. Notch and EGFR pathway interaction regulates neural stem cell number and self-renewal. Nature. 2010;467(7313):323-7.

[432] Dave RK, Ellis T, Toumpas MC, Robson JP, Julian E, Adolphe C, et al. Sonic hedgehog and notch signaling can cooperate to regulate neurogenic divisions of neocortical progenitors. PloS one. 2011;6(2):e14680.

[433] Conover JC, Doetsch F, Garcia-Verdugo JM, Gale NW, Yancopoulos GD, Alvarez-Buylla A. Disruption of Eph/ephrin signaling affects migration and proliferation in the adult subventricular zone. Nature neuroscience. 2000;3(11):1091-7.

[434] Khodosevich K, Watanabe Y, Monyer H. EphA4 preserves postnatal and adult neural stem cells in an undifferentiated state in vivo. Journal of cell science. 2011;124(Pt 8):1268-79.

[435] Theus MH, Ricard J, Bethea JR, Liebl DJ. EphB3 limits the expansion of neural progenitor cells in the subventricular zone by regulating p53 during homeostasis and following traumatic brain injury. Stem cells. 2010;28(7):1231-42.

[436] Doeppner TR, Bretschneider E, Doehring M, Segura I, Senturk A, Acker-Palmer A, et al. Enhancement of endogenous neurogenesis in ephrin-B3 deficient mice after transient focal cerebral ischemia. Acta neuropathologica. 2011;122(4):429-42.

[437] Depaepe V, Suarez-Gonzalez N, Dufour A, Passante L, Gorski JA, Jones KR, et al. Ephrin signalling controls brain size by regulating apoptosis of neural progenitors. Nature. 2005;435(7046):1244-50.

[438] Wu C, Qiu R, Wang J, Zhang H, Murai K, Lu Q. ZHX2 Interacts with Ephrin-B and regulates neural progenitor maintenance in the developing cerebral cortex. The Journal of neuroscience : the official journal of the Society for Neuroscience. 2009;29(23): 7404-12.

[439] Murai K, Qiu R, Zhang H, Wang J, Wu C, Neubig RR, et al. Galpha subunit coordinates with ephrin-B to balance self-renewal and differentiation in neural progenitor cells. Stem cells. 2010;28(9):1581-9.

[440] Qiu R, Wang J, Tsark W, Lu Q. Essential role of PDZ-RGS3 in the maintenance of neural progenitor cells. Stem cells. 2010;28(9):1602-10.

[441] Qiu R, Wang X, Davy A, Wu C, Murai K, Zhang H, et al. Regulation of neural progenitor cell state by ephrin-B. The Journal of cell biology. 2008;181(6):973-83.

[442] Islam O, Loo TX, Heese K. Brain-derived neurotrophic factor (BDNF) has proliferative effects on neural stem cells through the truncated TRK-B receptor, MAP kinase, AKT, and STAT-3 signaling pathways. Current neurovascular research. 2009;6(1): 42-53.

[443] Llado J, Haenggeli C, Maragakis NJ, Snyder EY, Rothstein JD. Neural stem cells protect against glutamate-induced excitotoxicity and promote survival of injured motor neurons through the secretion of neurotrophic factors. Molecular and cellular neurosciences. 2004;27(3):322-31.

[444] Wang F, Kameda M, Yasuhara T, Tajiri N, Kikuchi Y, Liang HB, et al. GDNF-pretreatment enhances the survival of neural stem cells following transplantation in a rat model of Parkinson's disease. Neurosci Res. 2011;71(1):92-8.

[445] Nguyen L, Malgrange B, Belachew S, Rogister B, Rocher V, Moonen G, et al. Functional glycine receptors are expressed by postnatal nestin-positive neural stem/ progenitor cells. The European journal of neuroscience. 2002;15(8):1299-305.

[446] Pitas RE, Boyles JK, Lee SH, Foss D, Mahley RW. Astrocytes synthesize apolipoprotein E and metabolize apolipoprotein E-containing lipoproteins. Biochimica et biophysica acta. 1987;917(1):148-61.

[447] Nathan BP, Bellosta S, Sanan DA, Weisgraber KH, Mahley RW, Pitas RE. Differential effects of apolipoproteins E3 and E4 on neuronal growth in vitro. Science. 1994;264(5160):850-2.

[448] Li G, Bien-Ly N, Andrews-Zwilling Y, Xu Q, Bernardo A, Ring K, et al. GABAergic interneuron dysfunction impairs hippocampal neurogenesis in adult apolipoprotein E4 knockin mice. Cell stem cell. 2009;5(6):634-45.

[449] Yang CP, Gilley JA, Zhang G, Kernie SG. ApoE is required for maintenance of the dentate gyrus neural progenitor pool. Development (Cambridge, England). 2011;138(20):4351-62.

[450] Gan HT, Tham M, Hariharan S, Ramasamy S, Yu YH, Ahmed S. Identification of ApoE as an autocrine/paracrine factor that stimulates neural stem cell survival via MAPK/ERK signaling pathway. Journal of neurochemistry. 2011;117(3):565-78.

[451] Tsai PT, Ohab JJ, Kertesz N, Groszer M, Matter C, Gao J, et al. A critical role of erythropoietin receptor in neurogenesis and post-stroke recovery. The Journal of neuroscience : the official journal of the Society for Neuroscience. 2006;26(4):1269-74.

[452] Noguchi CT, Asavaritikrai P, Teng R, Jia Y. Role of erythropoietin in the brain. Critical reviews in oncology/hematology. 2007;64(2):159-71.

[453] Liu X, Wang Q, Haydar TF, Bordey A. Nonsynaptic GABA signaling in postnatal subventricular zone controls proliferation of GFAP-expressing progenitors. Nature neuroscience. 2005;8(9):1179-87.

[454] Tham M, Ramasamy S, Gan HT, Ramachandran A, Poonepalli A, Yu YH, et al. CSPG Is a Secreted Factor that Stimulates Neural Stem Cell Survival Possibly by Enhanced EGFR Signaling. Plos One. 2010;5(12):e15341.

[455] Jones LL, Yamaguchi Y, Stallcup WB, Tuszynski MH. NG2 is a major chondroitin sulfate proteoglycan produced after spinal cord injury and is expressed by macrophages and oligodendrocyte progenitors. The Journal of neuroscience : the official journal of the Society for Neuroscience. 2002;22(7):2792-803.

[456] Babcock AA, Kuziel WA, Rivest S, Owens T. Chemokine expression by glial cells directs leukocytes to sites of axonal injury in the CNS. The Journal of neuroscience : the official journal of the Society for Neuroscience. 2003;23(21):7922-30.

[457] Owens T, Babcock AA, Millward JM, Toft-Hansen H. Cytokine and chemokine interregulation in the inflamed or injured CNS. Brain research Brain research reviews. 2005;48(2):178-84.

[458] Krathwohl MD, Kaiser JL. Chemokines promote quiescence and survival of human neural progenitor cells. Stem cells. 2004;22(1):109-18.

[459] Imitola J, Raddassi K, Park KI, Mueller FJ, Nieto M, Teng YD, et al. Directed migration of neural stem cells to sites of CNS injury by the stromal cell-derived factor 1alpha/CXC chemokine receptor 4 pathway. Proceedings of the National Academy of Sciences of the United States of America. 2004;101(52):18117-22.

[460] Reiss K, Mentlein R, Sievers J, Hartmann D. Stromal cell-derived factor 1 is secreted by meningeal cells and acts as chemotactic factor on neuronal stem cells of the cerebellar external granular layer. Neuroscience. 2002;115(1):295-305.

[461] Krumbholz M, Theil D, Cepok S, Hemmer B, Kivisakk P, Ransohoff RM, et al. Chemokines in multiple sclerosis: CXCL12 and CXCL13 up-regulation is differentially linked to CNS immune cell recruitment. Brain : a journal of neurology. 2006;129(Pt 1): 200-11.

[462] Hill WD, Hess DC, Martin-Studdard A, Carothers JJ, Zheng J, Hale D, et al. SDF-1 (CXCL12) is upregulated in the ischemic penumbra following stroke: association with bone marrow cell homing to injury. Journal of neuropathology and experimental neurology. 2004;63(1):84-96.

[463] Villeda SA, Luo J, Mosher KI, Zou B, Britschgi M, Bieri G, et al. The ageing systemic milieu negatively regulates neurogenesis and cognitive function. Nature. 2011;477(7362):90-4.

[464] Moriyama M, Fukuhara T, Britschgi M, He Y, Narasimhan R, Villeda S, et al. Complement receptor 2 is expressed in neural progenitor cells and regulates adult hippocampal neurogenesis. The Journal of neuroscience : the official journal of the Society for Neuroscience. 2011;31(11):3981-9.

[465] Aloisi F, Ria F, Adorini L. Regulation of T-cell responses by CNS antigen-presenting cells: different roles for microglia and astrocytes. Immunology today. 2000;21(3): 141-7.

[466] Storch A, Paul G, Csete M, Boehm BO, Carvey PM, Kupsch A, et al. Long-term proliferation and dopaminergic differentiation of human mesencephalic neural precursor cells. Experimental neurology. 2001;170(2):317-25.

[467] Riaz SS, Theofilopoulos S, Jauniaux E, Stern GM, Bradford HF. The differentiation potential of human foetal neuronal progenitor cells in vitro. Brain research Developmental brain research. 2004;153(1):39-51.

[468] Ben-Hur T, Ben-Menachem O, Furer V, Einstein O, Mizrachi-Kol R, Grigoriadis N. Effects of proinflammatory cytokines on the growth, fate, and motility of multipotential neural precursor cells. Molecular and cellular neurosciences. 2003;24(3):623-31.

[469] Pluchino S, Muzio L, Imitola J, Deleidi M, Alfaro-Cervello C, Salani G, et al. Persistent inflammation alters the function of the endogenous brain stem cell compartment. Brain : a journal of neurology. 2008;131(Pt 10):2564-78.

[470] Barnabe-Heider F, Wasylnka JA, Fernandes KJ, Porsche C, Sendtner M, Kaplan DR, et al. Evidence that embryonic neurons regulate the onset of cortical gliogenesis via cardiotrophin-1. Neuron. 2005;48(2):253-65.

[471] Butovsky O, Ziv Y, Schwartz A, Landa G, Talpalar AE, Pluchino S, et al. Microglia activated by IL-4 or IFN-gamma differentially induce neurogenesis and oligodendrogenesis from adult stem/progenitor cells. Molecular and cellular neurosciences. 2006;31(1):149-60.

[472] Guan Y, Jiang Z, Ciric B, Rostami AM, Zhang GX. Upregulation of chemokine receptor expression by IL-10/IL-4 in adult neural stem cells. Experimental and molecular pathology. 2008;85(3):232-6.

[473] Gomez-Nicola D, Valle-Argos B, Pallas-Bazarra N, Nieto-Sampedro M. Interleukin-15 regulates proliferation and self-renewal of adult neural stem cells. Molecular biology of the cell. 2011;22(12):1960-70.

[474] Iosif RE, Ekdahl CT, Ahlenius H, Pronk CJ, Bonde S, Kokaia Z, et al. Tumor necrosis factor receptor 1 is a negative regulator of progenitor proliferation in adult hippocampal neurogenesis. The Journal of neuroscience : the official journal of the Society for Neuroscience. 2006;26(38):9703-12.

[475] Cacci E, Ajmone-Cat MA, Anelli T, Biagioni S, Minghetti L. In vitro neuronal and glial differentiation from embryonic or adult neural precursor cells are differently affected by chronic or acute activation of microglia. Glia. 2008;56(4):412-25.

[476] Williams CA, Lavik EB. Engineering the CNS stem cell microenvironment. Regenerative medicine. 2009;4(6):865-77.

[477] Craig CG, Tropepe V, Morshead CM, Reynolds BA, Weiss S, van der Kooy D. In vivo growth factor expansion of endogenous subependymal neural precursor cell populations in the adult mouse brain. The Journal of neuroscience : the official journal of the Society for Neuroscience. 1996;16(8):2649-58.

[478] Kaneko N, Kako E, Sawamoto K. Enhancement of ventricular-subventricular zone-derived neurogenesis and oligodendrogenesis by erythropoietin and its derivatives. Frontiers in cellular neuroscience. 2013;7:235.

[479] Yoshimura S, Takagi Y, Harada J, Teramoto T, Thomas SS, Waeber C, et al. FGF-2 regulation of neurogenesis in adult hippocampus after brain injury. Proceedings of the National Academy of Sciences of the United States of America. 2001;98(10): 5874-9.

[480] Teramoto T, Qiu J, Plumier JC, Moskowitz MA. EGF amplifies the replacement of parvalbumin-expressing striatal interneurons after ischemia. The Journal of clinical investigation. 2003;111(8):1125-32.

[481] Pencea V, Bingaman KD, Wiegand SJ, Luskin MB. Infusion of brain-derived neurotrophic factor into the lateral ventricle of the adult rat leads to new neurons in the parenchyma of the striatum, septum, thalamus, and hypothalamus. The Journal of neuroscience : the official journal of the Society for Neuroscience. 2001;21(17): 6706-17.

[482] Benraiss A, Chmielnicki E, Lerner K, Roh D, Goldman SA. Adenoviral brain-derived neurotrophic factor induces both neostriatal and olfactory neuronal recruitment from endogenous progenitor cells in the adult forebrain. The Journal of neuroscience : the official journal of the Society for Neuroscience. 2001;21(17):6718-31.

[483] Zhang ZG, Zhang L, Jiang Q, Zhang R, Davies K, Powers C, et al. VEGF enhances angiogenesis and promotes blood-brain barrier leakage in the ischemic brain. The Journal of clinical investigation. 2000;106(7):829-38.

[484] Surget A, Tanti A, Leonardo ED, Laugeray A, Rainer Q, Touma C, et al. Antidepressants recruit new neurons to improve stress response regulation. Molecular psychiatry. 2011;16(12):1177-88.

[485] Malberg JE, Eisch AJ, Nestler EJ, Duman RS. Chronic antidepressant treatment increases neurogenesis in adult rat hippocampus. The Journal of neuroscience : the official journal of the Society for Neuroscience. 2000;20(24):9104-10.

[486] Wang HD, Dunnavant FD, Jarman T, Deutch AY. Effects of antipsychotic drugs on neurogenesis in the forebrain of the adult rat. Neuropsychopharmacology : official

publication of the American College of Neuropsychopharmacology. 2004;29(7):1230-8.

[487] Bessa JM, Ferreira D, Melo I, Marques F, Cerqueira JJ, Palha JA, et al. The mood-improving actions of antidepressants do not depend on neurogenesis but are associated with neuronal remodeling. Molecular psychiatry. 2009;14(8):764-73, 39.

[488] Halim ND, Weickert CS, McClintock BW, Weinberger DR, Lipska BK. Effects of chronic haloperidol and clozapine treatment on neurogenesis in the adult rat hippocampus. Neuropsychopharmacology : official publication of the American College of Neuropsychopharmacology. 2004;29(6):1063-9.

[489] Selemon LD, Lidow MS, Goldman-Rakic PS. Increased volume and glial density in primate prefrontal cortex associated with chronic antipsychotic drug exposure. Biological psychiatry. 1999;46(2):161-72.

[490] Green W, Patil P, Marsden CA, Bennett GW, Wigmore PM. Treatment with olanzapine increases cell proliferation in the subventricular zone and prefrontal cortex. Brain research. 2006;1070(1):242-5.

[491] Wang DD, Krueger DD, Bordey A. GABA depolarizes neuronal progenitors of the postnatal subventricular zone via GABAA receptor activation. The Journal of physiology. 2003;550(Pt 3):785-800.

[492] Nguyen L, Malgrange B, Breuskin I, Bettendorff L, Moonen G, Belachew S, et al. Autocrine/paracrine activation of the GABA(A) receptor inhibits the proliferation of neurogenic polysialylated neural cell adhesion molecule-positive (PSA-NCAM+) precursor cells from postnatal striatum. The Journal of neuroscience : the official journal of the Society for Neuroscience. 2003;23(8):3278-94.

[493] Bolteus AJ, Bordey A. GABA release and uptake regulate neuronal precursor migration in the postnatal subventricular zone. The Journal of neuroscience : the official journal of the Society for Neuroscience. 2004;24(35):7623-31.

[494] Wu X, Castren E. Co-treatment with diazepam prevents the effects of fluoxetine on the proliferation and survival of hippocampal dentate granule cells. Biological psychiatry. 2009;66(1):5-8.

[495] Petrus DS, Fabel K, Kronenberg G, Winter C, Steiner B, Kempermann G. NMDA and benzodiazepine receptors have synergistic and antagonistic effects on precursor cells in adult hippocampal neurogenesis. The European journal of neuroscience. 2009;29(2):244-52.

[496] Cossetti C, Alfaro-Cervello C, Donega M, Tyzack G, Pluchino S. New perspectives of tissue remodelling with neural stem and progenitor cell-based therapies. Cell and tissue research. 2012;349(1):321-9.

[497] Marchetti B, Pluchino S. Wnt your brain be inflamed? Yes, it Wnt! Trends in molecular medicine. 2013;19(3):144-56.

[498] Uccelli A, Moretta L, Pistoia V. Mesenchymal stem cells in health and disease. Nature reviews Immunology. 2008;8(9):726-36.

[499] Chu K, Kim M, Park KI, Jeong SW, Park HK, Jung KH, et al. Human neural stem cells improve sensorimotor deficits in the adult rat brain with experimental focal ischemia. Brain research. 2004;1016(2):145-53.

[500] Ziv Y, Avidan H, Pluchino S, Martino G, Schwartz M. Synergy between immune cells and adult neural stem/progenitor cells promotes functional recovery from spinal cord injury. Proceedings of the National Academy of Sciences of the United States of America. 2006;103(35):13174-9.

[501] Gerdoni E, Gallo B, Casazza S, Musio S, Bonanni I, Pedemonte E, et al. Mesenchymal stem cells effectively modulate pathogenic immune response in experimental autoimmune encephalomyelitis. Annals of neurology. 2007;61(3):219-27.

[502] Bacigaluppi M, Pluchino S, Peruzzotti-Jametti L, Kilic E, Kilic U, Salani G, et al. Delayed post-ischaemic neuroprotection following systemic neural stem cell transplantation involves multiple mechanisms. Brain : a journal of neurology. 2009;132(Pt 8): 2239-51.

[503] Pluchino S, Zanotti L, Rossi B, Brambilla E, Ottoboni L, Salani G, et al. Neurosphere-derived multipotent precursors promote neuroprotection by an immunomodulatory mechanism. Nature. 2005;436(7048):266-71.

[504] Calabrese C, Poppleton H, Kocak M, Hogg TL, Fuller C, Hamner B, et al. A perivascular niche for brain tumor stem cells. Cancer cell. 2007;11(1):69-82.

[505] Pluchino S, Gritti A, Blezer E, Amadio S, Brambilla E, Borsellino G, et al. Human neural stem cells ameliorate autoimmune encephalomyelitis in non-human primates. Annals of neurology. 2009;66(3):343-54.

[506] Pluchino S, Zanotti L, Brambilla E, Rovere-Querini P, Capobianco A, Alfaro-Cervello C, et al. Immune regulatory neural stem/precursor cells protect from central nervous system autoimmunity by restraining dendritic cell function. Plos One. 2009;4(6):e5959.

[507] Pluchino S, Martino G. The therapeutic plasticity of neural stem/precursor cells in multiple sclerosis. J Neurol Sci. 2008;265(1-2):105-10.

[508] Jeong SW, Chu K, Jung KH, Kim SU, Kim M, Roh JK. Human neural stem cell transplantation promotes functional recovery in rats with experimental intracerebral hemorrhage. Stroke; a journal of cerebral circulation. 2003;34(9):2258-63.

[509] Fujiwara Y, Tanaka N, Ishida O, Fujimoto Y, Murakami T, Kajihara H, et al. Intravenously injected neural progenitor cells of transgenic rats can migrate to the injured spinal cord and differentiate into neurons, astrocytes and oligodendrocytes. Neuroscience letters. 2004;366(3):287-91.

[510] Muller FJ, Snyder EY, Loring JF. Gene therapy: can neural stem cells deliver? Nat Rev Neurosci. 2006;7(1):75-84.

[511] Soares S, Sotelo C. Adult neural stem cells from the mouse subventricular zone are limited in migratory ability compared to progenitor cells of similar origin. Neuroscience. 2004;128(4):807-17.

[512] Doetsch F, Petreanu L, Caille I, Garcia-Verdugo JM, Alvarez-Buylla A. EGF converts transit-amplifying neurogenic precursors in the adult brain into multipotent stem cells. Neuron. 2002;36(6):1021-34.

[513] Maslov AY, Barone TA, Plunkett RJ, Pruitt SC. Neural stem cell detection, characterization, and age-related changes in the subventricular zone of mice. The Journal of neuroscience : the official journal of the Society for Neuroscience. 2004;24(7):1726-33.

[514] Mi R, Luo Y, Cai J, Limke TL, Rao MS, Hoke A. Immortalized neural stem cells differ from nonimmortalized cortical neurospheres and cerebellar granule cell progenitors. Experimental neurology. 2005;194(2):301-19.

[515] Pluchino S, Quattrini A, Brambilla E, Gritti A, Salani G, Dina G, et al. Injection of adult neurospheres induces recovery in a chronic model of multiple sclerosis. Nature. 2003;422(6933):688-94.

[516] Mueller FJ, Serobyan N, Schraufstatter IU, DiScipio R, Wakeman D, Loring JF, et al. Adhesive interactions between human neural stem cells and inflamed human vascular endothelium are mediated by integrins. Stem cells. 2006;24(11):2367-72.

[517] Darsalia V, Kallur T, Kokaia Z. Survival, migration and neuronal differentiation of human fetal striatal and cortical neural stem cells grafted in stroke-damaged rat striatum. The European journal of neuroscience. 2007;26(3):605-14.

[518] Lie DC, Song H, Colamarino SA, Ming GL, Gage FH. Neurogenesis in the adult brain: new strategies for central nervous system diseases. Annual review of pharmacology and toxicology. 2004;44:399-421.

[519] Emsley JG, Mitchell BD, Kempermann G, Macklis JD. Adult neurogenesis and repair of the adult CNS with neural progenitors, precursors, and stem cells. Progress in neurobiology. 2005;75(5):321-41.

[520] Benedetti S, Pirola B, Pollo B, Magrassi L, Bruzzone MG, Rigamonti D, et al. Gene therapy of experimental brain tumors using neural progenitor cells. Nature medicine. 2000;6(4):447-50.

[521] Sun C, Zhang H, Li J, Huang H, Cheng H, Wang Y, et al. Modulation of the major histocompatibility complex by neural stem cell-derived neurotrophic factors used for regenerative therapy in a rat model of stroke. Journal of translational medicine. 2010;8:77.

[522] Cusimano M, Biziato D, Brambilla E, Donega M, Alfaro-Cervello C, Snider S, et al. Transplanted neural stem/precursor cells instruct phagocytes and reduce secondary

tissue damage in the injured spinal cord. Brain : a journal of neurology. 2012;135(Pt 2):447-60.

[523] Cao QL, Howard RM, Dennison JB, Whittemore SR. Differentiation of engrafted neuronal-restricted precursor cells is inhibited in the traumatically injured spinal cord. Experimental neurology. 2002;177(2):349-59.

[524] Ben-Hur T. Immunomodulation by neural stem cells. J Neurol Sci. 2008;265(1-2): 102-4.

[525] Lee HJ, Kim KS, Kim EJ, Choi HB, Lee KH, Park IH, et al. Brain transplantation of immortalized human neural stem cells promotes functional recovery in mouse intracerebral hemorrhage stroke model. Stem cells. 2007;25(5):1204-12.

[526] Lee HJ, Kim KS, Park IH, Kim SU. Human neural stem cells over-expressing VEGF provide neuroprotection, angiogenesis and functional recovery in mouse stroke model. Plos One. 2007;2(1):e156.

[527] Lindvall O, Kokaia Z. Stem cells in human neurodegenerative disorders--time for clinical translation? The Journal of clinical investigation. 2010;120(1):29-40.

[528] Einstein O, Friedman-Levi Y, Grigoriadis N, Ben-Hur T. Transplanted neural precursors enhance host brain-derived myelin regeneration. The Journal of neuroscience : the official journal of the Society for Neuroscience. 2009;29(50):15694-702.

[529] Capone C, Frigerio S, Fumagalli S, Gelati M, Principato MC, Storini C, et al. Neurosphere-derived cells exert a neuroprotective action by changing the ischemic microenvironment. Plos One. 2007;2(4):e373.

[530] Daadi MM, Davis AS, Arac A, Li Z, Maag AL, Bhatnagar R, et al. Human neural stem cell grafts modify microglial response and enhance axonal sprouting in neonatal hypoxic-ischemic brain injury. Stroke; a journal of cerebral circulation. 2010;41(3): 516-23.

[531] Lalancette-Hebert M, Gowing G, Simard A, Weng YC, Kriz J. Selective ablation of proliferating microglial cells exacerbates ischemic injury in the brain. The Journal of neuroscience : the official journal of the Society for Neuroscience. 2007;27(10): 2596-605.

[532] Lu P, Jones LL, Snyder EY, Tuszynski MH. Neural stem cells constitutively secrete neurotrophic factors and promote extensive host axonal growth after spinal cord injury. Experimental neurology. 2003;181(2):115-29.

[533] Einstein O, Karussis D, Grigoriadis N, Mizrachi-Kol R, Reinhartz E, Abramsky O, et al. Intraventricular transplantation of neural precursor cell spheres attenuates acute experimental allergic encephalomyelitis. Molecular and cellular neurosciences. 2003;24(4):1074-82.

[534] Fainstein N, Vaknin I, Einstein O, Zisman P, Ben Sasson SZ, Baniyash M, et al. Neural precursor cells inhibit multiple inflammatory signals. Molecular and cellular neurosciences. 2008;39(3):335-41.

[535] Knight JC, Scharf EL, Mao-Draayer Y. Fas activation increases neural progenitor cell survival. Journal of neuroscience research. 2010;88(4):746-57.

[536] Wang L, Shi J, van Ginkel FW, Lan L, Niemeyer G, Martin DR, et al. Neural stem/progenitor cells modulate immune responses by suppressing T lymphocytes with nitric oxide and prostaglandin E2. Experimental neurology. 2009;216(1):177-83.

[537] Kim HM, Hwang DH, Lee JE, Kim SU, Kim BG. Ex vivo VEGF delivery by neural stem cells enhances proliferation of glial progenitors, angiogenesis, and tissue sparing after spinal cord injury. Plos One. 2009;4(3):e4987.

[538] Kim SY, Cho HS, Yang SH, Shin JY, Kim JS, Lee ST, et al. Soluble mediators from human neural stem cells play a critical role in suppression of T-cell activation and proliferation. Journal of neuroscience research. 2009;87(10):2264-72.

[539] Ricci-Vitiani L, Lombardi DG, Signore M, Biffoni M, Pallini R, Parati E, et al. Human neural progenitor cells display limited cytotoxicity and increased oligodendrogenesis during inflammation. Cell death and differentiation. 2007;14(4):876-8.

[540] Lee ST, Chu K, Jung KH, Kim SJ, Kim DH, Kang KM, et al. Anti-inflammatory mechanism of intravascular neural stem cell transplantation in haemorrhagic stroke. Brain : a journal of neurology. 2008;131(Pt 3):616-29.

[541] Einstein O, Fainstein N, Vaknin I, Mizrachi-Kol R, Reihartz E, Grigoriadis N, et al. Neural precursors attenuate autoimmune encephalomyelitis by peripheral immunosuppression. Annals of neurology. 2007;61(3):209-18.

[542] Yoo J, Kim HS, Hwang DY. Stem cells as promising therapeutic options for neurological disorders. Journal of cellular biochemistry. 2013;114(4):743-53.

[543] Tsuji O, Miura K, Fujiyoshi K, Momoshima S, Nakamura M, Okano H. Cell therapy for spinal cord injury by neural stem/progenitor cells derived from iPS/ES cells. Neurotherapeutics : the journal of the American Society for Experimental NeuroTherapeutics. 2011;8(4):668-76.

[544] Ben-Hur T. Cell therapy for multiple sclerosis. Neurotherapeutics : the journal of the American Society for Experimental NeuroTherapeutics. 2011;8(4):625-42.

[545] Willerth SM. Neural tissue engineering using embryonic and induced pluripotent stem cells. Stem cell research & therapy. 2011;2(2):17.

[546] Ellis-Behnke RG, Liang YX, You SW, Tay DK, Zhang S, So KF, et al. Nano neuro knitting: peptide nanofiber scaffold for brain repair and axon regeneration with functional return of vision. Proceedings of the National Academy of Sciences of the United States of America. 2006;103(13):5054-9.

[547] Guo J, Su H, Zeng Y, Liang YX, Wong WM, Ellis-Behnke RG, et al. Reknitting the injured spinal cord by self-assembling peptide nanofiber scaffold. Nanomedicine : nanotechnology, biology, and medicine. 2007;3(4):311-21.

[548] Christopherson GT, Song H, Mao HQ. The influence of fiber diameter of electrospun substrates on neural stem cell differentiation and proliferation. Biomaterials. 2009;30(4):556-64.

[549] Discher DE, Janmey P, Wang YL. Tissue cells feel and respond to the stiffness of their substrate. Science. 2005;310(5751):1139-43.

[550] Hynes SR, Rauch MF, Bertram JP, Lavik EB. A library of tunable poly(ethylene glycol)/poly(L-lysine) hydrogels to investigate the material cues that influence neural stem cell differentiation. Journal of biomedical materials research Part A. 2009;89(2): 499-509.

[551] Soen Y, Mori A, Palmer TD, Brown PO. Exploring the regulation of human neural precursor cell differentiation using arrays of signaling microenvironments. Molecular systems biology. 2006;2:37.

[552] Silva GA, Czeisler C, Niece KL, Beniash E, Harrington DA, Kessler JA, et al. Selective differentiation of neural progenitor cells by high-epitope density nanofibers. Science. 2004;303(5662):1352-5.

[553] Tysseling-Mattiace VM, Sahni V, Niece KL, Birch D, Czeisler C, Fehlings MG, et al. Self-assembling nanofibers inhibit glial scar formation and promote axon elongation after spinal cord injury. The Journal of neuroscience : the official journal of the Society for Neuroscience. 2008;28(14):3814-23.

[554] Saha K, Irwin EF, Kozhukh J, Schaffer DV, Healy KE. Biomimetic interfacial interpenetrating polymer networks control neural stem cell behavior. Journal of biomedical materials research Part A. 2007;81(1):240-9.

[555] Nakajima M, Ishimuro T, Kato K, Ko IK, Hirata I, Arima Y, et al. Combinatorial protein display for the cell-based screening of biomaterials that direct neural stem cell differentiation. Biomaterials. 2007;28(6):1048-60.

Section 5

Stem Cell Niches and Cancer

Chapter 9

The Role of the "Cancer Stem Cell Niche" in Cancer Initiation and Progression

Jacqueline E. Noll, Kate Vandyke and
Andrew C.W. Zannettino

Additional information is available at the end of the chapter

http://dx.doi.org/10.5772/58598

1. Introduction

1.1. Cancer stem cells: An introduction

Adult stem cells, also known as progenitor cells, have two major ascribed functions: (1) to replenish tissues throughout normal growth and development and (2) to repair tissues following damage by disease or injury. Classically, a stem cell is defined as possessing the capacity for self-renewal and potency. Self-renewal requires a stem cell to be able to divide in such a way as to maintain the pool of stem cells in an undifferentiated state, while potency (commonly referred to as pluri- or multi-potency) requires a stem cell to retain the capacity to differentiate into an array of specialized cell types. Cancer stem cells (CSC) represent a subset of tumour cells that exhibit the same properties as normal adult stem cells; namely the ability to self-renew, undergo asymmetric cell division and differentiate into a diverse range of cell types. In addition, the CSC population has the capacity to initiate tumours and has also been implicated in metastatic spread and in resistance to conventional anti-cancer therapies.

The majority of cells within a tumour are unable to sustain tumour growth and cannot initiate tumour establishment at secondary locations. The small population of cells that are inherently tumourigenic and commonly have a metastatic phenotype are referred to as CSC. CSC were initially identified and characterised in acute myeloid leukaemia (AML) [1]. In this study, a minority of the total tumour cell population (0.01-1%) was identified as having the unique capacity to induce leukemia following transplantation into immunodeficient mice implying that these cells alone had a tumour-initiating capacity. Following from this study, the ability of cells to cause tumour development in immune-compromised animals has been used both as proof-of-principle for the existence of CSC and a means of identifying CSC or tumour-initiating cells. Since the initial discovery of CSC, our knowledge and understanding of this area has increased exponentially, with CSC identified in an array of solid tumours beginning with breast cancer [2] and glioma [3, 4] and subsequently extending to prostate [5], pancreatic [6], melanoma [7], liver [8] and head and neck [9] cancers, among others.

As discussed throughout this chapter, the reliance of the CSC on their specific niche is a current area of research focus. Importantly, as our understanding of the signalling pathways and specific interactions that modulate and maintain CSC function (and subsequently tumour initiation and development) increases, we will become better placed to formulate novel treatment modalities that will not only target the tumour cells for destruction but also focus on abolishing integral factors within the supportive CSC niche. Novel treatments such as these will aim to eradicate residual CSC, resulting in a decrease in tumour recurrence following therapy and better overall prognosis for cancer patients.

2. CSC markers

The isolation of CSC from total tumour cell populations has been made possible due to the identification of CSC-specific markers. The use of these markers to classify and identify tumour-initiating cells and circulating tumour cells has allowed the investigation of the importance of CSC in the developing tumour, particularly in metastasis, drug resistance and patient prognosis. However, due to the similarities in cell surface phenotype and marker expression between CSC and normal adult stem cells, further research is needed to aid in our ability to distinguish between malignant stem cells and those required for normal tissue regeneration processes. Ideally, in the future, CSC-specific markers will be used to therapeutically target CSC for eradication. Notably, in-roads have already been made in this area.

CD44 has been identified as a CSC marker and its expression has been associated with high levels of metastasis, tumour recurrence and poor outcome in breast cancer, all factors associated with CSC sub-populations within tumours. Although it has been well characterised, there is some controversy regarding the specificity of CD44 to CSC as the full-length CD44 protein is widely expressed. However, recent studies have identified CSC-specific expression of a particular splice variant of CD44 [10, 11]. Aldehyde dehydrogenase isoform 1 (ALDH1) is another commonly used marker of CSC from a range of cancer types [12, 13], however similarly to CD44, ALDH1 expression is also associated with normal haematopoietic stem cells and therefore can be used as a marker of both normal and malignant stem cells [14]. A range of other markers have also been identified, with some CSC markers, including CD133, CD44 and ALDH1, remaining consistent across a number of tumour types. These are summarised in Table 1.

Tumour Type	CSC Phenotypic Markers	Selected References
Breast	$CD44^{+} CD21^{-/low} Lineage^{-} ALDH1^{+} CD133^{+} \alpha_6$-integrin	[2, 13, 15, 16]
AML	$CD34^{+} CD38^{-}$	[1]
Liver	$CD133^{+} CD49f^{+} CD90^{+} CD44^{+}$	[8, 17, 18]
Lung	$CD133^{+}$ $CD90^{+}$ $ALDH1^{+}$	[19-21]
Glioma	$CD133^{+}$ $Nestin^{+}$ $CD90^{+}$ α_6-integrin	[3, 4, 22-24]
Colon	$CD133^{+}$ $CD44^{+}$ $CD24^{+}$	[25, 26]
Prostate	$CD133^{+}CD44^{+}$ $\alpha 2\beta 1^{high}$ $ALDH1^{+}$	[5, 27]
Pancreatic	$CD44^{+} CD24^{+}$ $CD133^{+}$ $ALDH1^{+}$ $EpCAM^{+}$ Nestin $^{+}ABCG2^{high}$	[6, 28-30]
Melanoma	$CD20^{+}$ $CD166^{+} CD133^{+} Nestin^{+}$ $CD271^{+}$	[7, 31, 32]
Head and Neck	$CD44^{+}$ $CD133^{+} ALDH^{+}$	[9, 33]

Table 1. Representative cell surface phenotypic markers for human CSCs.

3. CSC niche components and function

The function and maintenance of stem cells is highly reliant on the specific anatomical and physiological location in which these cells reside. This highly specialised microenvironment in which stem cells are located is commonly referred to as the stem cell niche. The niche is made up of stromal support cells, soluble factors, blood vessels and extracellular matrix proteins that have both direct and indirect effects on stem cell number, proliferation, self-renewal and fate determination. The niche is essential for the maintaining the balance between pro- and anti-proliferative signals to ensure a controlled stem cell environment. In fact, stem cells outside of their defined niche are reported to have limited function as they are highly reliant on cell-cell interactions and signals within their local microenvironment for their basic proliferative and tissue renewal properties [34]. Importantly, in the context of CSC, the niche plays an important role in maintenance of stem cell function, tumour initiation and protection against chemotherapeutic agents. The reliance of CSC on the niche is becoming increasingly evident as studies have shown that while CSC-like cells can be isolated from cancer cell lines *in vitro*, these cell populations are difficult to maintain in an *in vitro* setting [35, 36]. In contrast, xenograft models have proven to be effective in faithfully recapitulating the characteristics of the CSC as found in the original tumour [37]. This is likely due to the ability of the CSC to grow within a supportive tumour microenvironment, or niche, and as such respond to cellular interactions and local signalling pathways.

The CSC niche may be derived through one of two mechanisms; either the CSC specifically manufacture the niche through the production of numerous factors that signal between stromal cells of the local microenvironment and the CSC themselves or, conversely, the CSC utilise the pre-existing, tissue-specific stem cell niche. The latter option results in the CSC "hijacking" the niche that would normally modulate the normal growth and development of local stem cells. It is clear that in both normal stem cell and CSC development and maintenance, there is a mutual dependence between the stem cells and their niche.

As discussed in detail throughout this chapter, both the CSC themselves and the stromal cells of the CSC niche secrete a variety of factors and signalling molecules to modulate pathways that are normally involved in growth and development. Together, these factors function to maintain CSC, support the maintenance of the CSC niche and promote tumour development. In addition, physiological consequences of tumour formation, such as the establishment of a hypoxic microenvironment and the subsequent induction of angiogenesis and neovascularisation, can result in a positive feed-forward effect on CSC. Furthermore, not only do CSC rely on factors expressed within the niche to ensure the maintenance of the CSC population, but the CSC themselves can regulate the composition and function of the niche [38, 39]. In this manner, the CSC and the niche function together to create a positive signalling loop to maintain a state conducive to the growth and further development of tumours.

4. CSC and metastasis

The metastatic potential of tumour cells relies not only on genetic alterations and expression of factors from the tumour cell itself, but also on interactions with structural, soluble and cellular components of the extracellular matrix (ECM) and stromal tissue compartment. These signals from the stromal microenvironment are required for the formation of what is termed the "pre-metastatic niche" and similarly to the CSC niche at the site of primary tumour establishment, this niche is a highly specialised microenvironment that aids in the migration, homing and colonisation of tumour cells, as well as subsequently enhancing tumour cell proliferation and disease development.

4.1. Epithelial-mesenchymal transition

Epithelial-mesenchymal transition (EMT) was originally described as an important process in early embryonic development, and has also been described as a significant feature in a number of cancers. Molecular markers of EMT include an increase in the expression of N-cadherin and vimentin coupled with decreased expression of E-cadherin, increased accumulation of β-catenin within the nucleus, increased secretion of matrix metalloproteinases (MMPs) and increased expression and activity of a number of transcription factors including SLUG, SNAIL and TWIST. Cells that have undergone an EMT exhibit a loss of epithelial cell polarity and intracellular adhesion accompanied by reorganisation of the cytoskeleton, which together results in an increased capacity for migration, invasion and cell scattering (reviewed in [40]). EMT in tumourigenesis has been described in detail in recent years with a large number of studies supporting a role for EMT in tumour progression, particularly the process of metastasis. In addition, the EMT process has been closely associated with the formation, and subsequent behaviour, of CSC populations. A number of different stimuli regulate and drive EMT both in development and disease. These include: signalling through specific pathways (e.g. TGFβ, Wnt, Notch and Hedgehog), direct cellular interactions and exposure to hypoxic conditions – all factors that also play key roles in the maintenance of CSC (as discussed throughout this chapter).

The process of EMT and the maintenance of CSC are closely linked, with the induction of EMT enriching for a CSC-like population. This has been observed by multiple means, including the observation that metastatic cells exhibit both an EMT- and CSC-like phenotype. Experimental evidence has demonstrated that circulating tumour cells in both prostate and breast cancers express both CSC and mesenchymal markers [41, 42]. In addition, induction of EMT in human mammary epithelial cells results not only in the acquisition of mesenchymal markers, but also the expression of a CSC phenotype, for example $CD44^{+}/CD24^{-/low}$ in breast cancer stem cells. The opposite is also seen to be true, with CSC-like cells expressing markers similar to those found on cells that have undergone an EMT, including decreased expression of E-cadherin and increased expression of N-cadherin, vimentin and Twist [43]. Notably, within a non-CSC population, the activation of EMT can in fact revert cells back to a CSC-like state, suggesting that this transition is a key player in modulating CSC function. It is likely, therefore, that the CSC population does, in fact, make up a large proportion of the tumour cells that have undergone EMT and as such can be found in the peripheral circulation and therefore play an integral role in the establishment of secondary/metastatic tumours. This has been demonstrated specifically in prostate cancer, with the vast majority of circulating tumour cells (those likely to be responsible for colonisation of distant metastatic sites) expressing both the stem cell marker CD133, as well as a range of mesenchymal-associated proteins [42].

4.2. Migration and invasion

Recent studies suggest that the CSC population within tumours, although comprising a very small percentage of the total tumour, are critical for metastatic colonisation and have been shown to initiate tumour growth at secondary sites [44, 45]. The ability of cells to migrate and invade is closely linked with their intrinsic ability to metastasise and colonise secondary tumour locations. The mechanisms utilised by CSC to modulate their invasive capacity are similar to those employed by normal stem cells for homing and/or mobilisation. Within the haematopoietic stem cell (HSC) niche, MMPs play a role in HSC mobilisation through the proteolysis of ECM components [46]. In addition, stem cell migration within both the haematopoietic and neural systems displays a reliance on integrins [47, 48]. Specific roles for MMPs and integrins in cancer cell metastasis have also been described [49, 50]. The CXCL12/CXCR4 axis (described in detail below) is also a key player in both normal and cancer stem cell migration and invasion [51, 52].

There have been a number of recent discoveries that support a direct link between CSC and the development of metastasis in cancer, as common pathways (including EMT) have been identified as integral for both regulation of CSC and driving metastasis [53]. Furthermore, CSCs have been demonstrated to be directly involved in metastasis in xenograft models of cancer [54, 55], due largely to the enhanced invasive capacity of CSC. This can be attributed, at least in part, to increased MMP secretion by CSC [56]. Furthermore, comparison of the genetic profiles of breast CSCs and normal breast epithelium has identified an "invasive" gene signature associated with the CSC population, which has been demonstrated to correlate with a metastatic phenotype in breast cancer patients [53, 57].

5. Extracellular matrix components of the CSC niche

The ECM has defined roles in cellular proliferation, differentiation and migration as well as tumour angiogenesis and protection from chemotherapy [58, 59] and represents an important component of the CSC niche. Receptors expressed within the ECM allow stem cells to anchor to specific locations within the niche and therefore maintain signalling and contact with other cells residing within the niche [58]. Loss of ECM contact with resident stem cells, through reducing and/or inhibiting the ECM glycoprotein components, results in a decrease in stem cell number [60, 61]. In addition, the ECM is also able to influence the behaviour of stromal cells within the niche, including endothelial cells, immune cells and fibroblasts [58]. Increased expression of specific ECM glycoproteins has been associated with various tumour types, including breast and pancreatic cancers [62-66], which in turn have been associated with poor prognosis [67]. The composition of the ECM is therefore critical in maintaining tissue and cellular homeostasis and in modulating the growth and proliferation of stem cells. Importantly, abnormal ECM dynamics compromise the role of the ECM as a physical barrier to tumour cell invasion, allowing for cellular migration and invasion. Modulation of the ECM, most commonly through enzymes such as MMPs that degrade the proteins making up the ECM, enhances cellular migration and subsequent colonisation of the pre-metastatic niche by CSC. High expression of Tenascin-C, a glycoprotein expressed within the ECM, has been linked to metastasis in a range of different cancer types [68-70]. Another component of the ECM, periostin (POSTN), is expressed by the stroma of primary tumours. Infiltrating tumour cells (i.e. cells undergoing metastasis) induce POSTN expression within the secondary target organ to initiate colonisation, a function that can be successfully inhibited by blocking POSTN [71]. The processes of cellular migration and invasion, and hence metastasis, therefore require significant modification of the ECM.

6. Stromal cells within the CSC niche

The CSC niche is composed of a range of specific cell types including fibroblasts, endothelial cells, mesenchymal stem cells (MSC) and immune cells. The "stemness" of CSC is in fact a dynamic quality that can be mediated by extrinsic cues derived from the local microenvironment. Direct cell-cell interactions between the stromal cell compartment and CSC, as well as signalling pathways mediated through the expression and secretion of a range of growth factors and cytokines play a role in the maintenance of the CSC population within the niche and in overall tumour growth.

http://dx.doi.org/10.5772/58598

6.1. Fibroblasts

The major cellular component of the tumour microenvironment is the stromal fibroblasts, commonly referred to as carcinoma-associated fibroblasts (CAFs). Factors that are secreted by fibroblasts within the tumour environment are able to revert differentiated tumour cells to a CSC-like phenotype, thereby maintaining the CSC population. CAFs have specifically been shown to alter the function of CSC through the expression of elevated levels of chemokine C-X-C motif ligand 12 (CXCL12) and MMP-1. The expression of CXCL12 by stromal fibroblasts within the tumour microenvironment adds to the supportive role of the niche, playing a role in the promotion of EMT in primary tumours ([72]; discussed in detail below). Coupled with the expression of MMPs, CAFs may directly stimulate the migration of CSC and therefore play a role in driving metastasis.

Tumour cells located in close proximity to stromal fibroblasts also exhibit increased expression of Wnt pathway regulated genes (as discussed below). The association between the tumour cells and their stromal environment is critical in modulating pathways and gene expression, both from the stromal cells (to create a permissive environment for CSC) and by the tumour cells (to further tumour development and CSC maintenance).

6.2. Immune cells

Tumour-associated macrophages (TAMs) represent the major immune component of stromal cells in tumour microenvironments. TAMs are identified as M2-type macrophages and as such express a range of factors that regulate matrix components resulting in remodelling of the ECM, activation of angiogenic pathways, suppression of adaptive immunity and enhancement of tumour cell proliferation and survival. Together, these functions cooperate to promote tumour development (Reviewed in [73]). TAMs are recruited to sites of tumour formation due to the expression of chemokines by the tumour cells. It has recently been suggested that the CSC population plays a significant role in the recruitment and modulation of TAM within the CSC niche. TAM recruitment factors, including soluble colony-stimulating factor 1 (sCSF-1), transforming growth factor beta (TGFβ) and macrophage inhibitory cytokine 1 (MIC-1), are specifically expressed by tumour-derived CSC [74]. In glioma, the presence of TGFβ-expressing TAM at the invasive tumour front has been correlated with the presence of glioma CSC and these CSC are shown to have a greater invasive potential in the presence of the TAMs [75]. In addition, CSC are able to promote the expression of IL-6 by TAMs [76], hence further enhancing the pro-growth signalling for CSC within the local environment. The specific depletion of TAMs in pancreatic cancers results in a concomitant decrease in the number of CSC [77]. It is apparent that the immune cell component of the CSC niche is important in the regulation of CSC number and activity in tumours.

6.3. Endothelial cells

The CSC population is commonly located in what has been termed a "perivascular" niche [78-80], which can best be defined by its anatomical/physical location directly adjacent to the endothelial cells comprising the blood vessels. Indeed a number of studies have shown that the removal of endothelial cells from the perivascular niche results in a reduction in CSC number, suggesting an absolute reliance on the cellular composition of the niche for maintenance and survival of CSC. Vascular endothelial cells interact directly with CSC and secrete a range of factors that influence both tumour growth and CSC maintenance. A number of studies have identified a range of secretory molecules that are produced by endothelial cells and subsequently support the growth, self-renewal and migratory capacity of CSC, including Bmi-1, Jagged-1, vascular endothelial growth factor (VEGF), interleukin 6 (IL-6), interleukin 8 (IL-8, also known as CXCL8) and epidermal growth factor (EGF) [78, 80-84].

6.4. Mesenchymal stem cells

Mesenchymal stem cells (MSC) are multi-potent stromal cells with the capacity to differentiate into osteoblasts, adipocytes and chondrocytes to make up bone, fat and cartilage tissues respectively. MSC are most commonly found within the bone marrow compartment; although, they have also been shown to reside in other tissue types. MSC have been implicated with a role in tumour progression. This has been described in haematological cancers, with an increase in MSC within the bone marrow being associated with myeloma disease [85]. In addition, a role for MSC in tumour development has been described in a number of solid cancers, including breast [86, 87], colon [88], lung [89] and prostate [90]. MSC injected into tumour-bearing mice exhibit preferential homing to the site of cancer formation and the presence of MSC within the local tumour environment was shown to accelerate tumour growth [86, 87]. The presence of MSC within the local tumour microenvironment was specifically shown to promote an increase in the CSC population, which is likely due to the expression of pro-proliferative factors by MSC, including CXCL12, IL-8 and IL-6 [87, 91-93]. These studies suggest that signalling between resident CSC and the MSC located within the niche results in enhanced proliferative and self-renewal signals.

7. Angiogenesis and hypoxia

Angiogenesis is a key player in the development of tumours and is defined by the proliferation of blood vessels within and around areas of established tumour, and aids in the supply of nutrients and oxygen to the malignant cells. In addition, as discussed above, vascular endothelial cells play an important role in maintaining the CSC niche. Exposure to hypoxic conditions, as are present in many solid tumours, results in the stabilisation of hypoxia inducible transcription factors (HIFs), which in turn mediate angiogenic pathways and stimulate neovascularisation within the tumour and at distant, metastatic sites. Angiogenesis is further regulated by a number of growth factors including VEGF and EGF.

The CSC population within the tumour, which (as previously discussed) has the potential to initiate metastatic growth at a distant site, will undergo an EMT in response to the hypoxic conditions present in the tumour microenvironment [94]. HIF-1α has multiple roles within the tumour microenvironment, functioning as a master regulator of angiogenesis in hypoxic conditions, maintaining the CSC niche and directly modulating the CSC population. HIF-1α

promotes EMT in a number of human cancers through the increased expression of EMT drivers (e.g. SNAIL) and the concurrent reduction of epithelial markers (e.g. E-Cadherin) [95]. As has already been discussed, the CSC population has distinct similarities to cells that have undergone an EMT, and as such it is plausible that this hypoxia-mediated EMT may also directly affect the CSC population. Indeed, the CSC population isolated from tumours following continuous cycles of hypoxia and re-oxygenation exhibit both stem-like and EMT-like phenotypes [94]. In addition, hypoxia is a critical physiological component of the CSC niche which functions to increase CSC number and maintain their stem-like state allowing for subsequent growth and metastasis [94, 96]. These effects are likely due to the aforementioned stabilisation of HIF-1α, which has a direct effect on CSC, enhancing their self-renewal capacity, promoting cellular proliferation and increasing the tumorigenic properties of these cells [94, 96, 97]. Together, these studies outline an important role for the hypoxic microenvironment in establishing and maintaining the CSC niche.

In addition to the hypoxic environment and subsequent angiogenic induction having a pro-growth effect on CSC, CSC are also able to promote angiogenesis. This can occur via both direct and indirect mechanisms. Firstly, CSC differentiate into cells of the vascular endothelium, thereby functioning to create their own niche [39, 98]. Secondly, CSC secrete factors that promote angiogenic induction. The CSC component of tumours exhibit a much greater expression of the pro-angiogenic factor VEGF compared to the non-CSC population and transplantation of these cells results in the formation of highly vascularised tumours *in vivo* [38, 39]. Furthermore, expression of VEGF by vascular endothelial cells acts directly on CSC to promote proliferation and stemness [78, 99]. VEGF expression is also regulated by EMT and as such provides a possible link between the VEGF-mediated angiogenesis and EMT-induced cancer stemness that is commonly observed in relation to CSC [100].

8. WNT/Notch/Hedgehog pathways

The Hedgehog (Hh), Wnt and Notch signalling pathways are involved in normal development and have also been implicated in tumour biology. These pathways have specific roles in CSC maintenance and differentiation, as discussed in detail below. The CSC niche modulates expression of these pathways, highlighting the importance of extrinsic effects of the microenvironment on signalling pathways within the CSCs.

Increased activity of the Hh pathway is associated with normal adult stem cell development, with down-regulation of the pathway components common following stem cell differentiation. Similarly, increased expression and activation of the Hh pathway is a common feature of CSC and has been described as an essential signalling component in CSC self-renewal and tumour-initiating properties [101-105]. These studies outline a role for the Hh signalling pathway in increasing CSC number and regulating the expression of CSC-related genes. Furthermore, a reliance on Hh signalling has been demonstrated for the recurrence and metastasis of xenograft tumours, most likely through the induction of EMT [105].

Activation of the Wnt signalling pathway is common throughout numerous developmental processes and is usually associated with the promotion of cell growth. This occurs through ligands binding to specific cell surface receptors, including Frizzled and Lrp5/6, which in turn mediates the inhibition of GSK3β and the subsequent accumulation of β-catenin within the nucleus. Therein, β-catenin functions as a transcription factor, modulating the expression of a range of genes that promote cellular growth and proliferation, including MYC [106], IL-8 [107] and Cyclin D [108, 109]. It therefore follows that abnormal or constitutive activation of the Wnt pathway will result in continuous signalling which promotes cell proliferation. Consequently, the Wnt inhibitors DKK1 and sFRP play a key role in regulating this pathway to ensure a balance in proliferative signals is maintained.

Wnt signalling is known to play a role in promoting normal adult stem cell activation and expansion, a function which is kept in check by the presence of specific Wnt inhibitors [110, 111]. In addition, bone morphogenic proteins (BMPs), which in general function to inhibit cell growth, work in concert with the Wnt signalling pathway to provide a balance between anti-and pro-growth signals to regulate stem cell self-renewal. However, in the context of cancer, this balance can be disrupted, through both loss of BMP signalling or aberrant activation of Wnt signalling, leading to uncontrolled proliferation of CSC [112]. The increase in Wnt signalling activity in tumour cells has been associated with cells in close proximity to the stromal fibroblasts, suggesting that modulation of this pathway may be through extrinsic factors and as such a niche-dependent function [113].

The Notch signalling pathway has also been associated with increased growth and tumourigenicity of CSC. High levels of Notch are observed in CSC populations, which have been demonstrated to lead to EMT and increased capacity to form tumours upon transplantation into immune-compromised animals [114-116]. Silencing Notch pathway signalling in CSC results in reduced growth, migration and invasion as well as enhanced apoptotic induction [117]. Furthermore, inhibition of Notch signalling has been demonstrated to inhibit tumour growth and, more specifically, reduce the number of CSC in a number of cancer types including colon and brain [115, 118, 119]. In addition, expression of hairy and enhancer of split 1 (Hes1), a target of the Notch signalling pathway, correlates with the expression of stem cell markers in colon cancer and is associated with self-renewal and increased tumourigenicty [120]. This recent study suggests that activation of the Notch signalling pathway and subsequent expression of target genes is an important component of maintaining CSC function.

The CSC population exhibits reliance on expression and deregulation of these developmental pathways that are modulated by stromal cells of the CSC niche. Together, these findings substantiate a reliance of the CSC on specific components of the niche. Specifically, these pathways provide novel opportunities for targeting the niche and signals emanating from it in an attempt to reduce CSC number *in vivo* and hence reduce the incidence of tumour recurrence following treatment.

9. The role of the CXCL12-CXCR4 axis in CSC

CXCL12 (also known as SDF-1) is a strong chemo-attractant initially identified with a role in the attraction of lymphocytes during haematopoiesis. In addition, CXCL12 has been demonstrated to play a role in immune signalling and inflammation and hence is an important regulator of key physiological processes in both development and disease. However in recent years, knowledge of its role has greatly diversified, particularly in the area of stem cell regulation and homing. The cognate receptor for CXCL12, CXCR4, is highly expressed on cells of the haematopoietic lineage, thereby promoting the localisation of HSC within their specific niche [121, 122]. Importantly, CXCR4 is also expressed on a range of stem cells including endothelial, haematopoietic, neural and embryonic stem cells, enabling these cells to respond to gradients of CXCL12, which are generated following tissue damage and irregular physiological events such as hypoxia [123]. This results in the efficient migration and homing of CXCR4 positive stem cells to areas of high CXCL12 expression, making the CXCL12-CXCR4 axis critical for normal development and tissue regeneration and repair.

In the context of human cancers, CXCR4 is highly expressed in many tumour types [124] and has been shown to not only regulate invasion of cancer cells to metastatic sites, but to be a marker of the CSC population of multiple tumour types [125, 126]. The CXCL12-CXCR4 axis has been implicated in metastatic processes in a range of cancers, including breast, prostate, lung, colon, kidney, melanoma, brain, leukaemia and myeloma, with expression of CXCL12 highest in tissues that are common sites of metastasis, such as liver, bone marrow and lungs [127], suggesting that tumour cells hijack this mechanism of normal tissue regeneration and stem cell localisation to aid the metastatic process. In all, it is clear that the CXCL12-CXCR4 axis plays an integral role in the development of metastasis and the definition of a pre-metastatic niche. Importantly, it is the CSC population within the tumour that is likely to utilise this axis for the development of distant metastases.

10. Cytokines and CSC regulation

Cytokines are a group of small molecules that are secreted by a broad range of cells, including immune cells, endothelial cells and fibroblasts. In general, following release by cells, cytokines act through binding their specific receptors on target cells and modulating immune responses as well as regulating cell growth and behaviour. A number of cytokines have been implicated in CSC maintenance and therefore are important secreted factors present within the CSC niche.

IL-6 is a pro-inflammatory cytokine which normally functions to stimulate the immune response following injury or infection. Importantly, IL-6 is expressed by a range of cells within the CSC niche, including macrophages and stromal fibroblasts, as well as CSC themselves, and has been demonstrated to directly regulate CSC behaviour. IL-6 signalling specifically through the STAT3 pathway has been identified as a key mechanism for the modulation of CSC. STAT3 is highly expressed in CSC derived from liver, bone, cervical and brain cancers

[128]. Furthermore, inhibition of the STAT3 pathway, through use of a specific STAT3 inhibitor, has been demonstrated to reduce the formation of glioblastoma CSC [129]. IL-6 signalling is a direct regulator of breast cancer CSC self-renewal, leading to subsequent increased expression and secretion of IL-6 and a resultant feed-forward loop that functions to further enhance CSC function [130]. In addition, glioblastoma CSC are reliant on increased expression of IL-6 to mediate growth, invasion and anti-apoptotic functions [131, 132]. Conversely, the loss of IL-6 in mouse models results in reduction in gastric tumour formation [133]. These mechanistic studies provide an explanation as to why patients with advanced or metastatic cancers exhibit increased serum expression of IL-6, which in turn is correlated with poor prognosis [134, 135].

IL-6 function within the CSC niche is not strictly limited to the direct maintenance of CSC self-renewal and growth, rather IL-6 also has an indirect effect on CSC through modulation of the stromal cell composition of the CSC niche itself. In breast cancer, the production of IL-6 by CSC acts as an attractant for MSC, resulting in the migration of MSC directly to sites of breast tumour growth [87]. This is particularly important as MSC have been described to play a major supportive role for CSC within their local microenvironment, promoting both tumour growth and angiogenic induction (as described above).

Aside from IL-6, a number of other interleukins play important roles in the regulation of CSC. IL-17 is expressed by cells within the tumour microenvironment and promotes the self-renewal of CSC in ovarian cancer [136]. High levels of IL-1 have been associated with advanced metastatic disease [137, 138], suggesting that it may also play a role in modulating CSC and the metastatic niche. Notably, increased IL-1 production by TAMs has been shown to increase angiogenesis and tumour cell growth as well as enhance metastasis [139]. Similarly, increased levels of IL-8 in serum are also correlated with more aggressive cancer and subsequently poor prognosis [140, 141]. This is likely due to the ability of IL-8 to modulate CSC self-renewal, and hence tumour growth, as the IL-8 receptor (CXCR1) is found to be highly expressed on CSC [142]. IL-8 in the tumour microenvironment is secreted by endothelial cells located within the perivascular niche, providing another mechanism through which the vasculature can induce CSC growth. In addition, the expression of IL-8 can also increase the expression of the CXCR1 receptor on CSC, making the CSC even more responsive to further IL-8 signalling [81].

11. Prognosis

The identification of CSC within the bulk tumour, as made possible through the use of phenotypic markers (see Table 1) has been shown to correlate with patient prognosis and survival. The expression of CD44, a common CSC marker, has been associated with high levels of metastasis, increased incidence of tumour recurrence following treatment, and as a result, poor patient outcome. This is likely due to the key role the CSC sub-population plays in the growth and development of tumours, both at the primary site and for establishment within the metastatic niche. An increase in the CSC isolated from breast cancer patients using the

CD44^{high}/CD24$^{-/low}$ phenotype was shown to correlate with more aggressive forms of disease and poor patient prognosis [143, 144]. A similar correlation was also noted in pancreatic cancer, with patients exhibiting increased expression of CD44 presenting with a median survival of only 10 months, compared to 43 months in patients with low levels of CD44 [145]. High ALDH1 expression was also demonstrated to correlate with poor prognosis in breast cancer [13], rectal adenocarcinoma [146] and colorectal cancer [147]. Furthermore, ALDH1 expression has also been shown to be indicative of tumour recurrence following treatment [148]. However, there remains some controversy in this area as high stromal cell expression of ALDH1 within the tumour microenvironment has been associated with improved patient outcome in breast cancer [149, 150], suggesting that use of stem cell markers as prognostic indicators must be approached with some caution. CD133 expression was correlated with metastasis in colorectal cancer patients, despite having no significant effect on overall survival in the same study [147]. However, high levels of CD133 have been demonstrated to correlate with poor clinical outcome in other studies [151, 152]. Other less commonly described markers of CSC, such as Nestin, have also been demonstrated to correlate with tumour size and lymph node metastasis in non-small cell lung carcinoma and poor patient prognosis in adenocarcinoma [153]. The analysis of CSC within a patient's tumour, through the use of defined CSC markers, may aid clinicians in defining more precise prognostic factors.

12. Resistance to therapy

Resistance to standard treatments, including radiation and chemotherapy regimens, frequently results in tumour relapse following a clinical response, due to the presence of residual tumour cells. One of the key attributes of the CSC, which highlights their importance in tumour development and maintenance, is the ability to resist chemotherapy-and radiation-mediated treatment strategies. This has been observed in breast cancer patients, with the remaining tumour cells following standard chemotherapy exhibiting both stem-and EMT-like gene expression profiles [154]. The CSC-enriched component from breast cancer cell lines also display a decreased sensitivity to radiation [155]. In addition, following radiation treatment in glioblastoma, the recurring tumours tend to exhibit a distinct nodular pattern, which is consistent with the theory of the recurring tumour arising from a clonal or sub-clonal source [156], likely due to the expansion of radio-resistant CSC. More recent studies have identified a specific increase in the percentage of CSC within the radio-resistant tumours in glioblastoma and breast cancer both in patients and in culture [38, 155, 157]. Furthermore, the CSC population is specifically involved in the generation of new tumours following treatment. This, coupled with the identification of both CSC and non-CSC cells within the recurring tumour, is supportive of a specific subpopulation of cells, namely the CSC, being responsible for the recurrent tumours.

The mechanism of therapy resistance in CSC is still largely unknown, however a number of signalling pathways and molecules have been implicated in this process. Notably, resistance is mediated by both extrinsic and intrinsic functions of the CSC, with extrinsically mediated resistance governed by the CSC niche. The tumour environment, or niche, is well-documented to mediate drug-resistance. This can be mediated both through the release of soluble factors and via cell adhesion mechanisms, with the latter commonly referred to as cell adhesion-mediated drug resistance (CAM-DR) (Reviewed by [158]). Both forms of niche-mediated drug resistance are well established in haematological malignancies. The bone marrow represents the most common site of metastases in these tumours and is characterised as an environment rich in IL-6 and fibronectin, with both of these factors being demonstrated to contribute to the acquisition of drug resistance [159]. As described earlier, the CXCL12-CXCR4 axis is an integral component in CSC homing to the niche and the colonisation of the pre-metastatic niche. This axis has also been shown to be a key player in tumour drug-resistance, with CXCR4 positive cells responding to stromal cell-derived CXCL12, resulting in increased activation of Akt/PKB and ERK signalling pathways which, in turn, mediate resistance by inducing anti-apoptotic pathways [160-162]. Direct cellular interactions mediated by adhesion between integrins and their receptors (often components of the ECM, including collagen and fibronectin), have also been directly associated with a decrease in drug-induced apoptosis [163, 164]. These studies show that CAM-DR is a key feature of chemotherapy resistance in tumours that is modulated by the niche. Due to the expression of CXCR4 and a range of integrins by CSC, coupled with the key interactions observed between CSC, the ECM and the CSC niche, it is likely that these mechanisms may be involved in CSC-specific resistance pathways.

In addition, activation of signalling pathways and increased expression of key molecules within the CSC themselves, provide an intrinsic means of modulating CSC resistance. The Wnt and Notch pathways that are commonly involved in regulating stem cell growth and behaviour have shown increased activation in response to irradiation and may confer radio-resistance on the CSC population. Inhibiting the Notch pathway or specific knockdown of Notch1 or Notch2 in glioma CSC sensitises these cells to radiation [165]. Furthermore, radiation-resistant CSC exhibit increased levels of stabilised β-catenin [157, 166], which suggests abnormal activation of the Wnt signalling pathway. Further to these signalling pathways, a population of stem-like cells identified in AML patients have been shown to exhibit a higher degree of drug efflux than non-stem cells isolated from the tumour [167], suggesting a mechanism by which the CSC component of tumours may be able to escape the effects of chemotherapeutic agents. This phenomenon is likely due to the increased expression of drug transporter genes ABCG2 and ABCA3 on CSC [168]. A further mechanism through which CSCs have been demonstrated to exhibit radiation-resistance is through enhanced activation of DNA damage checkpoint responses. A study utilising glioblastoma-derived CSC, prospectively isolated by their CD133+phenotype, showed that although DNA damage was initiated equally in both $CD133^-$ and $CD133^+$ cells following radiation, the $CD133^+$ cells (representing the enriched CSC population) were better able to repair the damage and hence displayed a lower rate of apoptosis [38]. It was subsequently shown that the DNA damage checkpoint proteins CHK1 and CHK2 showed preferential activation in the $CD133^+$ cells. In addition, these cells also showed a basal level of activation of another component of the DNA damage checkpoint, rad17. These data suggest that the CSC component of the tumour is specifically primed to respond to DNA damage caused by external stress stimuli, such as radiation, and as such exhibit increased survival following treatment. Furthermore, these studies suggest that radiation and chemotherapy resistance may be mediated by enhanced Wnt and/or Notch signalling, enhanced DNA damage responses and differential expression of drug transporter enzymes.

The observations detailed above explain why current cancer therapies are generally ineffective – or at least why tumour recurrence is extremely common in aggressive tumour types. Coupled with differential expression of key genes and pathways that can mediate cell survival following radiation and chemotherapy, the CSC are generally slow cycling and are not targeted by conventional therapies. In addition, CSC homing to the niche and subsequent adhesion provides important support for the CSC and in turn mediates drug resistance. Therefore, surviving CSC following treatment regimens continue to grow and differentiate and are able to re-populate the tumour.

13. Targeting CSC and the CSC niche in cancer therapeutics

Tumour regression following treatment does not always correlate with patient survival, and this is due largely to the remaining treatment-resistant CSC (as described above). Therefore, directly targeting the CSC, in conjunction with existing therapeutics, may provide a novel treatment strategy to eradicate the residual CSC and hence prevent tumour recurrence. When investigating suitable target pathways, it is also important to factor in both the targeting of the CSC population directly, as well as the CSC niche that supports their growth and survival. As has been discussed throughout this chapter, the niche provides essential support for CSC and disruption of these supportive processes presents an attractive focus for the development of new therapies.

CSC rely on increased expression of the developmental signalling pathways – namely Wnt, Notch and Hedgehog – to regulate their increased growth and survival. These pathways therefore represent key targets for the development of novel treatment strategies that may significantly inhibit the growth of the CSC population. Targeting the Notch pathway as a means of specifically targeting the CSC component of tumours is reviewed in detail by Pannuti *et al.* [169]. Inhibition of the Notch signalling pathway has been demonstrated to be effective in reducing the frequency of CSC derived from colon, medulloblastoma, glioblastoma and breast cancer [115, 118, 119, 170]. Importantly, inhibition of Notch signalling prevented breast cancer metastases [170] and when used in combination with a common chemotherapeutic agent delayed tumour recurrence [119]. As the Hedgehog pathway represents an essential signalling component in CSC to modulate self-renewal and tumour-initiating properties, targeting the Hedgehog pathway is another feasible therapeutic option (reviewed by [171]). Proof-of-principle for Hedgehog as a therapeutic target is demonstrated as inhibition of the hedgehog pathway results in a reduction in the tumourigenic capacity of human gliomas in immune-compromised mice [101]. The resident CSC population is required for the inherent tumourigenicty of glioma cells; therefore this data supports a role for inhibition of hedgehog in directly affecting the CSC component of the tumour. Specific inhibition of the hedgehog signalling pathway using cyclopamine is effective in preventing cancer growth as well as invasion and metastasis (another feature attributed to the CSC population) [172, 173]. Cyclopamine has also been shown to specifically reduce the number of CSC in pancreatic cancer [174], providing evidence that inhibition of the Hedgehog pathway can directly affect the CSC population. In addition to targeting the Notch and Hedgehog signalling pathways, as described here, inhibitors of Wnt signalling have also been developed in recent years and have been shown to be effective in reducing tumour cell growth [175-177]. Further development of these inhibitors and specific investigation of the effect of treatment on the CSC population is warranted.

Further studies have demonstrated a possible role for targeting a range of small molecules and proteins that play an integral role in the survival and growth of CSC. For example, the CXCR1/CXCR2 inhibitor repertaxin functions to inhibit IL-8 responses, and as such depletes the CSC component of breast cancer xenografts, resulting in a reduction in tumour growth and metastases [142]. In addition, CXCR4 antagonists have been established and are under investigation for clinical efficacy in treating leukaemia [91, 178]. Due to the high expression of CXCR4 on CSC and the role of the CXCL12-CXCR4 axis in mediating metastasis and drug resistance, these antagonists may also have a therapeutic role in targeting CSC and modulating their ability to colonise the pre-metastatic niche. The use of CSC phenotypic markers as possible additions to anti-tumour therapies is also beginning to be developed. A proof-of-principle study has demonstrated that use of an anti-CD44 antibody was able to reduce the number of tumour-initiating cells, both *in vitro* and in a xenograft model. In addition, this reduction of CSC resulted in decreased growth, metastasis and post-radiation tumour recurrence in mice [145]. In addition, targeting CD44 in AML has shown promise in eradicating the CSC population and reducing the tumorigenic capacity of cells *in vivo* [179]. The resistance of CSC to conventional therapeutics is perhaps the most critical function of CSC that must be overcome in the identification of novel therapeutic strategies. To this end, targeting critical enzymes in the DNA damage response pathway, CHK1 and CHK2, has been shown to reverse the radio-resistance observed in CD133 positive glioma CSC [38]. In addition, pharmacological inhibition or targeted knockdown of ABC drug transporter enzymes may provide useful tools to reverse chemoresistance of CSC [180, 181].

Perhaps as important as targeting the CSC specifically, is targeting the CSC niche. Disruption of critical components of the CSC niche has the potential to inhibit CSC growth and subsequently reduce tumour development and metastases. As tumour angiogenesis is supportive of CSC survival, anti-angiogenic treatments, when used in combination with cytotoxic chemotherapies, can reduce the CSC population [182]. In particular, VEGF has been shown to be expressed directly by CSC and by the tumour microenvironment and is an important molecule in regulating vessel formation as well as CSC growth. Targeting VEGF specifically with bevacizumab has been demonstrated to disrupt the CSC niche, and as such result in a reduction in the number of CSC and subsequently tumour growth [79, 183]. Specific disruption of cellular interactions between the CSC and stromal cells that comprise the CSC-and pre-metastatic-niche may also contribute to the inhibition of CSC growth and metastasis. For example, the fibronectin receptor VLA-4 is required for interactions between tumour cells and stromal cells of the pre-metastatic niche and antibodies targeting this receptor are able to prevent this association and reduce tumour burden following treatment [163, 184]. In addition, targeting this and similar interactions involving integrins in breast cancer cells can restore cells to an epithelial-like state [185, 186] and therefore may be useful in reversing the EMT which enhances the ability of malignant cells, specifically CSC, to metastasise.

14. Concluding remarks

CSC display a dependence on their niche for growth and survival. Signalling through various developmental pathways and via a range of cytokines and growth factors not only modulates CSC growth and function, but can also alter the composition of the CSC niche. The ability of tumours to metastasise is a feature of CSC, due to their enhanced capacity for migration and invasion. This process is reliant on the formation and regulation of a pre-metastatic niche, which is conducive to the colonisation of CSC and subsequent tumour growth at a secondary site. Metastasis accounts for almost 90% of all cancer-associated deaths. A better understanding of what drives tumour cells to metastasise and the changes in the host tissue and local microenvironments that may accommodate the establishment and colonisation of circulating tumour cells will increase our capacity to develop novel therapeutic agents that may inhibit or delay metastasis and as such have a significant impact on patient outcome. In addition, understanding the radio-and chemo-resistant properties of CSC has allowed us to ascertain possible mechanisms for tumour recurrence and poor patient outcome following initial tumour regression.

Described in this chapter are but a few of the possible novel therapies that target CSC-specific factors and/or components of the CSC niche. Further investigation of these, and others, as well as the development of novel agents, is still required to better understand the feasibility of direct targeting of CSC and/or their niche in cancer therapy. Furthermore, reliable biomarkers for the CSC population are required in order to accurately determine the efficacy of these treatments. Importantly, the development of these novel therapeutics would not replace existing effective treatment such as radiotherapy and cytotoxic chemotherapy, but rather would be used in conjunction with these known and proven agents. Combination therapies, such as these, as has already been demonstrated in a small number of *in vivo* studies, would better enable us to effectively inhibit the CSC and the functionality of the CSC niche, and subsequently lead to improved overall survival and reduced incidence of tumour recurrence.

Author details

Jacqueline E. Noll, Kate Vandyke and Andrew C.W. Zannettino

Myeloma Research Laboratory, School of Medical Sciences, Faculty of Health Science,
University of Adelaide, Adelaide, Australia

References

[1] Bonnet, D. and J.E. Dick, Human acute myeloid leukemia is organized as a hierarchy that originates from a primitive hematopoietic cell. Nat Med, 1997. 3(7): p. 730-7.

[2] Al-Hajj, M., et al., Prospective identification of tumorigenic breast cancer cells. Proc Natl Acad Sci U S A, 2003. 100(7): p. 3983-8.

[3] Singh, S.K., et al., Identification of a cancer stem cell in human brain tumors. Cancer Res, 2003. 63(18): p. 5821-8.

[4] Singh, S.K., et al., Identification of human brain tumour initiating cells. Nature, 2004. 432(7015): p. 396-401.

[5] Collins, A.T., et al., Prospective identification of tumorigenic prostate cancer stem cells. Cancer Res, 2005. 65(23): p. 10946-51.

[6] Li, C., et al., Identification of pancreatic cancer stem cells. Cancer Res, 2007. 67(3): p. 1030-7.

[7] Fang, D., et al., A tumorigenic subpopulation with stem cell properties in melanomas. Cancer Res, 2005. 65(20): p. 9328-37.

[8] Ma, S., et al., Identification and characterization of tumorigenic liver cancer stem/progenitor cells. Gastroenterology, 2007. 132(7): p. 2542-56.

[9] Prince, M.E., et al., Identification of a subpopulation of cells with cancer stem cell properties in head and neck squamous cell carcinoma. Proc Natl Acad Sci U S A, 2007. 104(3): p. 973-8.

[10] Jijiwa, M., et al., CD44v6 regulates growth of brain tumor stem cells partially through the AKT-mediated pathway. PLoS One, 2011. 6(9): p. e24217.

[11] Snyder, E.L., et al., Identification of CD44v6(+)/CD24-breast carcinoma cells in primary human tumors by quantum dot-conjugated antibodies. Lab Invest, 2009. 89(8): p. 857-66.

[12] Douville, J., R. Beaulieu, and D. Balicki, ALDH1 as a functional marker of cancer stem and progenitor cells. Stem Cells Dev, 2009. 18(1): p. 17-25.

[13] Ginestier, C., et al., ALDH1 is a marker of normal and malignant human mammary stem cells and a predictor of poor clinical outcome. Cell Stem Cell, 2007. 1(5): p. 555-67.

[14] Ma, I. and A.L. Allan, The role of human aldehyde dehydrogenase in normal and cancer stem cells. Stem Cell Rev, 2011. 7(2): p. 292-306.

[15] Cariati, M., et al., Alpha-6 integrin is necessary for the tumourigenicity of a stem cell-like subpopulation within the MCF7 breast cancer cell line. Int J Cancer, 2008. 122(2): p. 298-304.

[16] Schwab, L.P., et al., Hypoxia-inducible factor 1alpha promotes primary tumor growth and tumor-initiating cell activity in breast cancer. Breast Cancer Res, 2012. 14(1): p. R6.

[17] Rountree, C.B., et al., Expansion of liver cancer stem cells during aging in methionine adenosyltransferase 1A-deficient mice. Hepatology, 2008. 47(4): p. 1288-97.

[18] Yang, Z.F., et al., Significance of CD90+cancer stem cells in human liver cancer. Cancer Cell, 2008. 13(2): p. 153-66.

[19] Eramo, A., et al., Identification and expansion of the tumorigenic lung cancer stem cell population. Cell Death Differ, 2008. 15(3): p. 504-14.

[20] Sullivan, J.P., et al., Aldehyde dehydrogenase activity selects for lung adenocarcinoma stem cells dependent on notch signaling. Cancer Res, 2010. 70(23): p. 9937-48.

[21] Yan, X., et al., Identification of CD90 as a marker for lung cancer stem cells in A549 and H446 cell lines. Oncol Rep, 2013. 30(6): p. 2733-40.

[22] He, J., et al., CD90 is identified as a candidate marker for cancer stem cells in primary high-grade gliomas using tissue microarrays. Mol Cell Proteomics, 2012. 11(6): p. M111 010744.

[23] Lathia, J.D., et al., Integrin alpha 6 regulates glioblastoma stem cells. Cell Stem Cell, 2010. 6(5): p. 421-32.

[24] Zhang, M., et al., Nestin and CD133: valuable stem cell-specific markers for determining clinical outcome of glioma patients. J Exp Clin Cancer Res, 2008. 27: p. 85.

[25] O'Brien, C.A., et al., A human colon cancer cell capable of initiating tumour growth in immunodeficient mice. Nature, 2007. 445(7123): p. 106-10.

[26] Yeung, T.M., et al., Cancer stem cells from colorectal cancer-derived cell lines. Proc Natl Acad Sci U S A, 2010. 107(8): p. 3722-7.

[27] van den Hoogen, C., et al., High aldehyde dehydrogenase activity identifies tumor-initiating and metastasis-initiating cells in human prostate cancer. Cancer Res, 2010. 70(12): p. 5163-73.

[28] Hermann, P.C., et al., Distinct populations of cancer stem cells determine tumor growth and metastatic activity in human pancreatic cancer. Cell Stem Cell, 2007. 1(3): p. 313-23.

[29] Kim, M.P., et al., ALDH activity selectively defines an enhanced tumor-initiating cell population relative to CD133 expression in human pancreatic adenocarcinoma. PLoS One, 2011. 6(6): p. e20636.

[30] Matsuda, Y., S. Kure, and T. Ishiwata, Nestin and other putative cancer stem cell markers in pancreatic cancer. Med Mol Morphol, 2012. 45(2): p. 59-65.

[31] Civenni, G., et al., Human CD271-positive melanoma stem cells associated with metastasis establish tumor heterogeneity and long-term growth. Cancer Res, 2011. 71(8): p. 3098-109.

[32] Klein, W.M., et al., Increased expression of stem cell markers in malignant melanoma. Mod Pathol, 2007. 20(1): p. 102-7.

[33] Krishnamurthy, S. and J.E. Nor, Head and neck cancer stem cells. J Dent Res, 2012. 91(4): p. 334-40.

[34] Scadden, D.T., The stem-cell niche as an entity of action. Nature, 2006. 441(7097): p. 1075-9.

[35] Lee, J., et al., Tumor stem cells derived from glioblastomas cultured in bFGF and EGF more closely mirror the phenotype and genotype of primary tumors than do serum-cultured cell lines. Cancer Cell, 2006. 9(5): p. 391-403.

[36] Zheng, X., et al., Most C6 cells are cancer stem cells: evidence from clonal and population analyses. Cancer Res, 2007. 67(8): p. 3691-7.

[37] Pandita, A., et al., Contrasting in vivo and in vitro fates of glioblastoma cell subpopulations with amplified EGFR. Genes Chromosomes Cancer, 2004. 39(1): p. 29-36.

[38] Bao, S., et al., Glioma stem cells promote radioresistance by preferential activation of the DNA damage response. Nature, 2006. 444(7120): p. 756-60.

[39] Ricci-Vitiani, L., et al., Tumour vascularization via endothelial differentiation of glioblastoma stem-like cells. Nature, 2010. 468(7325): p. 824-8.

[40] Polyak, K. and R.A. Weinberg, Transitions between epithelial and mesenchymal states: acquisition of malignant and stem cell traits. Nat Rev Cancer, 2009. 9(4): p. 265-73.

[41] Aktas, B., et al., Stem cell and epithelial-mesenchymal transition markers are frequently overexpressed in circulating tumor cells of metastatic breast cancer patients. Breast Cancer Res, 2009. 11(4): p. R46.

[42] Armstrong, A.J., et al., Circulating tumor cells from patients with advanced prostate and breast cancer display both epithelial and mesenchymal markers. Mol Cancer Res, 2011. 9(8): p. 997-1007.

[43] Mani, S.A., et al., The epithelial-mesenchymal transition generates cells with properties of stem cells. Cell, 2008. 133(4): p. 704-15.

[44] Baccelli, I. and A. Trumpp, The evolving concept of cancer and metastasis stem cells. J Cell Biol, 2012. 198(3): p. 281-93.

[45] Dalerba, P., R.W. Cho, and M.F. Clarke, Cancer stem cells: models and concepts. Annu Rev Med, 2007. 58: p. 267-84.

[46] Heissig, B., et al., Recruitment of stem and progenitor cells from the bone marrow niche requires MMP-9 mediated release of kit-ligand. Cell, 2002. 109(5): p. 625-37.

[47] Andressen, C., et al., Beta1 integrin deficiency impairs migration and differentiation of mouse embryonic stem cell derived neurons. Neurosci Lett, 1998. 251(3): p. 165-8.

[48] Hirsch, E., et al., Impaired migration but not differentiation of haematopoietic stem cells in the absence of beta1 integrins. Nature, 1996. 380(6570): p. 171-5.

[49] Crowe, D.L. and A. Ohannessian, Recruitment of focal adhesion kinase and paxillin to beta1 integrin promotes cancer cell migration via mitogen activated protein kinase activation. BMC Cancer, 2004. 4: p. 18.

[50] Curran, S. and G.I. Murray, Matrix metalloproteinases in tumour invasion and metastasis. J Pathol, 1999. 189(3): p. 300-8.

[51] Kang, H., et al., The elevated level of CXCR4 is correlated with nodal metastasis of human breast cancer. Breast, 2005. 14(5): p. 360-7.

[52] Lapidot, T. and O. Kollet, The essential roles of the chemokine SDF-1 and its receptor CXCR4 in human stem cell homing and repopulation of transplanted immune-deficient NOD/SCID and NOD/SCID/B2m(null) mice. Leukemia, 2002. 16(10): p. 1992-2003.

[53] Liu, R., et al., The prognostic role of a gene signature from tumorigenic breast-cancer cells. N Engl J Med, 2007. 356(3): p. 217-26.

[54] Liu, H., et al., Cancer stem cells from human breast tumors are involved in spontaneous metastases in orthotopic mouse models. Proc Natl Acad Sci U S A, 2010. 107(42): p. 18115-20.

[55] Pang, R., et al., A subpopulation of CD26+cancer stem cells with metastatic capacity in human colorectal cancer. Cell Stem Cell, 2010. 6(6): p. 603-15.

[56] Justilien, V., et al., Matrix metalloproteinase-10 is required for lung cancer stem cell maintenance, tumor initiation and metastatic potential. PLoS One, 2012. 7(4): p. e35040.

[57] Charafe-Jauffret, E., et al., Breast cancer cell lines contain functional cancer stem cells with metastatic capacity and a distinct molecular signature. Cancer Res, 2009. 69(4): p. 1302-13.

[58] Lu, P., V.M. Weaver, and Z. Werb, The extracellular matrix: a dynamic niche in cancer progression. J Cell Biol, 2012. 196(4): p. 395-406.

[59] Wong, G.S. and A.K. Rustgi, Matricellular proteins: priming the tumour microenvironment for cancer development and metastasis. Br J Cancer, 2013. 108(4): p. 755-61.

[60] Garcion, E., et al., Generation of an environmental niche for neural stem cell development by the extracellular matrix molecule tenascin C. Development, 2004. 131(14): p. 3423-32.

[61] Kollet, O., et al., Osteoclasts degrade endosteal components and promote mobilization of hematopoietic progenitor cells. Nat Med, 2006. 12(6): p. 657-64.

[62] Barth, P.J., R. Moll, and A. Ramaswamy, Stromal remodeling and SPARC (secreted protein acid rich in cysteine) expression in invasive ductal carcinomas of the breast. Virchows Arch, 2005. 446(5): p. 532-6.

[63] Kwon, Y.J., et al., Expression patterns of aurora kinase B, heat shock protein 47, and periostin in esophageal squamous cell carcinoma. Oncol Res, 2009. 18(4): p. 141-51.

[64] Prenzel, K.L., et al., Significant overexpression of SPARC/osteonectin mRNA in pancreatic cancer compared to cancer of the papilla of Vater. Oncol Rep, 2006. 15(5): p. 1397-401.

[65] Rocco, M., et al., Proteomic profiling of human melanoma metastatic cell line secretomes. J Proteome Res, 2011. 10(10): p. 4703-14.

[66] Wikman, H., et al., Identification of differentially expressed genes in pulmonary adenocarcinoma by using cDNA array. Oncogene, 2002. 21(37): p. 5804-13.

[67] Chiodoni, C., M.P. Colombo, and S. Sangaletti, Matricellular proteins: from homeostasis to inflammation, cancer, and metastasis. Cancer Metastasis Rev, 2010. 29(2): p. 295-307.

[68] Chiquet-Ehrismann, R., et al., Tenascins in stem cell niches. Matrix Biol, 2014.

[69] Kaariainen, E., et al., Switch to an invasive growth phase in melanoma is associated with tenascin-C, fibronectin, and procollagen-I forming specific channel structures for invasion. J Pathol, 2006. 210(2): p. 181-91.

[70] Midwood, K.S., et al., Advances in tenascin-C biology. Cell Mol Life Sci, 2011. 68(19): p. 3175-99.

[71] Malanchi, I., et al., Interactions between cancer stem cells and their niche govern metastatic colonization. Nature, 2012. 481(7379): p. 85-9.

[72] Jung, Y., et al., Recruitment of mesenchymal stem cells into prostate tumours promotes metastasis. Nat Commun, 2013. 4: p. 1795.

[73] Sica, A., et al., Tumour-associated macrophages are a distinct M2 polarised population promoting tumour progression: potential targets of anti-cancer therapy. Eur J Cancer, 2006. 42(6): p. 717-27.

[74] Wu, A., et al., Glioma cancer stem cells induce immunosuppressive macrophages/ microglia. Neuro Oncol, 2010. 12(11): p. 1113-25.

[75] Ye, X.Z., et al., Tumor-associated microglia/macrophages enhance the invasion of glioma stem-like cells via TGF-beta1 signaling pathway. J Immunol, 2012. 189(1): p. 444-53.

[76] Jinushi, M., et al., Regulation of cancer stem cell activities by tumor-associated macrophages. Am J Cancer Res, 2012. 2(5): p. 529-39.

[77] Mitchem, J.B., et al., Targeting tumor-infiltrating macrophages decreases tumor-initiating cells, relieves immunosuppression, and improves chemotherapeutic responses. Cancer Res, 2013. 73(3): p. 1128-41.

[78] Beck, B., et al., A vascular niche and a VEGF-Nrp1 loop regulate the initiation and stemness of skin tumours. Nature, 2011. 478(7369): p. 399-403.

[79] Calabrese, C., et al., A perivascular niche for brain tumor stem cells. Cancer Cell, 2007. 11(1): p. 69-82.

[80] Krishnamurthy, S., et al., Endothelial cell-initiated signaling promotes the survival and self-renewal of cancer stem cells. Cancer Res, 2010. 70(23): p. 9969-78.

[81] Infanger, D.W., et al., Glioblastoma stem cells are regulated by interleukin-8 signaling in a tumoral perivascular niche. Cancer Res, 2013. 73(23): p. 7079-89.

[82] Lu, J., et al., Endothelial cells promote the colorectal cancer stem cell phenotype through a soluble form of Jagged-1. Cancer Cell, 2013. 23(2): p. 171-85.

[83] Neiva, K.G., et al., Cross talk initiated by endothelial cells enhances migration and inhibits anoikis of squamous cell carcinoma cells through STAT3/Akt/ERK signaling. Neoplasia, 2009. 11(6): p. 583-93.

[84] Zhu, T.S., et al., Endothelial cells create a stem cell niche in glioblastoma by providing NOTCH ligands that nurture self-renewal of cancer stem-like cells. Cancer Res, 2011. 71(18): p. 6061-72.

[85] Noll, J.E., et al., Myeloma plasma cells alter the bone marrow microenvironment by stimulating the proliferation of mesenchymal stromal cells. Haematologica, 2014. 99(1): p. 163-71.

[86] Karnoub, A.E., et al., Mesenchymal stem cells within tumour stroma promote breast cancer metastasis. Nature, 2007. 449(7162): p. 557-63.

[87] Liu, S., et al., Breast cancer stem cells are regulated by mesenchymal stem cells through cytokine networks. Cancer Res, 2011. 71(2): p. 614-24.

[88] Zhu, W., et al., Mesenchymal stem cells derived from bone marrow favor tumor cell growth in vivo. Exp Mol Pathol, 2006. 80(3): p. 267-74.

[89] Yu, J.M., et al., Mesenchymal stem cells derived from human adipose tissues favor tumor cell growth in vivo. Stem Cells Dev, 2008. 17(3): p. 463-73.

[90] Prantl, L., et al., Adipose tissue-derived stem cells promote prostate tumor growth. Prostate, 2010. 70(15): p. 1709-15.

[91] Burger, J.A. and A. Peled, CXCR4 antagonists: targeting the microenvironment in leukemia and other cancers. Leukemia, 2009. 23(1): p. 43-52.

[92] Rhodes, L.V., et al., Effects of human mesenchymal stem cells on ER-positive human breast carcinoma cells mediated through ER-SDF-1/CXCR4 crosstalk. Mol Cancer, 2010. 9: p. 295.

[93] Tsai, K.S., et al., Mesenchymal stem cells promote formation of colorectal tumors in mice. Gastroenterology, 2011. 141(3): p. 1046-56.

[94] Louie, E., et al., Identification of a stem-like cell population by exposing metastatic breast cancer cell lines to repetitive cycles of hypoxia and reoxygenation. Breast Cancer Res, 2010. 12(6): p. R94.

[95] Imai, T., et al., Hypoxia attenuates the expression of E-cadherin via up-regulation of SNAIL in ovarian carcinoma cells. Am J Pathol, 2003. 163(4): p. 1437-47.

[96] Heddleston, J.M., et al., Hypoxia inducible factors in cancer stem cells. Br J Cancer, 2010. 102(5): p. 789-95.

[97] Kim, Y., et al., Hypoxic tumor microenvironment and cancer cell differentiation. Curr Mol Med, 2009. 9(4): p. 425-34.

[98] Matsuda, S., et al., Cancer stem cells maintain a hierarchy of differentiation by creating their niche. Int J Cancer, 2013.

[99] Xu, C., X. Wu, and J. Zhu, VEGF promotes proliferation of human glioblastoma multiforme stem-like cells through VEGF receptor 2. ScientificWorldJournal, 2013. 2013: p. 417413.

[100] Fantozzi, A., et al., VEGF-mediated angiogenesis links EMT-induced cancer stemness to tumor initiation. Cancer Res, 2014.

[101] Clement, V., et al., HEDGEHOG-GLI1 signaling regulates human glioma growth, cancer stem cell self-renewal, and tumorigenicity. Curr Biol, 2007. 17(2): p. 165-72.

[102] Liu, S., et al., Hedgehog signaling and Bmi-1 regulate self-renewal of normal and malignant human mammary stem cells. Cancer Res, 2006. 66(12): p. 6063-71.

[103] Peacock, C.D., et al., Hedgehog signaling maintains a tumor stem cell compartment in multiple myeloma. Proc Natl Acad Sci U S A, 2007. 104(10): p. 4048-53.

[104] Takezaki, T., et al., Essential role of the Hedgehog signaling pathway in human glioma-initiating cells. Cancer Sci, 2011. 102(7): p. 1306-12.

[105] Varnat, F., et al., Human colon cancer epithelial cells harbour active HEDGEHOG-GLI signalling that is essential for tumour growth, recurrence, metastasis and stem cell survival and expansion. EMBO Mol Med, 2009. 1(6-7): p. 338-51.

[106] He, T.C., et al., Identification of c-MYC as a target of the APC pathway. Science, 1998. 281(5382): p. 1509-12.

[107] Masckauchan, T.N., et al., Wnt/beta-catenin signaling induces proliferation, survival and interleukin-8 in human endothelial cells. Angiogenesis, 2005. 8(1): p. 43-51.

[108] Shtutman, M., et al., The cyclin D1 gene is a target of the beta-catenin/LEF-1 pathway. Proc Natl Acad Sci U S A, 1999. 96(10): p. 5522-7.

http://dx.doi.org/10.5772/58598

[109] Tetsu, O. and F. McCormick, Beta-catenin regulates expression of cyclin D1 in colon carcinoma cells. Nature, 1999. 398(6726): p. 422-6.

[110] Lowry, W.E., et al., Defining the impact of beta-catenin/Tcf transactivation on epithelial stem cells. Genes Dev, 2005. 19(13): p. 1596-611.

[111] Tumbar, T., et al., Defining the epithelial stem cell niche in skin. Science, 2004. 303(5656): p. 359-63.

[112] Li, L. and W.B. Neaves, Normal stem cells and cancer stem cells: the niche matters. Cancer Res, 2006. 66(9): p. 4553-7.

[113] Vermeulen, L., et al., Wnt activity defines colon cancer stem cells and is regulated by the microenvironment. Nat Cell Biol, 2010. 12(5): p. 468-76.

[114] Bao, B., et al., Notch-1 induces epithelial-mesenchymal transition consistent with cancer stem cell phenotype in pancreatic cancer cells. Cancer Lett, 2011. 307(1): p. 26-36.

[115] Fan, X., et al., Notch pathway inhibition depletes stem-like cells and blocks engraftment in embryonal brain tumors. Cancer Res, 2006. 66(15): p. 7445-52.

[116] Harrison, H., et al., Regulation of breast cancer stem cell activity by signaling through the Notch4 receptor. Cancer Res, 2010. 70(2): p. 709-18.

[117] Suman, S., T.P. Das, and C. Damodaran, Silencing NOTCH signaling causes growth arrest in both breast cancer stem cells and breast cancer cells. Br J Cancer, 2013. 109(10): p. 2587-96.

[118] Fan, X., et al., NOTCH pathway blockade depletes CD133-positive glioblastoma cells and inhibits growth of tumor neurospheres and xenografts. Stem Cells, 2010. 28(1): p. 5-16.

[119] Hoey, T., et al., DLL4 blockade inhibits tumor growth and reduces tumor-initiating cell frequency. Cell Stem Cell, 2009. 5(2): p. 168-77.

[120] Gao, F., et al., Hes1 is involved in the self-renewal and tumourigenicity of stem-like cancer cells in colon cancer. Sci Rep, 2014. 4: p. 3963.

[121] Noll, J.E., et al., Tug of war in the haematopoietic stem cell niche: do myeloma plasma cells compete for the HSC niche? Blood Cancer J, 2012. 2: p. e91.

[122] Wright, D.E., et al., Physiological migration of hematopoietic stem and progenitor cells. Science, 2001. 294(5548): p. 1933-6.

[123] Martin, S.K., et al., The emerging role of hypoxia, HIF-1 and HIF-2 in multiple myeloma. Leukemia, 2011. 25(10): p. 1533-42.

[124] Balkwill, F., The significance of cancer cell expression of the chemokine receptor CXCR4. Semin Cancer Biol, 2004. 14(3): p. 171-9.

[125] Fusi, A., et al., Expression of chemokine receptors on circulating tumor cells in patients with solid tumors. J Transl Med, 2012. 10: p. 52.

[126] Muller, A., et al., Involvement of chemokine receptors in breast cancer metastasis. Nature, 2001. 410(6824): p. 50-6.

[127] Cojoc, M., et al., Emerging targets in cancer management: role of the CXCL12/CXCR4 axis. Onco Targets Ther, 2013. 6: p. 1347-1361.

[128] Cabarcas, S.M., L.A. Mathews, and W.L. Farrar, The cancer stem cell niche--there goes the neighborhood? Int J Cancer, 2011. 129(10): p. 2315-27.

[129] Sherry, M.M., et al., STAT3 is required for proliferation and maintenance of multipotency in glioblastoma stem cells. Stem Cells, 2009. 27(10): p. 2383-92.

[130] Sansone, P., et al., IL-6 triggers malignant features in mammospheres from human ductal breast carcinoma and normal mammary gland. J Clin Invest, 2007. 117(12): p. 3988-4002.

[131] Qiu, B., et al., Interleukin-6 is overexpressed and augments invasiveness of human glioma stem cells in vitro. Clin Exp Metastasis, 2013. 30(8): p. 1009-18.

[132] Wang, Z., et al., Emerging role of Notch in stem cells and cancer. Cancer Lett, 2009. 279(1): p. 8-12.

[133] Kinoshita, H., et al., Interleukin-6 mediates epithelial-stromal interactions and promotes gastric tumorigenesis. PLoS One, 2013. 8(4): p. e60914.

[134] Bachelot, T., et al., Prognostic value of serum levels of interleukin 6 and of serum and plasma levels of vascular endothelial growth factor in hormone-refractory metastatic breast cancer patients. Br J Cancer, 2003. 88(11): p. 1721-6.

[135] Zhang, G.J. and I. Adachi, Serum interleukin-6 levels correlate to tumor progression and prognosis in metastatic breast carcinoma. Anticancer Res, 1999. 19(2B): p. 1427-32.

[136] Xiang, T., et al., Interleukin-17 produced by tumor microenvironment promotes self-renewal of CD133 cancer stem-like cells in ovarian cancer. Oncogene, 2013.

[137] Elaraj, D.M., et al., The role of interleukin 1 in growth and metastasis of human cancer xenografts. Clin Cancer Res, 2006. 12(4): p. 1088-96.

[138] Pantschenko, A.G., et al., The interleukin-1 family of cytokines and receptors in human breast cancer: implications for tumor progression. Int J Oncol, 2003. 23(2): p. 269-84.

[139] Voronov, E., et al., IL-1 is required for tumor invasiveness and angiogenesis. Proc Natl Acad Sci U S A, 2003. 100(5): p. 2645-50.

[140] Benoy, I.H., et al., Increased serum interleukin-8 in patients with early and metastatic breast cancer correlates with early dissemination and survival. Clin Cancer Res, 2004. 10(21): p. 7157-62.

[141] Yao, C., et al., Interleukin-8 modulates growth and invasiveness of estrogen receptor-negative breast cancer cells. Int J Cancer, 2007. 121(9): p. 1949-57.

[142] Ginestier, C., et al., CXCR1 blockade selectively targets human breast cancer stem cells in vitro and in xenografts. J Clin Invest, 2010. 120(2): p. 485-97.

[143] Camerlingo, R., et al., The role of CD44+/CD24-/low biomarker for screening, diagnosis and monitoring of breast cancer. Oncol Rep, 2014. 31(3): p. 1127-32.

[144] Giordano, A., et al., Clinical relevance of cancer stem cells in bone marrow of early breast cancer patients. Ann Oncol, 2013. 24(10): p. 2515-21.

[145] Li, L., et al., Antibody Against CD44s Inhibits Pancreatic Tumor Initiation and Post-Radiation Recurrence in Mice. Gastroenterology, 2014.

[146] Deng, Y., et al., ALDH1 is an independent prognostic factor for patients with stages II-III rectal cancer after receiving radiochemotherapy. Br J Cancer, 2014. 110(2): p. 430-4.

[147] Zhou, F., et al., Expression and prognostic value of tumor stem cell markers ALDH1 and CD133 in colorectal carcinoma. Oncol Lett, 2014. 7(2): p. 507-512.

[148] Zhong, Y., et al., Expression of ALDH1 in breast invasive ductal carcinoma: an independent predictor of early tumor relapse. Cancer Cell Int, 2013. 13(1): p. 60.

[149] De Brot, M., et al., Prognostic impact of the cancer stem cell related markers ALDH1 and EZH2 in triple negative and basal-like breast cancers. Pathology, 2012. 44(4): p. 303-12.

[150] Resetkova, E., et al., Prognostic impact of ALDH1 in breast cancer: a story of stem cells and tumor microenvironment. Breast Cancer Res Treat, 2010. 123(1): p. 97-108.

[151] Lin, E.H., et al., Elevated circulating endothelial progenitor marker CD133 messenger RNA levels predict colon cancer recurrence. Cancer, 2007. 110(3): p. 534-42.

[152] Mehra, N., et al., Progenitor marker CD133 mRNA is elevated in peripheral blood of cancer patients with bone metastases. Clin Cancer Res, 2006. 12(16): p. 4859-66.

[153] Narita, K., et al., Nestin regulates proliferation, migration, invasion and stemness of lung adenocarcinoma. Int J Oncol, 2014. 44(4): p. 1118-30.

[154] Creighton, C.J., et al., Residual breast cancers after conventional therapy display mesenchymal as well as tumor-initiating features. Proc Natl Acad Sci U S A, 2009. 106(33): p. 13820-5.

[155] Phillips, T.M., W.H. McBride, and F. Pajonk, The response of CD24(-/low)/CD44+breast cancer-initiating cells to radiation. J Natl Cancer Inst, 2006. 98(24): p. 1777-85.

[156] Scheck, A.C., et al., Biological and molecular analysis of a low-grade recurrence of a glioblastoma multiforme. Clin Cancer Res, 1996. 2(1): p. 187-99.

[157] Woodward, W.A., et al., WNT/beta-catenin mediates radiation resistance of mouse mammary progenitor cells. Proc Natl Acad Sci U S A, 2007. 104(2): p. 618-23.

[158] Cukierman, E. and D.E. Bassi, The mesenchymal tumor microenvironment: a drug-resistant niche. Cell Adh Migr, 2012. 6(3): p. 285-96.

[159] Nefedova, Y., T.H. Landowski, and W.S. Dalton, Bone marrow stromal-derived soluble factors and direct cell contact contribute to de novo drug resistance of myeloma cells by distinct mechanisms. Leukemia, 2003. 17(6): p. 1175-82.

[160] Hattermann, K., et al., The chemokine receptor CXCR7 is highly expressed in human glioma cells and mediates antiapoptotic effects. Cancer Res, 2010. 70(8): p. 3299-308.

[161] Margolin, D.A., et al., Lymph node stromal cells enhance drug-resistant colon cancer cell tumor formation through SDF-1alpha/CXCR4 paracrine signaling. Neoplasia, 2011. 13(9): p. 874-86.

[162] Singh, S., et al., CXCL12-CXCR4 signalling axis confers gemcitabine resistance to pancreatic cancer cells: a novel target for therapy. Br J Cancer, 2010. 103(11): p. 1671-9.

[163] Matsunaga, T., et al., Interaction between leukemic-cell VLA-4 and stromal fibronectin is a decisive factor for minimal residual disease of acute myelogenous leukemia. Nat Med, 2003. 9(9): p. 1158-65.

[164] Weekes, C.D., C.A. Kuszynski, and J.G. Sharp, VLA-4 mediated adhesion to bone marrow stromal cells confers chemoresistance to adherent lymphoma cells. Leuk Lymphoma, 2001. 40(5-6): p. 631-45.

[165] Wang, J., et al., Notch promotes radioresistance of glioma stem cells. Stem Cells, 2010. 28(1): p. 17-28.

[166] Chen, M.S., et al., Wnt/beta-catenin mediates radiation resistance of Sca1+progenitors in an immortalized mammary gland cell line. J Cell Sci, 2007. 120(Pt 3): p. 468-77.

[167] Wulf, G.G., et al., A leukemic stem cell with intrinsic drug efflux capacity in acute myeloid leukemia. Blood, 2001. 98(4): p. 1166-73.

[168] Hirschmann-Jax, C., et al., A distinct "side population" of cells with high drug efflux capacity in human tumor cells. Proc Natl Acad Sci U S A, 2004. 101(39): p. 14228-33.

[169] Pannuti, A., et al., Targeting Notch to target cancer stem cells. Clin Cancer Res, 2010. 16(12): p. 3141-52.

[170] McGowan, P.M., et al., Notch1 inhibition alters the CD44hi/CD24lo population and reduces the formation of brain metastases from breast cancer. Mol Cancer Res, 2011. 9(7): p. 834-44.

[171] Merchant, A.A. and W. Matsui, Targeting Hedgehog--a cancer stem cell pathway. Clin Cancer Res, 2010. 16(12): p. 3130-40.

[172] Bar, E.E., et al., Cyclopamine-mediated hedgehog pathway inhibition depletes stem-like cancer cells in glioblastoma. Stem Cells, 2007. 25(10): p. 2524-33.

[173] Ramaswamy, B., et al., Hedgehog signaling is a novel therapeutic target in tamoxifen-resistant breast cancer aberrantly activated by PI3K/AKT pathway. Cancer Res, 2012. 72(19): p. 5048-59.

[174] Feldmann, G., et al., Blockade of hedgehog signaling inhibits pancreatic cancer invasion and metastases: a new paradigm for combination therapy in solid cancers. Cancer Res, 2007. 67(5): p. 2187-96.

[175] Chen, B., et al., Small molecule-mediated disruption of Wnt-dependent signaling in tissue regeneration and cancer. Nat Chem Biol, 2009. 5(2): p. 100-7.

[176] Fujii, N., et al., An antagonist of dishevelled protein-protein interaction suppresses beta-catenin-dependent tumor cell growth. Cancer Res, 2007. 67(2): p. 573-9.

[177] He, B., et al., Blockade of Wnt-1 signaling induces apoptosis in human colorectal cancer cells containing downstream mutations. Oncogene, 2005. 24(18): p. 3054-8.

[178] Konopleva, M., et al., Therapeutic targeting of microenvironmental interactions in leukemia: mechanisms and approaches. Drug Resist Updat, 2009. 12(4-5): p. 103-13.

[179] Jin, L., et al., Targeting of CD44 eradicates human acute myeloid leukemic stem cells. Nat Med, 2006. 12(10): p. 1167-74.

[180] Patil, Y., et al., Nanoparticle-mediated simultaneous and targeted delivery of paclitaxel and tariquidar overcomes tumor drug resistance. J Control Release, 2009. 136(1): p. 21-9.

[181] Sims-Mourtada, J., et al., Sonic Hedgehog promotes multiple drug resistance by regulation of drug transport. Oncogene, 2007. 26(38): p. 5674-9.

[182] Folkins, C., et al., Anticancer therapies combining antiangiogenic and tumor cell cytotoxic effects reduce the tumor stem-like cell fraction in glioma xenograft tumors. Cancer Res, 2007. 67(8): p. 3560-4.

[183] Burkhardt, J.K., et al., Orthotopic glioblastoma stem-like cell xenograft model in mice to evaluate intra-arterial delivery of bevacizumab: from bedside to bench. J Clin Neurosci, 2012. 19(11): p. 1568-72.

[184] Kaplan, R.N., et al., VEGFR1-positive haematopoietic bone marrow progenitors initiate the pre-metastatic niche. Nature, 2005. 438(7069): p. 820-7.

[185] Sandal, T., et al., Epigenetic reversion of breast carcinoma phenotype is accompanied by changes in DNA sequestration as measured by AluI restriction enzyme. Am J Pathol, 2007. 170(5): p. 1739-49.

[186] Weaver, V.M., et al., Reversion of the malignant phenotype of human breast cells in three-dimensional culture and in vivo by integrin blocking antibodies. J Cell Biol, 1997. 137(1): p. 231-45.